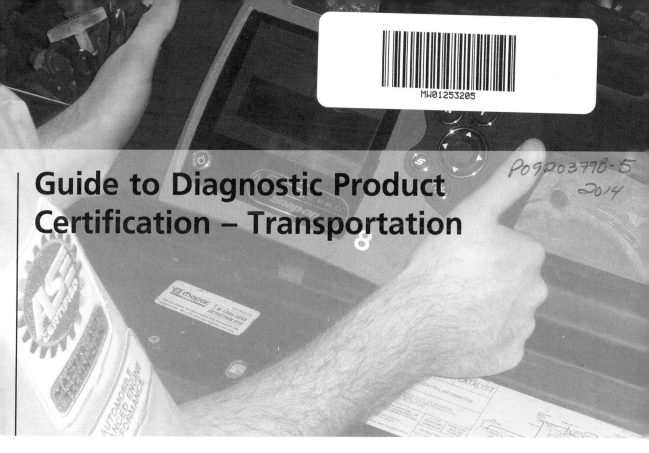

Guide to Diagnostic Product Certification – Transportation

Matthew Janisin

DELMAR
CENGAGE Learning

Australia • Brazil • Japan • Korea • Mexico • Singapore • Spain • United Kingdom • United States

Guide to Diagnostic Product
Certification – Transportation
Matthew Janisin

Vice President, Editorial: Dave Garza

Director of Learning Solutions: Sandy Clark

Executive Editor: Dave Boelio

Managing Editor: Larry Main

Senior Product Manager: Matthew Thouin

Editorial Assistant: Courtney Troeger

Vice President, Marketing: Jennifer Baker

Marketing Director: Deborah S. Yarnell

Marketing Manager: Erin Brennan

Associate Marketing Manager: Jillian Borden

Production Director: Wendy Troeger

Production Manager: Mark Bernard

Content Project Management: PreMediaGlobal

© 2013 Delmar, Cengage Learning

ALL RIGHTS RESERVED. No part of this work covered by the copyright herein may be reproduced, transmitted, stored, or used in any form or by any means graphic, electronic, or mechanical, including but not limited to photocopying, recording, scanning, digitizing, taping, Web distribution, information networks, or information storage and retrieval systems, except as permitted under Section 107 or 108 of the 1976 United States Copyright Act, without the prior written permission of the publisher.

> For product information and technology assistance, contact us at
> **Cengage Learning Customer & Sales Support, 1-800-354-9706**
> For permission to use material from this text or product,
> submit all requests online at **www.cengage.com/permissions**.
> Further permissions questions can be e-mailed to
> **permissionrequest@cengage.com**

Snap-on® is a trademark, registered in the United States and other countries, of Snap-on® Incorporated. Other marks are marks of their respective holders. Use of "Snap-on" in this publication does not constitute an endorsement by Snap-on Incorporated.

Library of Congress Control Number: 2012930061

ISBN-13: 978-1-4354-8379-8

ISBN-10: 1-4354-8379-0

Delmar
5 Maxwell Drive
Clifton Park, NY 12065-2919
USA

Cengage Learning is a leading provider of customized learning solutions with office locations around the globe, including Singapore, the United Kingdom, Australia, Mexico, Brazil, and Japan. Locate your local office at: **international.cengage.com/region**

Cengage Learning products are represented in Canada by Nelson Education, Ltd.

To learn more about Delmar, visit **www.cengage.com/delmar**

Purchase any of our products at your local college store or at our preferred online store **www.cengagebrain.com**

Notice to the Reader
Publisher does not warrant or guarantee any of the products described herein or perform any independent analysis in connection with any of the product information contained herein. Publisher does not assume, and expressly disclaims, any obligation to obtain and include information other than that provided to it by the manufacturer. The reader is expressly warned to consider and adopt all safety precautions that might be indicated by the activities described herein and to avoid all potential hazards. By following the instructions contained herein, the reader willingly assumes all risks in connection with such instructions. The publisher makes no representations or warranties of any kind, including but not limited to, the warranties of fitness for particular purpose or merchantability, nor are any such representations implied with respect to the material set forth herein, and the publisher takes no responsibility with respect to such material. The publisher shall not be liable for any special, consequential, or exemplary damages resulting, in whole or part, from the readers' use of, or reliance upon, this material.

Printed in the United States of America
1 2 3 4 5 6 7 16 15 14 13 12

Table of Contents

PREFACE. ix

CHAPTER 1
Snap-On Diagnostic Certification Program 1
    How to Use the Snap-On Certification Manual 3
    Snap-On Certification Tests 3

CHAPTER 2
Solus Pro 5
    Solus Pro Platform Overview. 6
    Solus Pro Platform 6
    Solus Pro Navigation 10
    System Tools 19
    Solus Pro Summary 22
    Review Questions. 23

CHAPTER 3
Vantage Pro 27
    Vantage Pro Overview 28
    Vantage Pro Platform 28
    Vantage Pro Navigation 34
    System Tools 44
    Vantage Pro Summary 49
    Review Questions. 49

CHAPTER 4
MODIS 53
    MODIS Platform Overview. 54
    MODIS Software, Navigation, and Tool Setup 60
    MODIS Main Menu. 61
    MODIS Utility Options 63
    Tool Setup Options 63
    Gas Bench Setup Options 71
    System Tools 71
    Legacy Software. 72
    MODIS Summary 77
    Review Questions. 78

CHAPTER 5
Scanner Introduction 85
    Scanner Overview 86
    Scanner Scenario 88
    Systematic Scanner Procedure 89
        Systematic Scanner Procedure 90
    Vehicle Connection 91
        General Motors 96
        Ford 96
        Chrysler 97
        OBDI 97
        OBDI Connection Steps 100
        OBDII 102
        OBDII Connection Steps 102
    Summary 107
    Review Questions 108

CHAPTER 6
Viewing and Interpreting Scan Data 111
    Viewing and Interpreting Data Overview 112
        Parameter Identifications (PIDs) 112
    Viewing Data 112
    Text View 113
        Version 9.2 and *earlier* Text View Functions 113
        Version 9.4 and *later* Text View Functions 121
    PID List View 126
        Freeze/Run 128
        Review 128
        Clear 129
        PID Sort 130
        PID List View Zoom 131
        PID List Custom Data View 133
        Save, Print, and Utilities in PID List View 139
    Graphing Data View 141
        Freeze/Run 145
        Reviewing Graphed Data and the Cursor Function 147
        Clearing the Data Memory Buffer 149
        Graphing PID Sort 150
        Graphing View Zoom 152
        Zoom Usage Example 155
        Snapshot 158
        Save, Print, and Utilities in Graphing View 158
        Custom Data View with Graphs 160
        Graphing PID Scaling Options 169
        Before Manual Scale 172
        After Manual Scale 172
    Custom Data List 173
    Custom Data List Summary 177
    Advanced Capturing Techniques 178
    Manual Snapshot 178

PID Trigger 180

PID Trigger Summary. 186

Viewing and Interpreting Data Summary 188

Summary Points. 189

Review Questions. 190

CHAPTER 7
Global OBD 199

Global OBDII 200

Global OBDII Navigation. 202
   Readiness Monitors 204
   MIL Status 206
   ($01) Display Current Data. 207
   ($02) Display Freeze Frame Data. 209
   ($03) Display Trouble Codes. 210
   ($04) Clear Emission Related Data 211
   ($05) Oxygen Sensor Monitoring 212
   ($06) Non-Cont. Monitored Systems 213
   ($07) DTCs Detected During Last Drive Cycle 215
   ($08) Request Control On-Board System 216
   ($09) Read Vehicle Identification. 217
   ($09) In-Use Performance Tracking 218
   ($0A) Display Permanent Trouble Codes 219

OBD Health Check 219

Generic Functions. 222

Global OBD Summary 223

Review Questions. 224

CHAPTER 8
Codes Menu. 227

Codes Menu Overview 228
   Diagnostic Trouble Codes (DTCs) 228
   Trouble Codes 230
   Clear Codes. 234
   Freeze Frame/Failure Records 235
   DTC Status 236

Summary. 237

Review Questions. 238

CHAPTER 9
Data Management 241

Data Management Overview. 242

Data Management Features. 243

Using the Edit Feature 244

Data Storage Devices 249

File Types 251

ShopStream Connect 253

Downloading, Installing, & Updating ShopStream Connect 254

Navigating ShopStream Connect 255

Reviewing a Scanner Movie File 258

Graph Properties 262

Custom Data Views 265

Custom Data List 267
    Reviewing a Lab Scope Movie File 270

Sample Lab Scope Navigation 272

Summary 278

Review Questions 278

CHAPTER 10
Functional Tests 283

Functional Tests Overview 284

Functional Test Descriptions 285
    Information Tests 285
    Toggle Tests 285
    Variable Control Tests 285
    Reset Tests 285
    CKP Variation Learn Test 287
    Injector Balance Test 287
    Output Controls 288
    Idle Air Control Test 289

Transmission Functional Tests 297

ABS Functional Tests 298

Airbag Functional Tests 299

Transfer Case Functional Tests 300

Body Control Module (BCM) Functional Tests 301

Instrument Panel Cluster (IPC) Functional Tests 302

Functional Test Training 304

Researching other Functional Tests 307

Functional Test Summary 314

Review Questions 315

CHAPTER 11
Fast-Track Troubleshooter 319

Overview 320

Fast-Track Troubleshooter Features 321

Troubleshooter in Detail 323

Code Tips 323

Symptom Tips 327

Tests & Procedures 329

Technical Assistance 330

Fast-Track Data Scan (Normal Values) 331

Fast-Track Troubleshooter Research 333

Summary 336

Review Questions 336

CHAPTER 12
Component Testing 341

    Component Test Meter Overview 342

    Component Testing Menu 342

    Component Test Meter Summary 374
        Online Info 374

    Power User Tests 375

    Fluid Pressures (Fuel, Transmission, Power Steering, ABS, etc) 377

    Compression Tests (Air/Gas/Vacuum Pressure) 378

    Features and Benefits 378

    A-Z Index 380

    How To 383

    Component Testing Summary 386

    Review Questions 387

CHAPTER 13
Lab Scope Operation 391

    Need for Lab Scopes 392

    Lab Scope Basics 392

    Labe Scope Channels and Connections 394

    Lab Scope Display Configurations 396

    Lab Scope Channel Configurations 406

    Trigger Configuration 421

    Lab Scope Presets 430

    Interpreting and Saving Lab Scope Data 435

    Ignition Scope Configuration 443

    Single Cylinder Ignition 448

    Viewing Multiple Ignition Scope Patterns 449

    Reviewing Ignition Scope Movies and Snapshots 458

    Finding Glitches Using a Lab Scope 462
        Steps to Finding a Glitch Using a Lab Scope 462
        Glitch Example #1: Bad Fuel Injector 466
        Glitch Example #2: Bad Wheel Speed Sensor 469
        Suspected Glitch 473

    Summary 473

    Review Questions 474

CHAPTER 14
Multimeter Operation 483

    Multimeter Overview 484

    Waveform Demonstrator Board 488

    Digital Meter Navigation 490
        Vantage Pro Specific 496
        MODIS Specific 497
        MODIS and Vantage Pro 500

Task: Resistance Testing Practice 501
    Vantage Pro Specific 502
    MODIS and Vantage Pro 504

Digital Meter Summary 506

Power Graphing Meter 507

Power Graphing Meter Connections 511

Power Graphing Meter Configuration 512
    Power Graphing Meter and Lab Scope Similar Configurations 512

PGM Channel Configurations 515

Practice 519

Summary 527

Review Questions 528

INDEX **537**

Preface

The *Guide to Diagnostic Product Certification – Transportation* is intended to give students the knowledge they need to more effectively use their Snap-On diagnostic tools. As vehicles have become more complex, so have the tools used to diagnose them. Although many automotive programs do a great job teaching the vehicle systems, few take the time to explain in detail the sophisticated diagnostic tools needed to diagnosis them. Current market research indicates that most technicians only use about 15% of their diagnostic tools' capabilities. The National Coalition of Certification Centers (NC3.net) was developed to address this situation. It is the goal of this manual, designed to accompany the NC3 curriculum, to support Snap-On Diagnostic Certification. This will allow students and technicians to become complete Power Users of their diagnostic equiptment. The manual begins with a physical overview of each of the tools and explains the various ports, connectors, buttons, and controls. In separate sections, each tool's utility menu is fully explained and all customizable features are discussed so the user will have full control to make the tool his or her own. Finally, every aspect of the tool, including the Scanner, Lab Scope, Component Test Meter, Troubleshooter, and all other options are explained in detail. The *Guide to Diagnostic Product Certification* will greatly increase a student's employability and a technician's productivity when faced with a complex diagnostic situation.

Acknowledgments

A project of this size requires the support and understanding of so many people, all of whom played a crucial role in its completion. I first have to thank Dr. Bryan Albrecht, president of Gateway Technical College, and Nick Pinchuk, chairman and CEO of Snap-On Incorporated, whose outstanding leadership skills combined to initiate and grow the certification program and then establish NC3 as a way to expand this model from coast to coast. Roger Tadajewski, executive director of NC3, has continued this expansion into the areas of transportation, energy, and aviation with great diligence and success. Without them, there would be no reason for this type of book. Next, I need to thank Bob Braun, automotive instructor and transportation division chair at Gateway Technical College, and Frederick Brookhouse, business and educational partnership manager at Snap-On Industrial, for their role as mentors. I could not be blessed with two more dedicated and knowledgeable individuals guiding me through both the educational and industrial sides of this endeavor. I would also like to thank DuWayne Jennings of Snap-On and Ken Dotzler of Gateway for all of their time and for their encouragement of this project. DuWayne, you have been one the keystones in the foundation of this entire certification program, and your hard work and long hours too often go unrecognized. Please know that I and the rest of the team at Gateway and Snap-On appreciate your commitment and hard work. Ken, thank you for always getting me to see the positive side of this challenge and helping me through the rough parts. It was great to know I could always count on your understanding. I also wish to thank the Cengage team, led by Dave Boelio and Matt Thouin, whose determination and dedication saw this project through to completion. Dave and Matt, thank you for your dedication and patience through all of the ups, downs, and sideways we traveled on this journey. It has been quite a ride. Finally, I wish to thank the people closest to me who have both helped and sacrificed to allow me the time to undertake and complete a project like this. To my parents, Ed and Mary Jo, and my in-laws, Ken and Jane, thank you for always being there and believing in me. I'm blessed to have two sets of parents that continue to teach and hold me to the highest standards. Ultimately my sincerest appreciation goes to my wife Brooke, who after giving away all of her patience to her eighth-grade English students each and every day, still found the time and tolerance to be my personal writing teacher and coach. I cannot thank you enough or put into words how much I love you.

Reviewers

The author and publisher wish to thank the following instructors for their feedback during the development of this text:

Benjamen Adams
Lakeshore Technical College
Cleveland, WI

Tom Berryman
Lawson State Community College
Bessemer, AL

Roger Blackburn
Aims Community College
Windsor, CO

Kenneth Dotzler
Gateway Technical College
Kenosha, WI

Eric Erskin
Ivy Tech Community College
Lafayette, IN

David E. Howell
Tidewater Community College
Chesapeake, VA

Tim Moy
Moraine Park Technical College
Fond du Lac, WI

About the Author

Matthew Janisin is the National Coalition of Certification Centers (NC3) instructor and coordinator at Gateway Technical College in Kenosha, Wisconsin. There he delivers a number of the certification courses available through NC3 and its partners, including the Snap-On Diagnostic Tool Certifications, to current Gateway students, local incumbent technicians, and fellow instructors from across the United States whose colleges have joined the NC3 team. Mr. Janisin holds both a Bachelor and Master of Science degree in industrial and technology education from the University of Wisconsin–Stout Polytechnic. He has been teaching automotive classes since 2003 and became the first official Snap-On Certified Diagnostician at the program's beginning in 2007. In addition to holding numerous NC3/Snap-On certifications, Mr. Janisin is also an ASE certified master technician with a number of other ASE endorsements, including the L1 advanced engine performance specialist, X1 undercar specialist, and the A9 light-duty diesel certification. He comes to Delmar/Cengage Learning with the strong backing of the faculty of Gateway Technical College, where their partnership with Snap-On, Inc., first developed the certification program idea, which eventually evolved into the NC3 educational model.

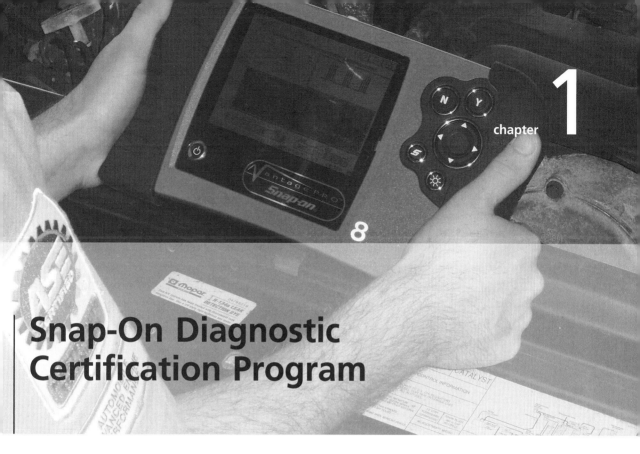

chapter 1

Snap-On Diagnostic Certification Program

The idea for a diagnostic tool certification program came from market research that indicated that most technicians only use about 15% of their diagnostic tools capabilities. A common example of this was a MODIS user who spent upwards of $7000 on that tool but really only feels comfortable using the scanner portion of it to read codes and to analyze a couple of data streams. This technician is, at best, only effectively using slightly over $1000 worth of that diagnostic tool's capabilities, a misuse of valuable capital. The fault lies with neither the tool nor the technician, but rather with both. To correct this problem, a formal training program was created that aims to combine the forces of the nation's technical colleges and Snap-On, Inc. This partnership creates a synergy in which the expertise of each group is maximized. The program combines the expertise of automotive educators who take complicated technical information and break it into manageable parts that are more easily learned by students and who have the proper facilities to deliver this information with the great depth of technical knowledge, application, and industry insight of Snap-On's employees. This raw technical knowledge and the full understanding of every aspect of a tool's capabilities is passed from the tool developer to the instructor, who then filters and sorts the information into a program of study that fits into the curriculum of the college. It is important to remember that, since the information comes directly from the tool's developer, no feature or function is left out or lost in transition and this information is delivered by trained instructors at training facilities designed for both classroom and live-application presentations. Expertise is focused in such a way that the certification program is of the highest quality for the technicians and students who go through the curriculum. This partnership of industry and education has expanded into the formation of the National Coalition of Certification Centers

(NC3). NC3 is continuing to establish new partnerships and expand to more schools to provide more comprehensive and better training coverage across the United States. To learn more about NC3 visit their website at: www.nc3.net. To find a certification center close to you go to: www.nc3.net/partners/certification-center-locations/.

In order for technicians to harness the full potential of their diagnostic tools capabilities, they need to become Power Users of that tool. A Power User is someone that can efficiently use 90+ percent of a tool's capability and is aware of and working towards incorporating the other 10%. However, the reality is that technicians are too busy on a normal day to ever be able to explore past what they already know, that initial 15%. In a work environment, you take what you know and apply it to what you need right now to fix the problem at hand. With real-world time constraints and complex diagnostic equipment, it is highly unlikely that a majority of users will ever achieve Power User status if left to learn it on their own. But a formal training course focused on specific diagnostic tools can make that goal a reality.

The vast amount of information that must be taught about vehicle systems in a typical automotive program combined with normal budget pressures often limits the amount of time spent teaching specific diagnostic tools. Diagnostic tools are usually covered in a broad sense, with little discussion of details about the the tool itself or how it can increase productivity in the shop. Snap-On Certified Diagnostic Training Centers offer this second avenue of specialized training that takes an average technician who has a sound understanding of vehicle systems and creates a more highly skilled diagnostician who understands how to efficiently gather needed data from the vehicle using the diagnostic tool. This diagnostic equipment certification, combined with the automotive degree earned from the representative school, provides graduates with proof to future employers and customers that they have all the necessary skills to quickly and accurately diagnose a vehicle, make the repair, and verify that repair.

In this training, you will become a Power User of each piece of equipment in the Snap-On diagnostic fleet. **See Figure 1-1.** This includes a scan tool with bidirectional control named Solus Pro and a dedicated lab scope and component test meter called the Vantage Pro. If a technician wants to have the power of both a scanner and lab scope in one unit, then the MODIS or

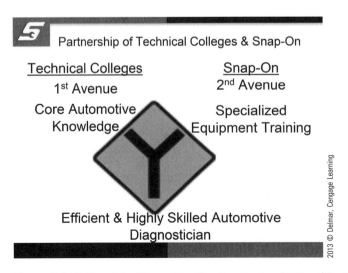

Figure 1-1 National Coalition of Certification Centers (NC3.net) Training Model

the newer Verus, which is not covered in this book, is the diagnostic tool of choice. All of these pieces of equipment share similarities, but all have very specific purposes and functions. This textbook will cover every aspect and function of each tool. It will also help technicians identify when and where to use each tool depending on the diagnostic problem at hand. Due to the length of this material, little if any time is spent on automotive terminology and the theory behind automotive systems. It is expected that the student or technician already has this knowledge or is getting this knowledge from a different class or training course. An example of this may be looking at Readiness Monitors under Global OBD. This textbook will show how to navigate to, change the view of, and print or archive the Readiness Monitor Data but it will not discuss what a Readiness Monitor is or how it fits into a diagnostic procedure.

How to Use the Snap-On Certification Manual

Although this textbook may resemble an instruction or user manual, the graphics and discussion will cover the features and benefits of each tool with much more depth. Due to the similarities between the different tools and the continuous updating of software, this manual has been strategically divided into various sections. The book can be used as a whole and read cover to cover, however, it is actually designed to allow readers to choose the sections that are most relevant to them. One will find a lot of repetition if reading the book straight through, but just enough to cover and reinforce the material if reading specific sections.

First, the textbook starts with a physical overview of each specific tool. These sections will introduce the reader to all necessary customization and optional settings, as well as general navigation and physical features. Then in subsequent chapters each function is covered in detail. These functions are not necessarily tool specific and will be applicable to a number of different tools. These sections are organized in the sequential way that users will encounter them on their diagnostic journey. This journey will start with basic scanner codes and data, progress to using diagnostic aids such as Fast-Track Troubleshooter and the performing of Functional Tests to further isolate the problem, and finish with using the Lab Scope or Component Test Meter to officially condemn a vehicle part before replacement.

Since students and technicians will have a wide range of experience using these features, the differentiated sections make it easier to skip right to the area where the most help and training for the specific individual is needed.

Within each section, step-by-step instructions to explain the various tool features are outlined by number. While at times this approach may seem unnecessary, remember that the textbook is also written for beginner diagnostic tool users. The numbered steps make it easier to quickly skip to the specific part of the book section that holds the information the reader is trying to find. To accommodate the beginner user, detailed information was included, but to help advanced users quickly skip through review material, a numbered system was used.

Snap-On Certification Tests

The review questions found at the end of each section are similar in scope to what one will find if taking a Snap-On Diagnostic Certification test. Snap-On's tests are completely tool specific and focus on navigation and tool set-up. The theory and diagnostic meaning behind the data is covered by your local technical college, ASE, or other automotive training program.

Snap-On is focusing on training and testing your ability to quickly gather data from the vehicle, organize it in an easily viewable way, and archive it for use in a future situation. As stressed earlier, both of these skill sets are required by today's technicians as vehicles become more and more complex, requiring increasingly sophisticated diagnostic tools and procedures. Use the review questions to understand the concepts that Snap-On will be testing before completing an actual certification test.

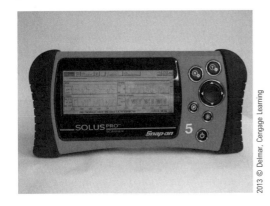

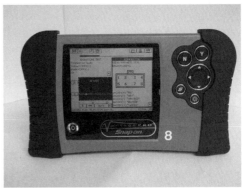

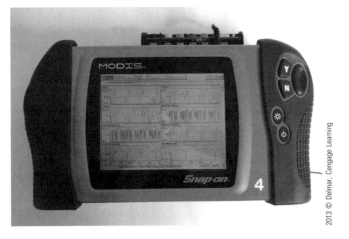

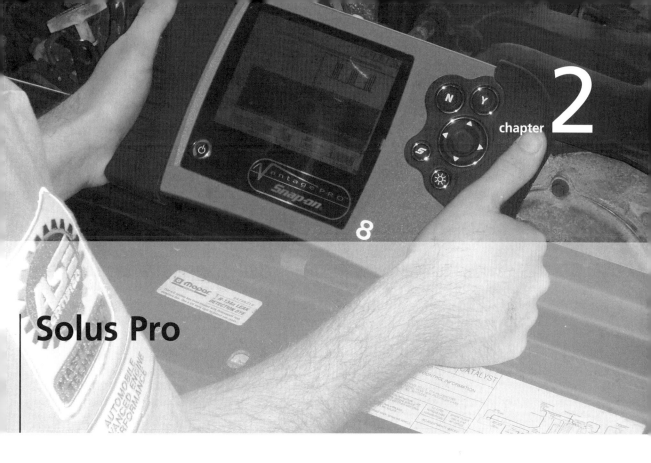

Solus Pro

Upon completion of the Solus Pro Platform module, you will be able to:

- Identify the various control buttons and explain their functions
- Locate the different connector ports and explain how each are used
- Locate the battery compartment and explain the different power sources' capabilities and how to maximize battery life while running the Solus Pro
- Locate and explain the use of the CF card memory slot
- Identify the main menu options
- Explain the time-saving capabilities of the "S" button
- Customize the Solus Pro to your specific shop environment and personal preferences to maximize efficiency
- Locate pertinent system information if needed
- Demonstrate how to connect the Solus Pro to a PC

Solus Pro Platform Overview

The Solus Pro hit the market in late 2007 and is an updated version of the Solus scanner, which came out around 2004. While there are some physical changes between the two units, the software navigation is almost identical. The Solus was really designed as an upgrade from the MT2500 or red brick scanner, and the original unit has a similar look to its predecessor. Recently a second generation scanner named the Solus Ultra was released and this tool has a completely different user interface or navigational structure. While this scanner is not covered in this textbook, many of the features and benefits discussed here can be found in the new scanner, so an in-depth understanding of the Solus Pro may, depending on the technician, extend its useful life a bit longer and will shorten the learning curve on the newer product if an upgrade is decided on in the future. The Solus Pro is more contoured and ergonomic and has some other physical differences that will be discussed in detail later in this section. Solus Pro is a multifunction scan tool capable of displaying 4 PID graphs at one time, pulling diagnostic trouble codes, gathering Global OBDII data, performing Functional Tests, and providing access to the Fast-Track Troubleshooter software. The scanner software is identical to that used on the MODIS, as the Solus Pro is marketed to technicians who want a more compact scanner separate from a lab scope. Solus Pro has a color 6½-inch display that is visible even in bright sunlight. The unit is run on an ARM processor using Windows CE as the operating system. The older Solus had two CF card slots, one of which was dedicated to the storage of the scanner software and the other dedicated to expanded storage and file-saving capabilities. This is no longer the case with the Solus Pro, as the software is stored internally on a larger chip memory and the remaining single CF card slot is for expanded storage, file saving, and updating capabilities. After updating to the internal storage chip the installation CF card can be removed so a storage card can be used once again. The Solus Pro is a very rugged, compact, and capable scan tool.

Solus Pro Platform

The back of the Solus Pro unit has a multi-position hanger/stand. It can be tucked away for easy hand held operations, propped out to allow the unit to stand on its own while on a bench (**See Figure 2-1**), flipped up so the unit can hang from overhead (**See Figure 2-2**), or put straight back to allow it to hang on a steering wheel while gathering data in the shop.

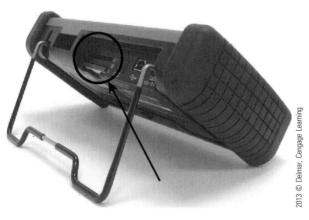

Figure 2-1 Older Solus Platform with 2 CF card slots

Solus Pro Platform 7

Figure 2-2 The hanger can be flipped to hang from overhead or put straight back to allow it to hang on a steering wheel.

The top of the Solus Pro unit has the following connectors (**See Figure 2-3**):

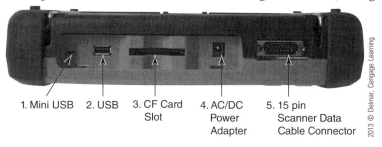

1. Mini USB   2. USB   3. CF Card Slot   4. AC/DC Power Adapter   5. 15 pin Scanner Data Cable Connector

Figure 2-3 Newer Solus Pro platform with 1 CF card slot.

1. **Mini USB:** This port is used to connect the Solus Pro directly to a personal computer (PC). It is very similar to many other computer peripherals such as a digital camera. Once properly connected, the Solus Pro will show up as a device under the My Computer icon of your PC, and then files can be cut, copied, pasted, clicked, and dragged to and from the computer or scan tool just as though it were part of your computer system. The free ShopStream Connect software download makes this process even easier. See the Data management section for more details.

2. **USB Port:** This USB port allows the Solus Pro to be connected to many PC type accessories, such as a printer, keyboard, or mouse. If you are using software version 7.2 or later you may also connect a USB storage device (jump drive) for even more storage capacity. The keyboard is very useful for entering Shop Information, renaming saved files, or adding notes under the "Edit" command under Data Management. With a mouse installed, a cursor will appear on the screen, but it has little function.

3. **CF Card Slot:** The top Compact Flash (CF) card slot has multiple purposes. Its common use is for extra storage, but it can also be used for software upgrades. This CF card can be "hot swapped," meaning that inserting or removing the CF card while the unit is on will not damage the card or the unit.

4. **AC/DC Power Adapter:** The external power sources will plug into this connector. This may be from the vehicle's battery or the AC/DC adapter cord.
5. **Scanner Data Cable Connector:** This is where the vehicle data link cable is connected to the Solus Pro.

There is a removable and replaceable plastic screen protector on the new Solus Pro. Use a small screwdriver to gently pry on the two plastic tabs at the top of the screen when it is time to replace the old, scratched-up screen cover. The right side hand grip is removable to allow access to the battery. See **Figure 2-4**. This is one of the very important upgrades to come out with the Solus Pro. The older Solus unit used six batteries that didn't last long, so it became expensive to continually replace them. The Solus Pro has a rechargeable Nickel-Metal Hydride battery similar to those found in many cordless tools. It has a much longer life and reduces operational costs with its recharging capability. The Solus Pro unit *cannot* charge the battery itself. The unit does come with an external charging stand and uses the same AC adapter cord to power the charger that it uses to power the scan tool itself. **See Figure 2-5**. Even with a dead battery, or no battery, the unit will function if plugged into the OBDII connector or external AC power source.

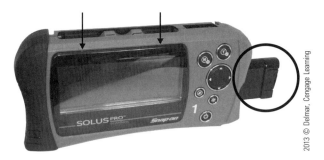

Figure 2-4 The Solus Pro unit will not charge the battery. The external charging stand must be used.

Figure 2-5 Newer Solus Pro unit with removable and replaceable plastic screen cover, as well as the new, rechargeable Nickel-Metal Hydride battery, accessible under the right hand grip.

The right front of the Solus Pro houses all of the buttons needed to control and navigate the tool. **See Figure 2-6.**

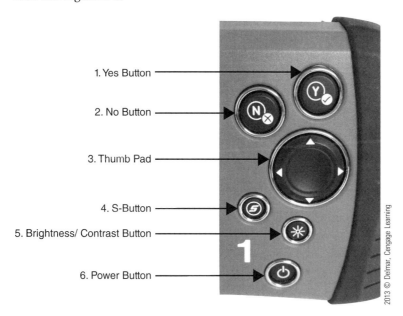

Figure 2-6 The basic controls of the Solus Pro are intuitive and similar to all Snap-On Diagnostic tools.

1. **Yes Button:** The Yes button is used for selecting functions on the screen. It is very similar to the Enter button on a PC. You can also think of it as a gas pedal that keeps you moving forward with the tool.

2. **No Button:** The No button is used for confirming selections made on the screen and then backing up in order to perform another task with the tool. It is very similar to the Esc button on a PC that will take you back to where you were or the brake pedal of a car that stops you before backing up. Remember that pressing No does not cancel selections, it just backs you out of a screen so you can perform different functions, but it still leaves everything the way you changed it.

3. **Thumb Pad:** This is the steering wheel for the scan tool. It allows you to move vertically and horizontally around the screen. With Easy Scroll (discussed later in this section) enabled, it can also function as the Yes and No buttons.

4. **S-Button:** This S-button is a programmable button: You set its function. It can save screenshots, print PID lists, save movies, and much more. Details on how to program this feature are discussed later in this section.

5. **Brightness/Contrast Button:** Pressing this button brings up the display properties box and allows the user to quickly adjust the brightness and contrast to fit the current lighting conditions and specific user preferences.

6. **Power Button:** Pressing this button will power up the unit. The button is purposely indented into the unit to protect it from being accidently pushed while working with the tool. Pressing the button again will turn the tool off.

Now that we have looked at all of the physical properties and external connections, let's power up the unit and begin to look at the software navigation and tool customization.

REMINDER: This section is written so the reader can follow along using the Solus Pro scan tool, if available. It is strongly encouraged that the reader follow along using the diagnostic tool, as this provides the hands-on knowledge necessary to pass the Certification Exam. The only required equipment is the Scanner unit itself and the AC power adapter cord. The demonstrations built into the Scanner software will provide the rest of the required material.

Solus Pro Navigation

1. Press the Power button to turn on the unit as indicated by the solid circle.
2. Once booted up, press the Brightness/Contrast Button as indicated by the dashed circle.

3. With the Brightness control highlighted by the blue box, use the Up and Down arrow buttons to adjust the brightness to your preference.

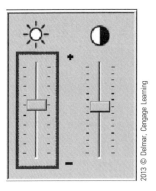

4. When finished, press the Right arrow to highlight the Contrast control and again use the Up and Down arrows to set the contrast to your preference.

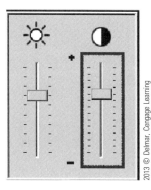

5. When the display is to your liking, press the Brightness/Contrast button (dashed circle on the earlier picture) to close the display properties box.

6. Next, we'll look at our Main Screen options.

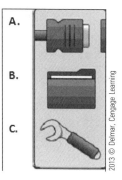

A: Scanner: The Scanner drawer is where all the vehicle communication software is located, including any optional software, such as European. This is where the technician will be able to access all of the back-door diagnostic information, such as trouble codes and data streams. The scanner and its functions are discussed in greater detail in its specific chapter.

B: Data Management: This is the drawer that allows users to access and manipulate all of the stored data files saved on the internal top CF card or any USB storage devices. Screenshots, snapshots, and movies from the scanner can be moved, copied, deleted, or have file names and other information edited using the USB keyboard option via this interface. Data Management is discussed in greater detail in its specific chapter.

C: Utilities: This menu allows technicians to customize the tool to meet their specific needs so that the tool's efficiency is maximized for the owner. Many of the customizable options found in this menu allow the user more flexibility to get the job done faster and ultimately get the most efficient use out of the tool and therefore squeezing every dollars worth of value out of it. Complete understanding of the utilities menu brings the user closer to utilizing 90+ percent of the tool's capabilities. Solus Pro specific utility options are discussed in the next section.

1. Scroll down to highlight the Utilities Menu (wrench icon).

2. Press the Right arrow button to highlight Tool Setup.

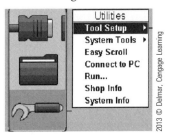

3. Press the Right arrow button again to highlight Power Management and press Yes.

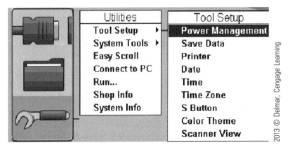

4. Press Yes to open the Backlight drop down menu options. The options available are Timer, On, or Off. The more often the backlight is used, the shorter the amount of time the battery power will last. On and Off are self explanatory. It is advisable to use the Timer feature, as this provides the best balance between visually display and battery life.

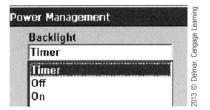

5. Scroll if necessary to highlight Timer and press Yes.
6. Press Down to move to Timer Settings. This allows a finer adjustment between display and battery life.
7. Press Yes to open the Timer Settings drop down menu.

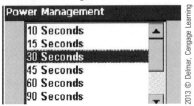

8. Scroll to your desired setting and press Yes. We are going to use 30 seconds for the demonstration.

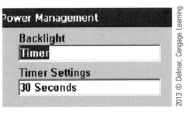

9. After the Power Management features have been set, press No to return to the Tool Setup options.
10. A warning box will appear explaining the relationship between the use of the backlight and battery life.

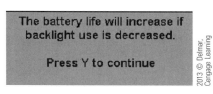

11. Press Yes to continue.
12. Scroll down to highlight Save Data and press Yes.

    Save Data: To fully understand the following options, a little background information is needed. To use the "% after trigger" option, one must realize that in the Scanner a Movie is *always* being recorded, and this information is being saved to the internal memory buffer. The amount of frames or data saved in the buffer will change with software upgrades, but for example purposes let's assume the buffer can hold 2000 frames of data. Once the buffer is full, the first frame of data that entered the buffer is erased and replaced with the 2001st frame of data. This continues so that most recent 2000 frames are always in the memory.

    When attempting to capture a glitch or event in the scanner data, the user can permanently save or archive the information temporarily stored in the memory buffer for later use by either saving a Movie or Snapshot. When a user saves a Movie, all of the data in the buffer is collected and saved to a single file. When the button is pushed to save the Movie, no new data is added to the buffer and only the current frames of data in the buffer are saved.

    A Snapshot is slightly different; it can be made up of both past data stored in the buffer and new data that is continually being collected into the buffer. The moment when the user activates or takes the Snapshot is called the trigger. The "% after trigger" option in the dialogue box allows the user to change how much old information is taken from the buffer and how much new information is collected. The default setting is typically 30% after trigger. Assuming a buffer of 2000 frames of data, this means that once the user triggers a Snapshot, the tool takes 1400 frames (70%) of the data from the buffer, which was automatically recorded before the trigger, and then continues to capture 600 frames (30%) of data after the trigger. These 2000 frames are bundled together and stored as a Snapshot file, with the triggering event located between the beginning and the end of the entire Snapshot file and at the user-specified percentage. If the user wants to see more of the effect of the triggering event, then the percent after trigger should be increased; if the user wants to see more of the cause of the triggering event, then the percent after trigger should be decreased.

    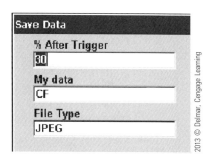

13. Press Yes to select "% After Trigger" and scroll to highlight your desired value. Press Yes again to select it. The Demonstration will use 30%.

14. Scroll down to My Data and press Yes. This chooses the destination of all saved data files. The Solus Pro does have some excess memory on the internal chip that can be used but also supports the Top CF card and USB jump drive storage devices. Again, scroll to your preferred storage device and press Yes. The demonstration will use CF card as the example.

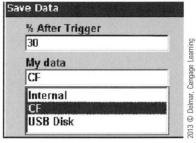

"My Data," allows the user to change where files are saved to. Users have three places to choose from for saving data: the Internal Storage, Top CF card, or a USB storage device. This selection also indicates what the active or viewed data source is when using the Data Management options.

15. Scroll down to File Type and press Yes. This allows you to choose between either a bitmap or jpeg image file when screenshots are taken.

The final option in this dialogue box is "File Type." This option refers to only Screenshots or pictures that can be taken of the screen at anytime. A Screenshot is simply a "what you see is what you get" picture file. Most of the pictures in this book, including

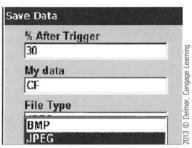

the ones in this section, were created using this feature. The File Type option allows users to choose between two different picture formats. The options are either a bitmap (bmp) or a Joint Photographic Experts Group (jpeg or jpg) format. Both file types are very PC friendly and can be viewed and edited using basic PC software. They differ mainly in their ability to compress the image size to save memory. A newer file type, jpeg, has higher compression and is more easily integrated into newer PC programs but either file type should work fine for the average technician wanting to post something online or e-mail a screenshot to a fellow technician. Refer to the Data Management section for more information.

16. Press No to save your settings and return to the Tool Setup menu.
17. Scroll down to highlight Printer and press Yes.

    Printer: Supported printers include most Hewlett-Packard (HP) printers that use the PCL 3 or higher driver standard, as well as Epson ESC/P2 and Stylus printer drivers that work on most Epson inkjet and Stylus printers. These drivers may allow the Solus Pro to print on other printers, but it works best on the brands just mentioned.

18. Using the same method, select the type of printer you wish to print to and where you are going to connect the printer. The Solus Pro only has one option, the USB port, so this Port box is no longer really used. The older Solus had an option to print wirelessly from an infrared port, and you would switch to that option here.

19. Press No to save your settings and return to the Tool Setup menu.

Solus Pro Navigation **15**

20. A dialogue box will appear, asking if you want to print a test page. Press No to exit.

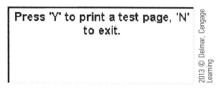

21. Scroll down to highlight Date and press Yes.

22. The date is set on the internal chip, and only the display format can be changed.

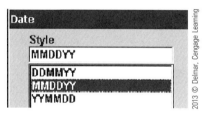

23. Press No to save your settings and return to the Tool Setup menu.
24. Scroll down to highlight Time and press Yes.

25. Set the hour, minutes, and type of clock (12 or 24 hour).

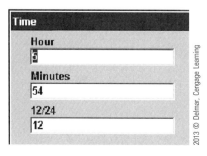

26. Press No to save your settings and return to the Tool Setup menu.
27. Scroll down to highlight Time Zone and press Yes.

28. Scroll and select your specified time zone and choose daylight savings time (DST) if applicable.

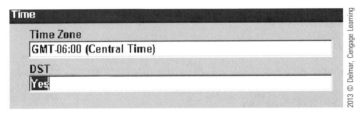

29. Press No to save your settings and return to the Tool Setup menu.
30. Scroll down to highlight the S-Button and press Yes.

Here the user can assign different functions to the S-button. Whatever option the user selects in this screen will then be activated from this point forward every time the S-Button is pressed. When used efficiently, this can be a big time saver for the technician. Here is an outline of what the S-Button can be programmed to do:

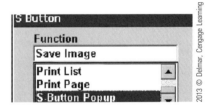

Save Image: This will make the S-Button function like a camera—a picture of the screen will be taken each time the button is pushed. The file will either be saved as a bitmap or jpeg depending on the user's selection under the Save Data options we discussed earlier. It is important to remember that these are just image files and can not be manipulated like Movies or Snapshots. What you see is what you get.

Freeze/Run: When viewing any screen that has the ability to be frozen or paused, this setting will allow this action to be done without navigating to that icon on the screen. Simply toggle the S-Button to freeze or run the data stream.

Save Frame: This option saves all of the frames seen on the graphing data screen, depending on the zoom level, for all of the PIDs being collected. Example: Assume that you are viewing graphing data at a 1x zoom level that allows you to see 500 frames of data across the screen. Saving the frame will then save these 500 frames of data for *all* the PIDs. This means that while reviewing the Frame, you will not be able to scroll left and right to see more data, but you will be able to scroll up and down to view all of the PIDs.

Print List: This is a shortcut to the print command when viewing either the PID list or Text view in the scanner. The full text list of all the PIDs will print when the S-Button is pushed when viewing either screen.

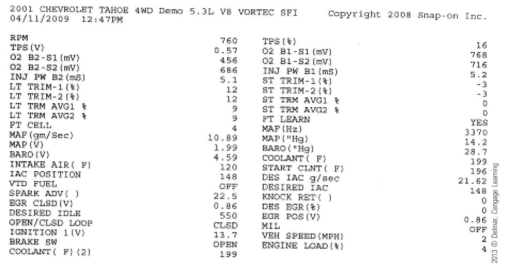

Print List Sample

Print Page: This option will print out exactly what is displayed on the screen in the form of a hard-copy screenshot.

S-Button Popup: This feature allows a quick way to perform any of the available tasks from this menu. When you press the S-Button, the menu appears but has a slightly different function. At this time, one highlights the required action and then presses Yes, at which time the action is performed. The next time you press the S-Button, the menu will again appear allowing you to choose either the same action or a different one by selecting it from the list and then pressing Yes. The benefit of this is that the user does not have to navigate around the screen as much to accomplish these actions—they are instantly brought up in a menu with a push of the S-button.

31. Our demonstration will use Save Image.
32. Press No to save your settings and return to the Tool Setup menu.
33. Scroll down to highlight Color Theme and press Yes.

34. This allows you to choose the background color of the main screen. Classic view will change the icon pictures back to the old style.

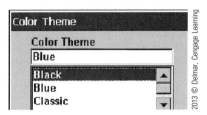

35. Press No to save your settings and return to the Tool Setup menu.
36. Scroll down to highlight Scanner View and press Yes.

    Scanner View: The default view when entering the Scanner is Text view. This is the view most commonly recognized by past users of the MT-2500 and resembles two basic columns of PIDs with their current values. Newer versions of software now put this information into a single column list. Newer scan tools have other options for viewing data such as the PID List and Graphs. This option allows the user to change the default view as the scanner is entered. If you prefer to see the Text first, then leave this option alone, but if you would rather go into the  PID List or Graph view first, then scroll and select to change view. There is also an option to have the last scanner view you used before exiting the scanner to become the default view. This gives multiple users of a piece of equipment the option of having the tool automatically change default view based on what view is being used most by the current technician.

37. The demonstration will use Text view.

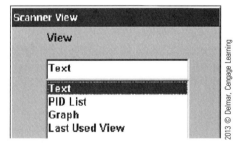

38. Press No to save your settings and return to the Tool Setup menu.
39. Press the Left arrow button to move back to the Utilities menu box.
40. Scroll down to highlight System Tools. As before, scroll back to the right to select each option.

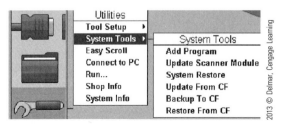

System Tools

The options found under the System Tools menu will most likely be used under the direction of a customer care representative or your sales representative.

Add Program: This allows the user to add optional software such as the European database.

Update Scanner Module: If files need to be added or updated to the scanner in the future, this is where a customer care or help desk person will instruct you to go.

System Restore: This is similar to the System Restore feature you may have used on your home or business PC. It allows the user to reinstall their original software files to reboot the machine after a major software failure. If after a period of time the tool seems to be running slower and slower, this feature will clear out the memory and thus increase the overall speed of the tool.

Update from CF: Use this option after you purchase an upgraded software bundle to install it onto the unit.

Backup to CF: You will need to insert a CF card with at least 1 GB of free space, after which the system information will be backed up to that card in case of a major failure in the future.

Restore from CF: If you have sustained a major failure and previously made a Backup CF card, as described in the last step, then use this option to restore the tool back to the good settings saved on the CF card.

41. Scroll down to highlight Easy Scroll.

    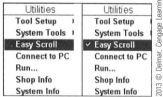

    Easy Scroll: This feature is toggled on and off by pressing the Yes button while highlighted in the Utilities menu. When Easy Scroll is enabled, a check mark will appear by it. The Easy Scroll feature makes it quicker for the user to navigate the menus. It simplifies the selection process by incorporating the Yes and No buttons with the thumb pad. While enabled, the up and down arrows will function as usual, but the right arrow can be used as the Yes button and the left arrow can be used as the No button. The user will not have to move his or her thumb off of the directional pad to navigate and make any selections. This is an intelligent feature that works differently depending on what module or screen the user is in. It will automatically be disabled on any screen in which the right and left arrows are needed to move or navigate around the screen. This does not disable the Easy Scroll feature in the Utilities menu but limits its use to certain screens. Once the user gets accustomed to this feature, it will save navigation time and allow for more efficient use of the tool.

42. It is recommended that during training you test-drive Easy Scroll. It takes a little getting used to but saves a lot of time in the long run.

43. Scroll down to highlight Connect to PC.

    Connect to PC: This command allows the Solus Pro to communicate with a PC. The user will have to physically connect the tool to their PC by using the mini USB port on top of the tool and a standard USB port on their PC. You will need a typical Mini-B USB cable, which probably came with your diagnostic tool but is also available from any office or electronics supply store to accomplish this. This is the same type of cable used to connect digital cameras and some external hard drives to PCs. To perform this task, from the Utilities menu select Connect to PC. A dialogue box will appear and indicate that the system will have to restart. After you press Yes, the tool will reboot in Connect to PC mode and the screen will have a different background. At this point, connect the mini USB end to the tool and the other end to your PC's USB port and then follow the on screen instructions. Any data storage device you have connected to the tool such as the internal storage or the top CF card will now be recognized as external drives on your PC. These can be viewed under the My Computer option when using Windows. From here, you can cut, copy, paste, or drag and drop files between your PC and diagnostic tool. To make the transfer of files even easier, Snap-On has developed a program that can be loaded on your PC. This program is called ShopStream Connect and it is a *free* download available on the Snap-On Diagnostics website located at: http://www1.snapon.com/diagnostics/us and then search "Shopstream Connect." It can also be found under the diagnostic software page located at: http://www1.snapon.com/diagnostics/us/Software.

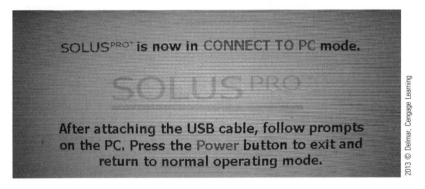

44. Scroll down to highlight Run....

    Run...: This feature is very similar to the Run feature on your home PC. It is used to run special application from CF cards. This is something you will be doing under the instruction of your sales or customer care representative.

45. Scroll down to highlight Shop Info.

Shop Info: Selecting Shop Info will bring up a dialogue box that enables users to enter their business information using a USB keyboard. After you are finished inputting your information, follow the directions at the bottom of the screen, which indicate to either press Esc on the keyboard or No if the keyboard is not connected. That will bring up another dialogue box giving the option of turning on or off the print header. If you press Yes, every page you print will include the information you just entered as a header on the sheet. By pressing No, you *do save* the information you just entered, but it will not be printed as a header. The information, once entered, is saved either way, but the option is whether you want it to be printed or not.

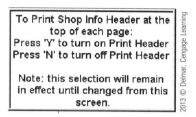

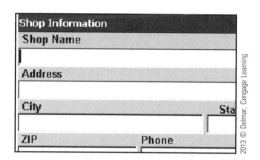

Example of the "Shop Info Header" when printed on a PID list

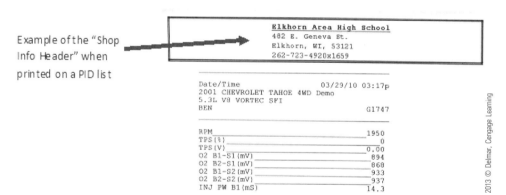

46. Scroll down to highlight System Info.

System Info: This option allows you to see the configuration information of your diagnostic equipment. This is normally used while working with the Technical Help line. Remember to scroll to the right (double arrow in the lower right corner) to view all of the information, as it may be necessary for you to find this information to help Snap-on

22 Chapter 2 Solus Pro

help you. It is a good idea to print all of these screens out and keep them with your user manuals for future reference and security reasons. If the tool is ever stolen, you can report the stolen serial number to customer care, and they will flag the number so if it ever shows up for service or they get a call for some assistance, they can track the tool and possibly locate it for you.

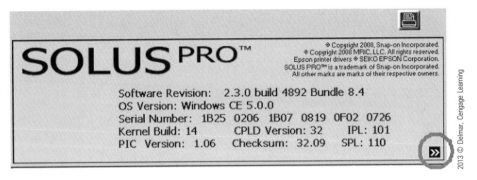

47. Press No to exit System Info.
48. Look at the lower right corner of the screen. You should see some information including Date, Time, and an icon. The icon will vary depending on your power connection.

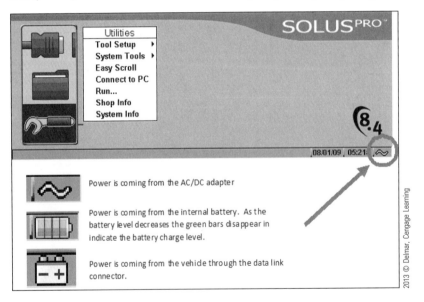

Solus Pro Summary

The Solus Pro is a dedicated scan tool with the capability of displaying 4 PID graphs at one time, pulling diagnostic trouble codes, gathering Global OBDII data, performing Functional Tests, and providing access to the Fast-Track Troubleshooter software. Because the Solus Pro is run from a Windows-based operating system, we can get PC-like functions from the scan tool, including the ability to use a USB keyboard to input data, a USB storage device (jump drive) to expand storage space, and the ability to connect it directly to a PC to cut, copy, paste, or drag and drop files between the PC and the diagnostic tool. The rechargeable battery and external charging stand offer an economical way to sustain power to the tool when it is not connected to a vehicle. Time saving features such as the S-Button and Easy Scroll are simple ways to improve diagnostic productivity and allow more focus to be placed on the vehicle and less on the tool. Likewise, using the Utilities Menu to customize the tool to fit specific needs and personal preferences can translate into more time being spent diagnosing the vehicle and less time figuring out or manipulating the diagnostic tool.

Review Questions

1. Technician A states that the Solus Pro is run on Windows CE. Technician B states that the Solus Pro has bi-directional or Functional Test capabilities. Who is correct?
    a. Tech A only
    b. Tech B only
    c. Both Tech A and Tech B
    d. Neither Tech A nor Tech B

2. Technician A states that the Solus Pro has two CF card slots on the top. Technician B states that you can access the Fast-Track Troubleshooter database through the Solus Pro. Who is correct?
    a. Tech A only
    b. Tech B only
    c. Both Tech A and Tech B
    d. Neither Tech A nor Tech B

3. Technician A states that the mini USB port is for connecting to printers. Technician B states that the large USB port is for transferring files to and from a PC. Who is correct?
    a. Tech A only
    b. Tech B only
    c. Both Tech A and Tech B
    d. Neither Tech A nor Tech B

4. Technician A states that it is all right to "hot swap" the top CF card while the tool is on. Technician B states that when powered by the AC/DC adapter the Solus Pro will charge the internal battery. Who is Correct?
    a. Tech A only
    b. Tech B only
    c. Both Tech A and Tech B
    d. Neither Tech A nor Tech B

24 Chapter 2 Solus Pro

5. Two technicians are discussing the screenshot below. Technician A states that this screen is accessed through the System Tools menu. Technician B states that pressing No now will reset the backlight to the "off" setting. Who is correct?

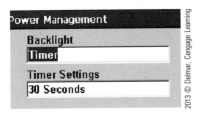

   a. Tech A only
   b. Tech B only
   c. Both Tech A and Tech B
   d. Neither Tech A nor Tech B

6. Two technicians are discussing the screenshot below. Technician A states that pressing Yes now will highlight the "My Data" box. Technician B states that any file saved while using the tool will be stored on the top CF card. Who is correct?

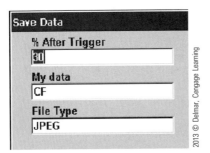

   a. Tech A only
   b. Tech B only
   c. Both Tech A and Tech B
   d. Neither Tech A nor Tech B

7. Technician A states that the date does *not* come set from the factory, so it has to be changed in the Utilities menu. Technician B states that the time zone can be changed through the Utilities menu. Who is correct?

   a. Tech A only
   b. Tech B only
   c. Both Tech A and Tech B
   d. Neither Tech A nor Tech B

8. Technician A states that to perform a System Restore you would have to navigate to the System Tools menu. Technician B states that, once connected to a computer, the Solus Pro will automatically connect and be able to transfer files. Who is correct?

   a. Tech A only
   b. Tech B only
   c. Both Tech A and Tech B
   d. Neither Tech A nor Tech B

9. Two technicians are discussing the screenshot below. Technician A states that Easy Scroll is enabled. Technician B states that, when navigating a screen that requires movement to the left or right, Easy Scroll automatically turns off. Who is correct?

   a. Tech A only
   b. Tech B only
   c. Both Tech A and Tech B
   d. Neither Tech A nor Tech B

10. Technician A states that filling in the Shop Info requires the use of a USB keyboard. Technician B states that pressing No to turn off the print header also erases the information from the Shop Info fields. Who is correct?

    a. Tech A only
    b. Tech B only
    c. Both Tech A and Tech B
    d. Neither Tech A nor Tech B

11. Two technicians are discussing the screenshot below. Technician A states that pressing Yes right now will allow you to view vehicle system information. Technician B states that System Info is sometimes used by customer care to help diagnose problems. Who is correct?

a. Tech A only
b. Tech B only
c. Both Tech A and Tech B
d. Neither Tech A nor Tech B

12. Two technicians are discussing the screenshot below. Technician A states that pressing Yes now will run the scanner program and allow for vehicle identification. Technician B states that this Solus Pro is being powered externally by the vehicle. Who is correct?

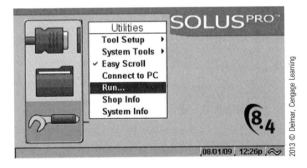

a. Tech A only
b. Tech B only
c. Both Tech A and Tech B
d. Neither Tech A nor Tech B

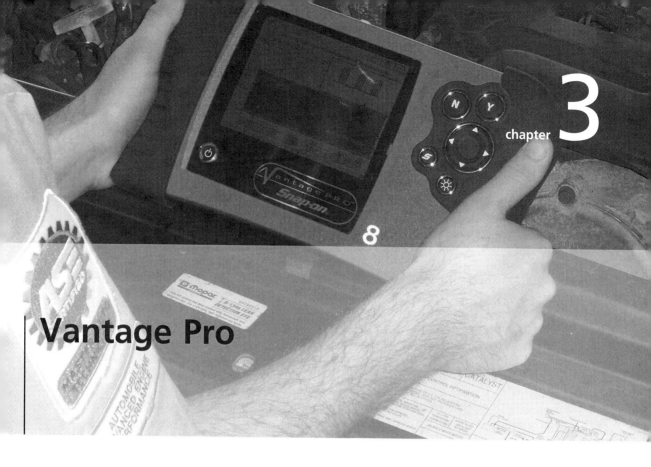

chapter 3

Vantage Pro

Upon completion of the Vantage Pro Platform module, you will be able to:

- Identify the various control buttons and explain their functions
- Locate the different connector ports and explain how each are used
- Locate the battery compartment and explain the different power sources capable of running the Vantage Pro and how to maximize battery life
- Locate and explain the use of the CF card memory slots
- Explain the various stand positions and functions
- Identify the main menu options
- Explain how to incorporate S-Button functions to save time
- Customize the Vantage Pro to your specific shop environment and personal preferences to maximize efficiency
- Respond to immediate questions using the Tool Help function
- Locate pertinent system information if needed
- Demonstrate how to connect the Vantage Pro to a PC

Vantage Pro Overview

The original Vantage meter has served the automotive industry for well over a decade. In 2005, it was upgraded to the Vantage Pro. The Vantage Pro is a 2 Channel lab scope that also has digital and graphing meter capabilities. With additional accessories such as the low amps probe, pressure transducers, and ignition adapters, there is no diagnostic situation that the Vantage Pro cannot help with. The Vantage Pro runs Windows CE and has identical software when compared to the MODIS lab scope and meter, but it comes in a more compact unit. The Vantage Pro has a 5.7" color display that can be read in bright sunlight, as well as a fast and reliable ARM processor. The Vantage Pro's real benefit to shop productivity is the Component Test Meter (CTM). Currently, the CTM has well over 45,000 vehicle configurations, and that equates to over 2 million individual component tests stretching back to 1979. The CTM instantly gives the technician the required VIN-specific information to test a component, including component locations, wiring diagrams, connector views, and a library of known good waveforms and some common faults to look out for. Step-by-step instructions—from component identification, to connecting the Vantage Pro, to viewing the waveform on the screen—are all given to the technician at the times when they are needed most. The Component Test Meter (CTM) is covered in greater detail later in this book, in its own specific chapter. Being Windows based, the Vantage Pro is friendly to use in a PC environment. The Vantage Pro can be connected directly to a computer, so a file scan be cut, copied, pasted, or clicked and dragged to and from the tool to the PC. The larger USB port supports printing, a keyboard, and a USB storage device or jump drive. Using the free ShopStream Connect download, data can be easily saved, archived, or e-mailed to a colleague for a second opinion. The data that is captured with the Vantage Pro lab scope is sure to help find the problem with its fast 8 mega samples per second. With the Vantage Pro, you do not need to guess and end up installing unneeded parts on the vehicle. Use the tool and the information found on it to test suspected components before paying for and installing new ones.

Vantage Pro Platform

The back of the Vantage Pro houses two key features. The first is a 10 amp mini automotive fuse that protects the tool from overload when testing amperage in series with the tool's shunt amp measurement capability, and is readily accessible, so changing the fuse does not require tool disassembly. **See Figure 3-1.** The second feature is the combination wire stand and hanging hook. When popped out at an angle, the stand allows the Vantage Pro to sit upright on a bench top, but if flipped to the top of the tool, the hook will allow it to be hung from above. See **Figure 3-2.**

Vantage Pro Platform 29

The top of the Vantage Pro contains the following connectors:

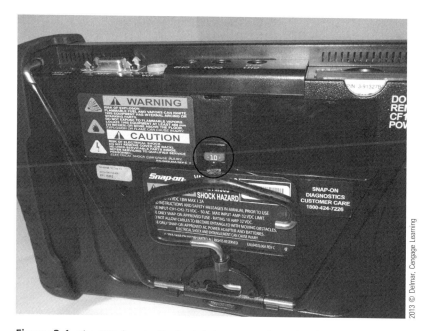

Figure 3-1 The 10A fuse in the back is to protect the unit when using the shunt Amp feature.

Figure 3-2 Use the multi-position stand and hanger to position the tool for comfortable and hands free viewing. It can be propped up on a bench (45 degree), slung from a steering wheel (90 degree), or hung overhead (180 degree).

30 Chapter 3 Vantage Pro

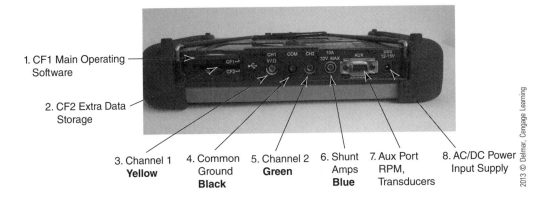

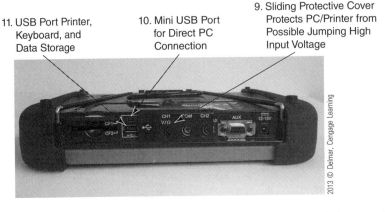

1. **CF 1:** Compact Flash (CF) card one is the main "hard drive" of the tool. CF1 stores the operating software. There is some space left over on it that can be used for general storage, but it is important to *never* pull the CF1 out of the slot with the tool turned on, as damage could result.

2. **CF2:** CF2 is open for additional storage and is used to add additional programs or software upgrades. CF2 can be removed while the Vantage Pro is turned on. This is commonly called "hot swapping" and will not damage the tool or card.

3. **Channel 1:** This is the first channel of the lab scope and is color coded yellow.

4. **Common Ground:** This is the common ground for both channels and is color coded black. Only the single flat-topped connector should be placed in here with the double-ended, piggy-back "flying" ground hanging off the side. It is to the flying ground that all other grounds should be attached. See the lab scope section for more information.

5. **Channel 2:** This is the second channel for the lab scope and is color coded green.
6. **Shunt Amps:** This is the fused testing port in which we can hook the tool into series with a circuit and test the amperage, similar to what is done with a standard digital multimeter. Be sure to keep the amperage under 10, so as not to blow the fuse. This is different from using the Low Amps probe, which measures amperage indirectly while connected in a circuit and is connected to the Vantage Pro via the ground and either channel one or two. See the lab scope section for more information.
7. **Aux Port:** Just as on the MODIS, we can attach the inductive RPM pickup and Split Lead Adapter for pressure transducers to the Aux port.
8. **AC/DC Power Input:** This is where we can connect the AC/DC power adapter to power the Vantage Pro. This will *not* charge the battery.
9. **Protective Sliding Cover:** This sliding cover protects the USB ports when the lab scope channels are open. In order to uncover the USB ports, the test leads must be disconnected. The two test leads and USB cables cannot be connected at the same time. This helps ensure that no high voltage coming in on a test lead could jump to the USB ports, which are commonly connected to a PC or printer. This potential high voltage could damage the PC and similar equipment.
10. **Mini USB Port:** This port is used to connect the Vantage Pro directly to a personal computer (PC). It is very similar to many other computer peripherals, such as digital cameras. Once properly connected, the Vantage Pro will show up as a device under the My Computer icon of your PC, and then files can be cut, copied, pasted, clicked, and dragged to and from the computer or scan tool just as it were part of your computer system. The free ShopStream Connect software download makes this process even easier. See the Data management section for more details.
11. **USB Port:** This USB port allows the Vantage Pro to be connected to many PC-type accessories, such as a printer, keyboard, or mouse. If you are using software version 7.2 or later, you may also connect a USB storage device (jump drive) for even more storage capacity. The keyboard is very useful in renaming saved files. With a mouse plugged in, a cursor will appear on the screen but has little function.

The right side hand grip is removable to allow access to the battery. **See Figure 3-3.** This is one of the very important upgrades to come out with the Vantage Pro. The Vantage Pro has a rechargeable Nickel-Metal Hydride battery similar to those found in many cordless tools and is the same one found in the Solus Pro. It has a much longer run time, almost five hours, and reduces costs with the recharging capability. The Vantage Pro unit cannot charge the battery itself. The unit does come with an external charging stand and uses the same AC adapter cord to power the charger as it uses to power the tool itself. **See Figure 3-4.** Having a second battery on hand in the charger will provide continuous power no matter how long the workday.

32 Chapter 3 Vantage Pro

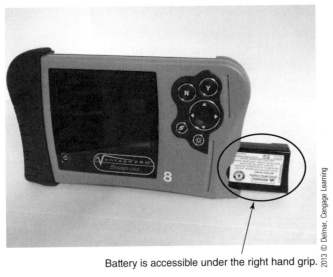

Battery is accessible under the right hand grip.

Figure 3-3 The battery is accessible under the right hand grip.

Figure 3-4 The Vantage Pro unit will not charge the battery. The external charging stand must be used.

The front of the Vantage Pro houses all of the buttons needed to control and navigate the tool.

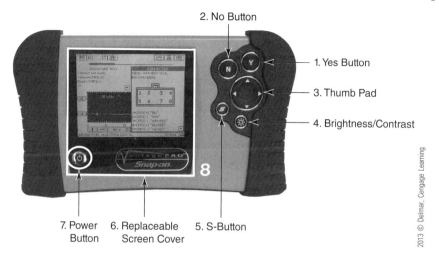

1. **Yes Button:** The Yes button is used for selecting functions on the screen. It is very similar to the Enter button on a PC. You can also think of it as a gas pedal that keeps you moving forward with the tool.

2. **No Button:** The No button is used for confirming selections made on the screen and then backing up in order to perform another task with the tool. It is very similar to the Esc button on a PC that will take you back to where you were or the brake pedal of a car that stops you before backing up. Remember that pressing No does not cancel selections, it just backs you out of a screen so you can perform different functions, but it still leaves everything the way you changed it.

3. **Thumb Pad:** This is the steering wheel for the scan tool and allows you to move vertically and horizontally around the screen. With the Easy Scroll, discussed later in this section, enabled it can also function as the Yes and No buttons.

4. **Brightness/Contrast Button:** Pressing this button brings up the display properties box and allows the user to quickly adjust the brightness and contrast to fit the current lighting conditions and specific user preferences.

5. **S-Button:** This S-button is a programmable button that you set the function of. It can save screen shots, print PID lists, save movies, and much more. Details on how to program this feature is discussed later in this section.

6. **Replaceable Screen Cover:** Since a working lab scope will get its fair share of abuse the Vantage Pro's screen is protected by a plastic cover than can easily be popped out and replaced using a small standard screwdriver and gently inserting it into the slot on the right side of the screen. This slot is just to the left of the S-button.

7. **Power Button:** Pressing this button will power up the unit. The button is purposely indented into the unit to protect it from being accidently pushed while working with the tool. Pressing the button again will turn the tool off.

Now that we have looked at all of the physical properties and external connections, let's power up the unit and begin to look at the software navigation and tool customization.

REMINDER: This section is written so the reader can follow along using the Vantage Pro if available. It is strongly encouraged that you follow along using the diagnostic tool, as this provides the hands-on knowledge necessary to pass the Certification Exam. The only required equipment is the unit itself and the AC power adapter cord.

Vantage Pro Navigation

1. Press the Power button to turn on the unit.
2. Once booted up, press the Brightness/Contrast Button.
3. With the Brightness control highlighted by the blue box, use the Up and Down arrow buttons to adjust the brightness to your preference.

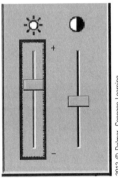

4. When finished, press the Right arrow to highlight the Contrast control and again use the Up and Down arrows to set the contrast to your preference.

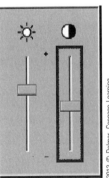

5. When the display is to your liking, press the Brightness/Contrast button again to close the display box.
6. Next, we'll look at our Main Screen options.

After you power up the Vantage Pro, you will see the main screen with the various menu options down the left side. There are many different tools that the Vantage Pro can bring to bear on a diagnostic problem. Just as your toolbox has many drawers of tools, so does the Vantage Pro. Snap-On would like you to think of the Vantage Pro main menu in the same way as you think of your toolbox. And after proper training, you'll be able to choose the best tool to complete the job, go directly to the proper drawer of the Vantage Pro, and efficiently diagnose the problem.

The main menu is the home screen that is shown after the Vantage Pro is finished booting up. There are five main "drawers," or options, that a technician can choose from. See **Figure 3-5**. The following will identify and briefly describe the main menu choices and each will be discussed in greater detail either in this specific Vantage Pro section or a future chapter dedicated to software features that pertain to more than one diagnostic tool.

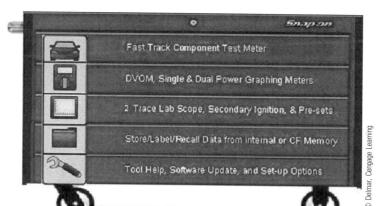

Figure 3-5 As you gain experience on the diagnostic tool each menu option will become as familiar to you as your toolbox drawers. Choosing the right tool (menu option) is the key to an efficient and successful diagnosis.

A. **Component Testing:** This drawer holds the Component Test Meter, which is a very powerful vehicle-specific tool that can display pin connections, component operation information, specific component tests, best test and hook locations, as well as specifications with which to compare your test results. This tool drawer provides just-in-time training for front-door diagnostics. The component test meter is discussed in greater detail in its specific chapter.

B. **Multimeter:** The Multimeter drawer holds both a digital volt/ohm/amp meter and a very powerful dual-channel graphing meter. The graphing meter is preconfigured for many basic readings such as volts, frequency, duty cycle, amps, dwell, vacuum, and pressure. The Multimeter drawer is discussed in greater detail in its specific chapter.

C. **Scope:** The scope drawer holds many different scope and channel configurations as well as a shortcut to some scope presets that come with the software. The two basic scope configurations are the 2-Channel Scope and Ignition Scope both of which can be modified and configured to your diagnostic situation. The lab scope is discussed in greater detail in its specific chapter.

D. **Saved Data:** This is the drawer that allows users to access and manipulate all of the stored data files saved on the internal and top CF cards, as well as any USB storage devices. Screen shots, snapshots, and movies from both the multimeter and lab scope can be moved, copied, deleted, or have the file name and other information edited using the USB keyboard option via this interface. Data Management is discussed in greater detail in its specific chapter.

E. **Utilities:** This menu allows technicians to customize the tool to meet their specific needs to maximize tool efficiency. Many of the customizable options found in this menu allow the user more flexibility to get the job done faster and ultimately get the most efficient use out of the tool and therefore maximizing a shop's profits. Complete understanding of the utilities menu brings the user closer to using 90+ percent of the tool's capabilities. Vantage Pro specific utility options are discussed in the next section.

7. Scroll down to highlight the Utilities Menu (wrench icon).

8. Press the Right arrow button to highlight Tool Setup.

9. Press the Right arrow button again to highlight Units and press Yes.

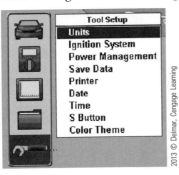

Units: Pressing Yes on Units will bring up the Units dialogue box in which one can switch between U.S. standard units and their metric counterparts. Remember, pressing Yes will open the drop-down menu so all the available options can be scrolled through. The last box to the right is labeled "Display As." This box allows the user to change how the units are displayed in the lab scope. When adjusting voltage in the lab scope using either the Factory Default or Full Scale option, the voltage reading will be that of the total scale. Since there are always 5 divisions on the lab scope screen, the total voltage chosen by the user will be overlaid on these 5 divisions so as if the user picks the 100 volt scale, each division will be worth 20 volts and the lab scope will read 20 volts per division. If the Display As setting is changed to Units/Division, then the voltage adjustment in the lab scope is set to increments of volts per division and not to the total voltage scale. As an example, to pick the 100 volt scale, one has to choose the 20 volts per division option. This gives the user flexibility to change the tool to his or her thinking style. If one usually thinks of units in per division increments, then switch this to the Units/Division setting. If one thinks of units as a total scale, then leave it at either Factory Default or the Full Scale Option.

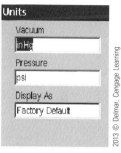

10. After the Units have been set, press No to return to the Tool Setup menu.
11. Scroll down to highlight Ignition System and press Yes.

Ignition System: The ignition system dialogue box allows the user to enter information about the specific ignition system they are working on. This, in turn, allows the Vantage Pro to automatically configure and label some of the waveforms that will later be analyzed by the technician. The more accurate the information that is entered here, the easier the diagnosis will be later. The user must choose the appropriate ignition type. Standard is the traditional distributor system; Wasted Spark is selected when coil packs are used to fire multiple cylinders; Direct is a Coil over Plug (COP) situation; and Other is selected if one is working on a 2-stroke engine or multiple strike engine where the RPM has to be compensated for. Any number of cylinders between 1 and 12 can be chosen, and the corresponding firing orders will then be listed. The last option will change depending on the selection of your ignition type. If Standard or Direct is chosen, then the user has to choose between either the Plug wire or Coil as the trigger. If the Wasted Spark option is chosen, then this last box changes to Cylinder Polarity, and the user must then identify the positively and negatively firing cylinders. See the lab scope section for more information.

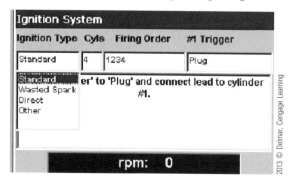

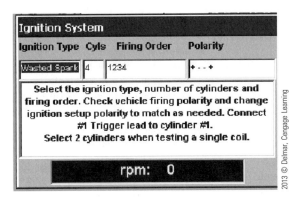

12. Press No to save the Ignition System data and return to the Tool Setup menu.
13. Scroll down to highlight Power Management and press Yes.

14. Press Yes to open the Backlight drop down menu options. To conserve battery power, the backlight will automatically turn off after a set period of time. Here is where you decide what the period of time will be. The less time that the backlight is on, the longer the battery will last.
15. Scroll to your desired setting and press Yes. For demonstration purposes we'll use the 10 minute setting.

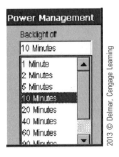

16. The next option is Stand By. To conserve battery power, the Vantage will go into a standby mode that after a set amount of time. For demonstration purposes we'll use the 30 minute setting.
17. The next option is Turn Off. To conserve battery power, the Vantage will turn off after a set amount of time. For demonstration purposes we'll use the 1 hour setting.

18. After the Power Management features have been set, press No to return to the Tool Setup options.

    Backlight Off: This timer begins from the last input activity you gave the tool. After the time period has elapsed, the backlight will turn off to conserve battery power.

    Stand By: The unit will enter Stand By mode after inactivity for the set amount of time. The benefit of Stand By is that all data recordings and snapshots are still being collected and can be saved.

    Turn Off: After entering Stand By mode this timer begins. Once the set time has elapsed, the unit will turn off. Remember to add the Stand By time. In this example, the unit will stay on for one hour and thirty minutes. If you are recording for long periods of time, you should connect to an external power source using the power input adapter. When connected to an external power source, the options under Power Management are ignored and the backlight will remain on. The tool will not go into Stand By or turn off. This would be important if performing an overnight parasitic draw test.

19. Scroll down to highlight Save Data, and press Yes.

    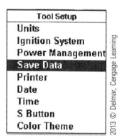

    Save Data: To fully understand the "% after trigger" option, one must realize that in the lab scope a movie is *always* being recorded and this information is being saved to the memory buffer. The amount of frames saved in the buffer will change depending on the scope sweep, but most often, 1048 frames of data will be saved. Once the buffer is full, the first frame of data that entered the buffer is erased and replaced with the 1049th frame of data. This continues so that the most recent 1048 frames are always in the memory.

    When you are attempting to capture a glitch, or event, in the lab scope, you can either save a Movie or create a Snapshot. When a user saves a Movie, all of the data in the buffer is collected and saved to a single file. When the button is pushed to save the Movie, no new data is added to the buffer, and only the current frames of data in the buffer are saved.

    A Snapshot is slightly different; it can be made up of both past data stored in the buffer and new data that is continually being fed into the buffer. The moment when the user activates or takes the Snapshot is called the trigger. The "% after trigger" option in the dialogue box allows the user to change how much old information is taken from the buffer and how much new information is collected. The default setting is typically 30% after trigger. Assuming a buffer of 1048 frames of data, this means that once the user triggers a Snapshot, the tool takes 734 frames (70%) of the data from the buffer that was automatically recorded before the trigger and then continues to capture 314 frames (30%) of data after the trigger. These 1048 frames are bundled together and stored as a Snapshot file, with the triggering event located between the beginning and the end of the entire Snapshot file and at the user-specified percentage. If the user wants to see more of the effect of the triggering event, then the percent after trigger should be increased. If the user wants to see more of the cause of the triggering event, then the percent after trigger should be decreased.

20. Press Yes to select "% After Trigger" and scroll to highlight your desired value. Press Yes again to select it. The Demonstration will use 30%.

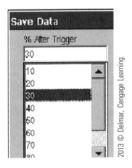

21. Scroll down to My Data and press Yes. This chooses the destination of all saved data files. Again, scroll to your preferred storage device and press Yes. The demonstration will use CF2 card as the example.

    The next option, "My Data," allows the user to change where files are saved to. Users have three places to choose from for saving data. The options are CF1, CF2, or a USB storage device. This selection also indicates what the active or viewed data source is when using the Data Management options.

22. Scroll down to File Type and press Yes. This allows you to choose between either a bitmap or jpeg image file when screen shots are taken.

    The final option in this dialogue box is "File Type." This option refers to screenshots, or pictures that can be taken of the screen at any time. A screenshot is simply a "what you see is what you get" picture file. Most of the pictures in this book, including the ones in this section, were created using this feature. The File Type option allows users to choose between two different picture formats. The options are either a bitmap (bmp) or Joint Photographic Experts Group (jpeg or jpg) format. Both file types are very PC friendly and can be viewed and edited using basic PC software. They differ mainly in their ability to compress the image size to save memory. Jpeg is a newer file type with higher compression and is more easily integrated into newer PC programs, but either file type should work fine for the average technician wanting to post something online or e-mail the screenshot to a fellow technician. Refer to the Data Management section for more information.

23. Press No to save your settings and return to the Tool Setup menu.
24. Scroll down to highlight Printer and press Yes.

    Printer: Supported printers include most Hewlett-Packard (HP) printers that use the PCL 3 or higher driver standard, as well as Epson ESC/P2 and Stylus printer drivers that work on most Epson inkjet and Stylus printers. These drivers may allow the Vantage Pro to print on other printers, but it works best on the brands mentioned.

25. Using the same method, select the type of printer you wish to print to and where you are going to connect the printer. The Vantage Pro only has one option, the USB port, so this Port box is no longer really used.

    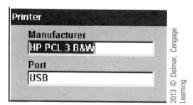

26. Press No to save your settings and return to the Tool Setup menu.
27. A dialogue box will appear asking if you want to print a test page. Press No to exit.

    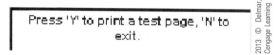

28. Scroll down to highlight Date and press Yes.

    Date: The date option allows the user to enter the correct date into the Vantage Pro. This is helpful when saving files, as the default file name includes a date and time stamp. This is also helpful because, in the upcoming options, the user will have the ability to print sheets with their business information as a print header that includes the date and time. This is discussed in greater detail under the Shop Info portion coming later in this Utilities section. Use the drop down menus to enter the date and the format style or arrangement of month, day, and year.

29. Press No to save your settings and return to the Tool Setup menu.
30. Scroll down to highlight Time and press Yes.

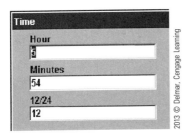

31. Set the hour, minutes, and type of clock (12 or 24 hour).

32. Press No to save your settings and return to the Tool Setup menu.
33. Scroll down to highlight S-Button and press Yes.

   This is where the user can assign different functions to the S-button. Whatever option the user selects in this screen will then be activated from this point forward every time the S-Button is pressed. When used efficiently, this can be a big time saver for the technician. Here is an outline of what the S-Button can be programmed to do.

   Save Image: This will make the S-Button function like a camera, as a picture of the screen will be taken each time the button is pushed. The file will either be saved as a bitmap or jpeg depending on the user's selection under the Save Data options we discussed earlier. It is important to remember that these are just image files and cannot be manipulated like Movies or Snapshots. What you see is what you get.

Freeze/Run: When you are viewing any screen that has the ability to be frozen or paused, this setting will allow the action to be done without navigating to that icon on the screen. Simply toggle the S-Button to freeze or run the data stream.

Save Frame: This option saves the current frame on the lab scope screen. This looks similar to a saved image, but remember that an image is not active, whereas frame is. Functions as such as cursors and zero offset are still active in a frame.

Print Page: This option will print out exactly what is displayed on the screen. It is a printed screen shot.

S-Button Popup: This feature allows a quick way to perform any of the available tasks from this menu. When you press the S-Button, the menu appears but has a slightly different function. At this time, one highlights the required action and then presses Yes, at which time the action is performed. The next time you press the S-Button, the menu will again appear, allowing you to choose either the same action or a different one by selecting it from the list and then pressing Yes. The benefit of this is that the user does not have to navigate around the screen as much to accomplish these actions. Instead, the actions are instantly brought up in a menu with a push of the S-button.

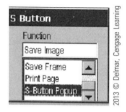

34. Our demonstration will use Save Image.
35. Press No to save your settings and return to the Tool Setup menu.
36. Scroll down to highlight Color Theme and press Yes.

37. This allows you to choose the background color of the main screen.

38. Press No to save your settings and return to the Tool Setup menu.
39. Press the Left arrow button to move back to the Utilities menu box.
40. Scroll down to highlight System Tools, and, as before, scroll back to the right to select each option.

System Tools

The options found under the System Tools menu will most likely be used under the direction of a Customer Care representative or your account manager.

Add Program: This allows the user to add upgraded software.

System Restore: This is similar to the System Restore feature you may have used on your home or business PC. It allows the user to reinstall original software files to reboot the machine after a major software failure.

41. Press the Left arrow button to move back to the Utilities menu box.
42. Scroll down to highlight Tool Help and press Yes.

    Tool Help: All of the user manuals that come with the Vantage Pro on its CD/DVD are also stored in electronic format on the Vantage Pro itself. These manuals are able to be printed out if needed and can be quickly browsed to find an answer without the inconvenience of having to dig out the actual user manual.

43. For practice, let's look up the next item we are going to talk about—the Easy Scroll feature. Notice that at first there is no blue box around the text display area. First, press Down to put a blue border around the lower display box.

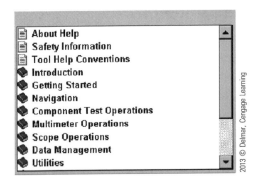

44. Next, continue to scroll down to highlight Utilities.

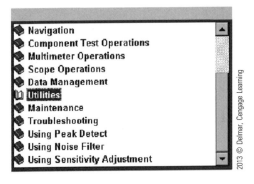

45. Press Yes to open the Utilities book and then scroll down to highlight Easy Scroll.

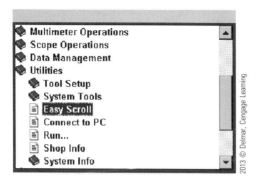

46. With Easy Scroll highlighted, press the Right button to read about Easy Scroll.
47. Use the Up and Down arrows to scroll the text.

48. When finished reading, press No to exit Easy Scroll.

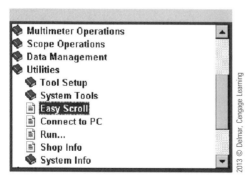

49. Next, scroll back up to Utilities and press Yes to close the book.

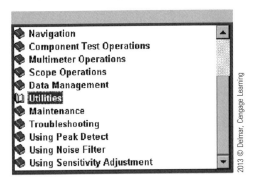

50. Next, press No twice to exit and return to the Utilities menu.

    NOTE: The printer icon is not visible at the top of the screen until an active printer is plugged into the USB port. This information can be printed, but a printer has to be connected to make the icon active and visible.

51. Scroll down to highlight Easy Scroll.

    Easy Scroll: This feature is toggled on and off by pressing the Yes button while highlighted in the Utilities menu. When Easy Scroll is enabled, a check mark will appear next to it. The Easy Scroll feature makes it quicker for the user to navigate the menus, simplifying the selection process by incorporating the Yes and No buttons with the thumb pad. While enabled, the up and down  arrows will function as normal, but the right arrow can be used as the Yes button and the left arrow can be used as the No button. The user will not have to move his or her thumb off of the directional pad to navigate and make any selections. Easy Scroll also allows you to more quickly adjust the lab scope, as all of the settings can be changed by scrolling to the feature button and pressing the Up and Down arrows to change the values. This is an intelligent feature that works differently depending on what module or screen the user is in. It will automatically be disabled on any screen in which the right and left arrows are needed to move or navigate around the screen. This does not disable the Easy Scroll feature in the Utilities menu but limits its use to certain screens. Once the user gets accustomed to this feature, it will save navigation time and allow for more efficient use of the tool.

52. It is recommended that, during training, you test-drive Easy Scroll. It takes a little getting used to, but it saves a lot of time in the long run.

53. Scroll down to highlight Connect to PC.

    Connect to PC: This command allows the Vantage Pro to communicate with a PC. The user will have to physically connect the tool to the PC by using the mini USB port on top of the tool and a standard USB port on their PC. To accomplish this, you will need a typical Mini-B USB cable, which probably came with your diagnostic tool, but is also available from any office or electronics supply store. This is the same type of cable used to connect digital cameras and

some external hard drives to PCs. To perform this task, from the Utilities menu select Connect to PC. At this time, a dialogue box will appear and indicate that the system will have to restart. After pressing Yes, the tool will reboot in Connect to PC mode and the screen will have a different background. At this point, connect the mini USB end to the tool and the other end to your PC's USB port and then follow the on screen instructions. Any data storage device you have connected to the tool, such as the internal storage or the top CF card, will now be recognized as external drives on your PC. These can be viewed under the My Computer option when using Windows. From here, you can cut, copy, paste, or drag and drop files between your PC and diagnostic tool. To make the transfer of files even easier, Snap-On has developed a program that can be loaded on your PC. This program is called ShopStream Connect, and it is a *free* download available on the Snap-On Diagnostics website at http://www1.snapon.com/diagnostics/us/Software.

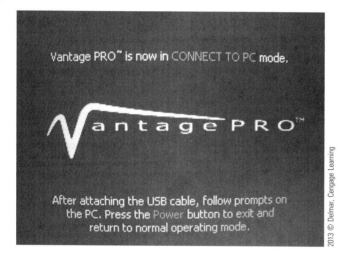

54. Scroll down to highlight Run....

    Run...: This feature is very similar to the Run feature on your home PC. It is used to run special applications from CF cards. This is something you will be doing under the instruction of your sales representative or a Customer Care person.

55. Scroll down to highlight Shop Info.

    Shop Info: Selecting Shop Info will bring up a dialogue box that enables the user to enter their business information using a USB keyboard. After you are finished typing in your information, follow the direction at the bottom of the screen that indicates to either press Esc on the keyboard or No if the keyboard is not connected. That will bring up another dialogue box that gives the option of turning on or off the print header. If you press Yes, every page you print will include the information you just entered as a header on the sheet. By pressing No, you *do save* the information you just entered, but it will not be printed as a header. The information, once entered, is saved either way, but the option is whether you want it to be printed or not.

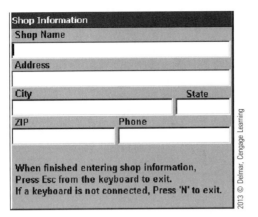

56. Scroll down to highlight System Info.

    System Info: This option allows you to see the configuration information of your diagnostic equipment. This is normally used while working with the Technical Help line. Remember to scroll to the right to view all of the information, as it may be necessary for you to find this information to help Snap-On help you. It is a good idea to print all of these screens out and keep them with your user manuals for future reference and security reasons. If the tool is ever stolen, you can report the stolen serial number to customer care, and they will flag the number, so if it ever shows up for service or they get a call for some assistance, they can track the tool and possibly locate it for you.

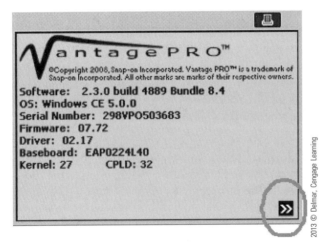

57. Press No to exit System Info.

58. Look at the lower right corner of the screen. You should see some information, including the Time and an icon. The icon will change depending on your power connection.

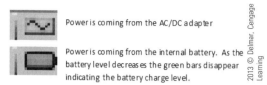

Vantage Pro Summary

The Vantage Pro is a dedicated lab scope and meter capable of displaying two channels simultaneously and with optional accessories can perform pressure, amps, and ignition testing. Even with all of this high-end capability, the tool is very user friendly as the Component Test Meter (CTM) brings together all of the required VIN specific information a technician will need to diagnosis a particular component including step-by-step instructions, component locations, wiring diagrams, connector views, and a library of known good waveforms and some bad examples to look out for. Currently, the CTM has well over 45,000 vehicle configurations. That equates to over 2 million individual component tests stretching back to 1979, so you can be confident that no matter what diagnostic problem comes through the door, you will be able to handle it. Once connected, the data-saving abilities of the Vantage Pro will allow you to easily freeze incoming data for immediate review, save a movie that includes all data frames collected by the memory buffer, or save a Snapshot that will capture both before and after trigger data. Capturing high-speed glitches has never been easier. Since the Vantage Pro is run from a Windows-based operating system, we can get PC-like functions from the diagnostic tool, including the ability to use a USB keyboard to input data, an USB storage device (jump drive) to expand storage space, and the ability to connect it directly to a PC to cut, copy, paste, or drag and drop files between the PC and the diagnostic tool. The rechargeable battery and external charging stand offer an economical way to sustain power to the tool when performing all-day diagnostics. Time saving features such as the S-Button and Easy Scroll are simple ways to improve diagnostic productivity and allow more focus to be placed on the vehicle and less on the tool. Likewise, using the Utilities Menu to customize the tool to fit specific needs and personal preferences can translate into more time being spent diagnosing the vehicle and less time figuring out or manipulating the diagnostic tool.

Review Questions

1. Technician A states that the Vantage Pro is run on Windows CE. Technician B states that the Vantage Pro has VIN specific component test capabilities. Who is correct?
    a. Tech A Only
    b. Tech B Only
    c. Both Tech A and Tech B
    d. Neither Tech A nor Tech B

2. Technician A states that the Vantage Pro has two CF card slots on the top. Technician B states that you can access vehicle-specific wiring connector information through the Vantage Pro. Who is correct?

   a. Tech A only
   b. Tech B only
   c. Both Tech A and Tech B
   d. Neither Tech A nor Tech B

3. Technician A states that the mini USB port is for connecting to printers. Technician B states that the large USB port is for transferring files to and from a PC. Who is correct?

   a. Tech A only
   b. Tech B only
   c. Both Tech A and Tech B
   d. Neither Tech A nor Tech B

4. Technician A states that the protective sliding cover is solely designed to prevent dirt and debris from contaminating the USB ports. Technician B states that to use a low amps probe you should connect it to the common ground and the shunt amps test ports. Who is correct?

   a. Tech A only
   b. Tech B only
   c. Both Tech A and Tech B
   d. Neither Tech A nor Tech B

5. Technician A states that there is a 10 amp fuse protecting the Vantage Pro, and it is easily accessible from the back of the tool. Technician B states that the vehicle data cable is commonly connected to the AUX port. Who is correct?

   a. Tech A only
   b. Tech B only
   c. Both Tech A and Tech B
   d. Neither Tech A nor Tech B

6. Technician A states that it is all right to "hot swap" the top CF2 card while the tool is on. Technician B states that when powered by the AC/DC adapter the Vantage Pro will charge the internal battery. Who is correct?

   a. Tech A only
   b. Tech B only
   c. Both Tech A and Tech B
   d. Neither Tech A nor Tech B

7. Two technicians are discussing the screen shot below. Technician A states that this screen is accessed through the Tool Setup menu. Technician B states that pressing Yes now will save the current settings and close the Power Management window. Who is correct?

    a. Tech A only
    b. Tech B only
    c. Both Tech A and Tech B
    d. Neither Tech A nor Tech B

8. Two technicians are discussing the screen shot below. Technician A states that pressing Yes now will highlight the "My Data" box. Technician B states that any file saved while using the tool will be stored on the top CF2 card. Who is correct?

    a. Tech A only
    b. Tech B only
    c. Both Tech A and Tech B
    d. Neither Tech A nor Tech B

9. Technician A states that the date does *not* come set from the factory, so it has to be changed in the Utilities menu. Technician B states that the time zone can be changed through the Utilities menu. Who is Correct?

    a. Tech A only
    b. Tech B only
    c. Both Tech A and Tech B
    d. Neither Tech A nor Tech B

10. Technician A states that to perform a System Restore you would have to navigate to the System Tools menu. Technician B states that once connected to a computer the Vantage Pro will *not* automatically connect and Connect to PC will have to be selected from the Utilities menu. Who is correct?

    a. Tech A only
    b. Tech B only
    c. Both Tech A and Tech B
    d. Neither Tech A nor Tech B

11. Two technicians are discussing the screen shot below. Technician A states that this is an example of the System Info feature. Technician B states that to view information about Easy Scroll you would have to press the Right arrow button now. Who is correct?

    a. Tech A only
    b. Tech B only
    c. Both Tech A and Tech B
    d. Neither Tech A nor Tech B

52 Chapter 3 Vantage Pro

12. Two technicians are discussing the screen shot below. Technician A states that Easy Scroll is enabled. Technician B states that when navigating a screen that requires movement to the left or right, Easy Scroll automatically turns off even if enabled. Who is correct?

    a. Tech A only
    b. Tech B only
    c. Both Tech A and Tech B
    d. Neither Tech A nor Tech B

13. Technician A states that filling in the Shop Info requires the use of an USB keyboard. Technician B states that pressing No to turn off the print header also erases the information from the Shop Info fields. Who is correct?

    a. Tech A only
    b. Tech B only
    c. Both Tech A and Tech B
    d. Neither Tech A nor Tech B

14. Two technicians are discussing the screen shot below. Technician A states that pressing Yes right now will allow you to view vehicle system information. Technician B states that System Info is sometimes used by Customer Care to help diagnosis problems. Who is correct?

    a. Tech A only
    b. Tech B only
    c. Both Tech A and Tech B
    d. Neither Tech A nor Tech B

15. Two technicians are discussing the screenshot at the right. Technician A states that pressing Yes now will run the component test meter and allow for vehicle identification. Technician B states that this Vantage Pro is being powered internally by the battery. Who is correct?

    a. Tech A only
    b. Tech B only
    c. Both Tech A and Tech B
    d. Neither Tech A nor Tech B

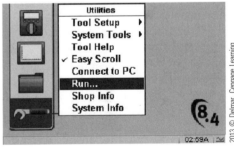

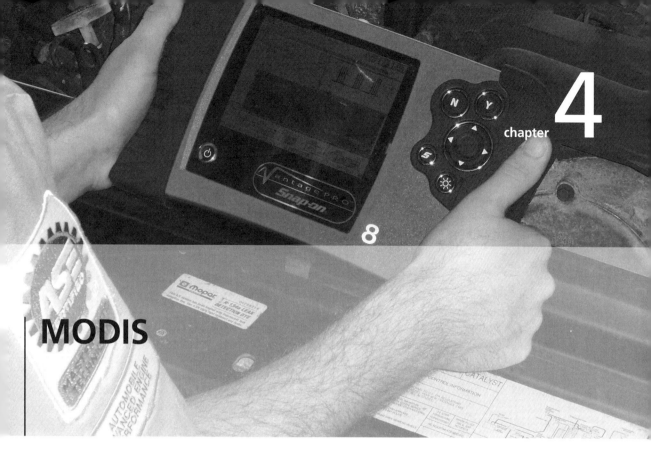

chapter 4

MODIS

MODIS Section Outline:

1. Platform Overview
2. Main Menu
3. Utility Options
4. Summary

Upon completion of the MODIS Platform module, the technician will be able to:

- Identify the various control buttons and explain their functions
- Locate the different connector ports and explain how each are used
- Locate the battery compartment and explain the different power sources capable of running the MODIS
- Locate and explain the use of the CF card memory slot
- Explain the various stand positions and functions
- Explain the reasons behind the MODIS's modular design
- Connect the MODIS to a larger monitor or LCD projector for larger viewing options

MODIS Platform Overview

The Modular Diagnostic Information System, or MODIS, is a rugged yet relatively compact piece of diagnostic equipment considering the vast functionality that is packed in its frame. MODIS has all the scanner capabilities of the Solus Pro and the lab scope and meter capabilities of the Vantage Pro, plus an additional two channels on the lab scope. MODIS is a multifunction scan tool capable of displaying 8 PID graphs at one time, pulling diagnostic trouble codes, gathering Global OBDII data, performing Functional Tests, and providing access to the Fast-Track Troubleshooter software. But it is also a lab scope and meter, capable of displaying four channels simultaneously and, with optional accessories, can perform pressure, amps, and ignition testing while providing access to the Component Test Meter (CTM), which brings together all of the required VIN specific information a technician will need to diagnosis a particular component, including step-by-step instructions, component locations, wiring diagrams, connector views, and a library of known good waveforms. Information is processed on a fast ARM Processor that is running a Windows CE–based platform. This ensures that the user will have fast access to data and easier future updates with a familiar computer operating system. All of this information is provided to the user on a seven-inch color LCD display. The scanner, as well as the lab scope module, is able to be removed from the MODIS unit for easier repair. The MODIS was first introduced around 2002 and has gone through one hardware revision in 2006. It is important to be able to physically recognize the difference between the two different tools, as some of the functions change depending on the hardware revision. Any MODIS produced in 2006 and after has a slightly raised border around the screen, as well as a mini USB port under the left hand grip. Each of these features is shown in the following pictures. It is best to check for the mini USB port as there are some units out there that have the new plastic case and raised border but never received the new internal board that would include the Mini USB port. Bottom line, if the unit does not have the Mini USB port, then it does not have the updated hardware.

Old Style (2002–05), No Raised Border

The top of the MODIS unit has the following connectors:

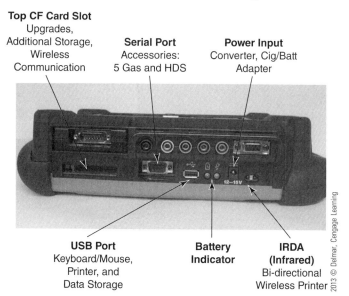

Top CF Card Slot: The top Compact Flash (CF) card slot has multiple purposes. Its main use is for extra storage, but it can also be used for software upgrades and wireless communication with a software revision older than 7.2 or the older MODIS hardware addition. The wireless CF card must be purchased separately and will allow your MODIS to operate on a network. This is not an option for the newer software updates.

Serial Port: This connector is dedicated for accessories such as the 5 gas analyzer and the Heavy Duty Vehicle Software connector (HDS).

USB Port: This USB port allows the MODIS to be connected to many PC type accessories such as a printer, keyboard, mouse, or USB storage device (jump drive) for more storage.

Battery Indicator: These LEDs show the user if the MODIS is being powered from either the internal battery or the AC adapter.

Power Input: The external power sources will plug into this connector. This may be from the vehicle's battery or the AC/DC adapter.

IRDA: Infrared printer port that allows the MODIS to print wirelessly. This option only works with a separate infrared printer adapter that is *not* sold by Snap-On. This adapter, Part number ACT-IR100MU, is sold directly through ACTiSYS www.actisys.com. It has been tested using USB HP DeskJet printers with the PCL3 printer protocol selected in the printer utility menu. This product is *not* supported or sold by Snap-On, Inc. It is much easier and more reliable to print using the USB connection, and it is suggested that this method be used rather than the IRDA.

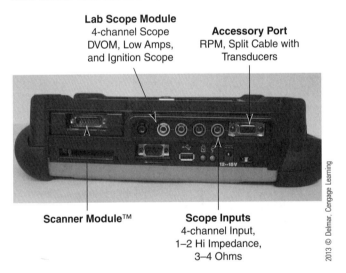

Scanner Module™: This is where the vehicle data link cable is connected to the MODIS. If you push on the release button at the back of the Scanner Module™ and pull up on the module itself, it will pull out of the MODIS. This will allow the Scanner Module™ to be replaced or upgraded without having to replace the entire device.

Lab Scope Module: This module is also referred to as the lab scope plug-in, or LSPI. It has four input channels that include a Digital Volt Ohm Meter (DVOM), an optional Low Amps Probe capabilities, and an Ignition Scope. Color coding helps eliminate confusion and provides visual recognition of what is being tested. The color code starts at the lab scope module, continues to the test leads and ends with the color-coded traces on the screen. Channels one and two are the high impedance channels and also have the fastest sampling rate. Channels three and four must be used to measure resistance, test diodes, and check for continuity.

MODIS Platform Overview **57**

Channel 1 Yellow: Channel one is the fastest sampling channel able to sample down to a 50 microseconds (50μs) or 50 millionths of a second. Your fastest component should always be tested on channel one.

Channel 2 Green: Channel two is the next fastest channel and also has a shielded ground like channel one.

Channel 3 Blue: Channel three can be used as a normal lab scope channel but also has a special designation as the negative input when performing ohms, continuity, or diode testing.

Channel 4 Red: Channel four is similar to channel three except that it is designated as the positive input when performing ohms, continuity, or diode testing.

AUX Port: The accessory port is used with the inductive RPM pick up clamp that collects the vehicles RPM and can also be used to trigger the lab scope and sync the ignition scope. The split lead adapter cable that is used with the pressure transducers also attaches to the accessory port.

Under the left handgrip you will find the following:

Side CF Card Slot
CF Slot for Software Upgrades, and Memory Storage

Mini-USB Connection
Direct Cable Connection for Connection to PC

2013 © Delmar, Cengage Learning

Left Side Panel: The left side rubber hand grip is removable. Gently peel it from the MODIS to uncover two features of interest.

Side CF Card Slot: This card holds the MODIS software and is removable to perform software upgrades. One should *never* remove this CF card with the MODIS turned on. Think of this as the internal hard drive on your PC, and without properly shutting down you can cause damage to the equipment. The MODIS software does not take up all of the available space on the CF card, so there is some extra storage for the user to save information such as movies, snapshots, and screen shots.

Mini-USB Connection: This connection is only found on the 2006 and newer MODISs. It uses a typical Mini-B USB cable. This is the same type of cable used to connect digital cameras and some external hard drives to PCs. This connection is currently used to directly connect the

MODIS to a PC. To perform this function, see the "Connect to PC" option under the Utilities menu covered later in this section.

The bottom of the MODIS has a VGA output and provides access to the battery pack and the internal fuse.

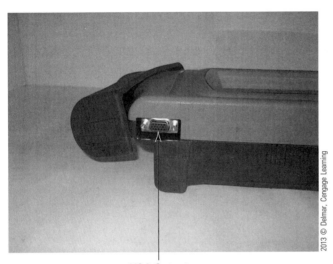

VGA Output
Direct VGA Interface
for Monitors or Projection Units

VGA Output: From the underside of the MODIS, with the left hand guard still removed, you can see the VGA Output. This is a MODIS exclusive and is excellent for drivability technicians who may be using the tool all day, as they can connect it to a larger monitor and not have to strain their eyes in front of a smaller screen. In addition, instructors can now project the MODIS screen using an LCD projector so all students can see the screen and more easily follow along.

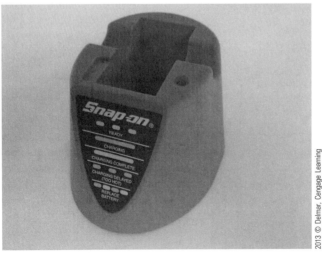

External Battery Charger for the MODIS

MODIS Platform Overview   **59**

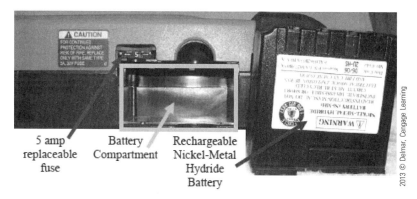

5 amp replaceable fuse — Battery Compartment — Rechargeable Nickel-Metal Hydride Battery

5 Amp Fuse: This protects the internal circuitry from overload. The fuse can only be accessed with the battery removed.

Battery: The battery is a rechargeable Nickel-Metal Hydride and can be charged by the MODIS unit but there also an external battery charger available if a spare battery is wanted to be maintained for a reserve. Finally, an important fact that is often misunderstood by technicians in the field is that the MODIS must be plugged in to an external power source and powered up (ON) for the battery to charge. The internal software controls the battery charger and will not function if the unit is powered down, even if the power adapter is plugged in. *The unit must be on for the battery to charge.*

The right front of the MODIS houses all of the buttons needed to control and navigate the tool.

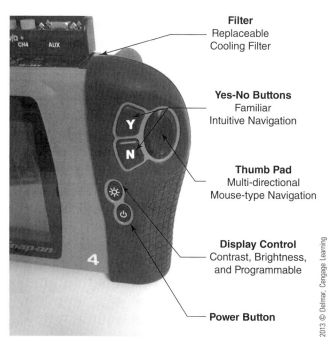

Filter
Replaceable Cooling Filter

Yes-No Buttons
Familiar Intuitive Navigation

Thumb Pad
Multi-directional Mouse-type Navigation

Display Control
Contrast, Brightness, and Programmable

Power Button

Filter: Just as any other type of mobile computer device generates heat, so does the MODIS, especially when charging the battery. The internal fan will start automatically if the internal temperature gets too high. To protect the internal electronics from dirt and other debris, incoming air is filtered. It is important to check the filter periodically to ensure that it is in place and not plugged with grease, dirt, or other debris.

Yes-No Buttons: This is a Snap-On standard when it comes to diagnostic equipment and provides familiar, intuitive navigation. An automotive illustration would be to compare the Yes button to a vehicle gas pedal that keeps you moving forward when pressed and the No button as the brake pedal that stops and backs you out of current selections. A PC example would be to think of the Yes button as the Enter key on a keyboard and the No button as the Esc key. Enter moves you forward, and Esc backs you up. Remember that No does *not* cancel selections, it just backs you out of a screen so you can perform different functions, but it still leaves everything the way you changed it.

Thumb Pad: This thumb pad is a multi-directional control that can be used as a mouse in certain applications. It is normally used in the traditional up, down, left, and right screen navigation, but it is not limited to this.

Display Control: By default, this button functions as the brightness and contrast control. Pushing it will bring up the display properties screen and slide bars to change the brightness and contrast settings. You may have noticed that the Solus, Solus Pro, and Vantage Pro have an S-button and that MODIS does not. This S-button is a programmable button: You set its function. It can save screenshots, print PID lists, save movies, and much more. The software in the MODIS allows us to turn the display control button into the S-button and program it to do what we want it to do. This is a great use of a button whose default setting is not used very often. Rather than perform another hardware revision and modify the MODIS case by adding another button, this is a quick yet effective software fix to a hardware shortfall. The S-button is programmed under the Utilities menu, which will be discussed later in this section.

Power Button: This button powers the unit on and off.

MODIS Software, Navigation, and Tool Setup

Upon completion of the MODIS Software module, you will be able to:

- Identify the main menu options
- Elaborate on the helpfulness of the component test meter
- Explain how to incorporate S-button functions with the brightness/contrast control
- Customize the MODIS to your specific shop environment and personal preferences to maximize efficiency
- Explain how to maximize battery life
- Respond to immediate questions using the Tool Help function
- Locate pertinent system information if needed
- Demonstrate how to connect the MODIS to a PC

REMINDER: This section is written so the reader can follow along using the MODIS scan tool, if available. It is strongly encouraged to follow along using the diagnostic tool as this provides the hands-on knowledge necessary to pass the Certification Exam. The only required equipment is the Scanner unit itself and the AC power adapter cord. The demonstrations built into the Scanner software will provide the rest of the required material.

1. Press the Power button to power up the MODIS unit

After you power up the MODIS you will see the main screen with the various menu options down the left side of the screen. There are many different tools that the MODIS can bring to bear on a diagnostic problem. Just as your toolbox has many drawers of tools so does the MODIS. Snap-On would like you to think of the MODIS main menu as you do your toolbox. After proper training, you'll be able to choose the best tool to complete the job, go directly to the proper drawer of the MODIS, and efficiently diagnose the problem.

MODIS Main Menu

The main menu is the home screen that is shown after the MODIS is finished booting up. There are seven main "drawers" or options that a technician can choose from. The following will identify and briefly describe the main menu choices, and each will be discussed in greater detail, either in this specific MODIS section or in a future chapter dedicated to software features that pertain to more than one diagnostic tool. Depending on your software version or color theme setting, which we will cover later, your main menu will look similar to one of the following screenshots. Newer software is defaulted to a more graphical interface, while the classic menu features both graphics and word label identification.

MODIS Menu Formats

Classic Gold (Word & Graphics)   Newer Version (Graphics Only)

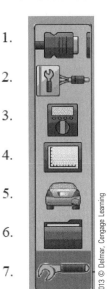

1.
2.
3.
4.
5.
6.
7.

1. **Scanner:** The Scanner drawer is where all the vehicle communication software is located, including any optional software such as European and Heavy Duty Truck. This is where the technician will be able to access all of the back door diagnostic information, such as trouble codes and data streams. The scanner is discussed in greater detail in its specific chapter.

2. **Info/Component Testing:** The Info drawer holds the Component Test Meter. This is a very powerful vehicle specific tool that can display pin connections, component operation information, specific component tests, best test and connection locations, as well as specifications to compare your test results. This tool drawer provides just-in-time training for front-door diagnostics. The component test meter is discussed in greater detail in its specific chapter. The other option in this drawer is Online Info, which is a shortcut to online service information that your shop has a subscription to, such as Shop Key. This option will only work assuming you have the wireless CF card option properly configured and connected to the Internet. This isn't an option for the newer MODIS updates.

3. **Multimeter:** The Multimeter drawer holds both a digital volt/ohm/amp meter and a very powerful dual channel graphing meter. The graphing meter is preconfigured for many basic readings such as volts, frequency, duty cycle, amps, dwell, vacuum, and pressure. The Multimeter drawer is discussed in greater detail in its specific chapter.

4. **Scope:** The scope drawer holds many different scope and channel configurations, as well as a shortcut to some scope presets that come with the software. The two basic scope configurations are the 4-Channel Scope and Ignition Scope, both of which can be modified and configured to your diagnostic situation. The lab scope is discussed in greater detail in its specific chapter.

5. **Gases:** This drawer holds the interface for the optional Flexible Gas Analyzer. This option allows the technician to graph out and record five gas values as the vehicle is running. The basic digital/graph interface, any needed zero/calibration options, and access to presets are found in this drawer.

6. **Saved Data:** This is the drawer that allows users to access and manipulate all of the stored data files saved on the internal and top CF cards, as well as any USB storage devices. Screenshots, snapshots, and movies from both the scanner and lab scope can be moved, copied, deleted, or have the file name and other information edited using the USB keyboard option via this interface. Data Management is discussed in greater detail in its specific chapter.

7. **Utilities:** This menu allows technicians to customize the tool to meet their specific needs in order to maximize tool efficiency. Many of the customizable options found in this menu allow the user more flexibility to get the job done faster and ultimately get the most efficient use out of the tool and thus maximize shop profit. Complete understanding of the utilities menu brings the user closer to using 90+ percent of the tool's capabilities. MODIS specific utility options are discussed in the next section.

MODIS Utility Options

When configuring the MODIS using the utilities menu, it is important to remember some basic tool-browsing procedures. The Yes button will open up options, and that can include drop-down menus. Scroll down to highlight your preferred choice, and then push Yes to confirm that choice. The confusing part for some people is that after you make your choice and confirm it by pushing Yes, you then need to push No to back out and continue using the tool. Pushing No at this point will not deselect your choice or default it back to the original setting. Your confirmed choice remains, but pushing No allows you to exit that menu and perform a different operation.

Tool Setup Options

1. Scroll down to highlight the utilities drawer.
2. Now scroll to the right to highlight Tool Setup and then scroll right one more time to Units. Your screen should now look similar to the following picture. Let's take a detailed look at each of the options under Tool Setup.

```
                Tool Setup
         1.  Units
         2.  Scanner Units
         3.  Ignition System
         4.  Power Management
         5.  Save Data
         6.  Printer
         7.  Date
         8.  Time
         9.  Brightness/Contrast Button
        10.  Color Theme
        11.  Scanner View
        12.  FGA Demo
```

Task: Please scroll to, highlight, and press Yes in response to the following menu options. Pushing the buttons, navigating the menu, and seeing the actual features are important in order to become a power user of the tool and will also help in passing the Certification Exam.

1. **Units:** Pressing Yes on Units will bring up the Units dialogue box, in which one can switch between U.S. standard units and their metric counterparts. Remember, pressing Yes will open the drop-down menu so that all the available options can be scrolled through. The last box is labeled "Display As." This box allows the user to change how the units are displayed in the lab scope. When adjusting voltage in the lab scope using either the Factory Default or Full Scale option, the voltage reading will be that of the total scale. Since there are always 10 divisions on the lab scope screen, the total voltage chosen by the user will be overlaid on these 10 divisions so if the user picks the 100 volt scale, each division will be worth 10 volts and the lab scope will read 10 volts per division. If the Display As setting is changed to Units/Division, then the voltage adjustment in the lab scope is set to increments of volts per division and not the total voltage scale. An example of this would be that to pick the 100 volt scale, one has to choose the 10 volts per division option. This gives the user flexibility to change the tool to their thinking style. If one usually thinks of units in per division increments, then switch this to the Units/Division setting. If one thinks of units as a total scale, then leave it at either Factory Default or the Full Scale Option. These units are specific to the lab scope and flexible gas analyzer. The units in the Scanner have their own menu option that we will look at next.

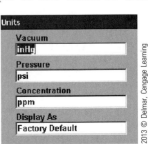

2. **Scanner Units:** Highlighting Scanner Units and pressing Yes will bring up a dialogue box that allows the user to customize the units of measure in the Scanner. The options are either the U.S. standard units or their metric counterparts. It is important to remember that many of the specifications found in today's service information are given in the metric units. Instead of calculating the U.S. equivalent, it is much faster to switch the tool to the metric units and then make the comparison between the spec and actual data.

3. **Ignition System:** The ignition system dialogue box allows the user to enter information about the specific ignition system they are working on, and this in turn allows the MODIS to automatically configure and label some of the waveforms that will later be analyzed by the technician. The more accurate the information that is entered here, the easier the diagnosis will be later. The user must choose their ignition type. Standard is the traditional distributor system, Wasted Spark is when coil packs are used to fire multiple cylinders, Direct is a Coil over Plug (COP) situation, and Other is selected if one is working on a 2-stroke engine or multiple strike engine where the RPM has to be compensated for. Any number of cylinders between 1 and 12 can be chosen and the corresponding firing orders will then be listed. The last option will change depending on the selection of your ignition type. If Standard or Direct is chosen then the user has to choose between either the Plug wire or Coil as the trigger. If the Wasted Spark option is chosen then this last box changes to Cylinder Polarity and the user must then identify the positive and negative firing cylinders. See the lab scope section for more information.

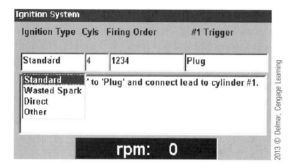

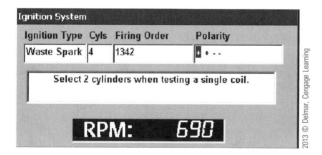

4. **Power Management:** The Power Management dialogue box allows the user to alter the battery charging mode. The two options are either a Quick-4 hour charge or a slower Trickle charge that will take 8 hours. There are advantages and disadvantages of each. Obviously, there is a charging time difference that is self-explanatory but it is highly recommended that one keeps the MODIS set to the 8-hour Trickle charge, as it will prolong the life of the battery pack. The amount of time the battery will power the MODIS is not greatly impacted by the charge time, but the number of charges and discharges that battery itself can endure over its lifespan will be increased if Trickle charge is used on a routine basis. This is also important in hot climates, as the internal battery charger will create a lot of heat and the battery itself will also become hot while charging. The heat in both situations is increased if the unit is set to the 4 hour Quick charge. The last important fact that is often misunderstood by technicians in the field is that the MODIS must be

powered up and ON for the battery to charge. The internal software controls the battery charger and will not function if the unit is powered down, even if the power adapter is plugged in. *The unit must be on for the battery to charge.*

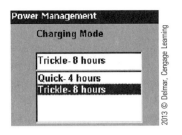

5. **Save Data:** To fully understand the following options, you will need a little background information. To utilize the "% after trigger" option, one must realize that in both the Scanner and lab scope a movie is *always* being recorded, and this information is being saved to the internal memory buffer. The amount of frames or data saved in the buffer will change with software upgrades, but for example purposes let's assume the buffer can hold 2000 frames of data. Once the buffer is full, the first frame of data that entered the buffer is erased and replaced with the 2001st frame of data. This continues so that most recent 2000 frames are always in the memory.

When attempting to capture a glitch, or event, in the scanner data, you can save or archive it for later use by either saving a Movie or Snapshot. When a user saves a Movie, all of the data in the buffer is collected and saved to a single file. When the button is pushed to save the Movie, no new data is added to the buffer and only the current frames of data in the buffer are saved.

A Snapshot is slightly different; it can be made up of both past data stored in the buffer and new data that is continually being collected into the buffer. The moment when the user activates or takes the Snapshot, it is called the trigger. The "% after trigger" option in the dialogue box allows the user to change how much old information is taken from the buffer and how much new information is collected. The default setting is typically 30% after trigger. Assuming a buffer of 2000 frames of data, this means that once the user triggers a Snapshot, the tool takes 1400 frames (70%) of the data from the buffer, which was automatically recorded before the trigger, and then continues to capture 600 frames (30%) of data after the trigger. These 2000 frames are bundled together and stored as a Snapshot file with the triggering event located between the beginning and the end of the entire Snapshot file and at the user specified percentage. If the user wants to see more of the effect of the triggering event, then the percent after trigger should be increased. If the user wants to see more of the cause of the triggering event, then the percent after trigger should be decreased.

1. Press Yes to select "% After Trigger" and scroll to highlight your desired value. Press Yes again to select it. The Demonstration will use 30%.

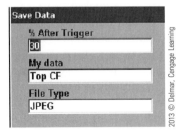

2. Scroll down to My Data and press Yes. This chooses the destination of all saved data files. The MODIS does have some excess memory on the side CF card that can be used, but it also supports the Top CF card and USB jump drive storage devices. Again, scroll to your preferred storage device and press Yes. The demonstration will use top CF card as the example.

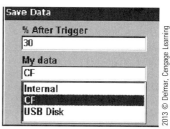

"My Data," allows the user to change where files are saved to. Users have three places to choose from for saving data: the Internal Storage, Top CF card, or a USB storage device. This selection also indicates what the active or viewed data source is when using the Data Management options.

3. Scroll down to File Type and press Yes. This allows you to choose between either a bitmap or jpeg image file when screen shots are taken.

The final option in this dialogue box is "File Type." This option refers to only screenshots, or pictures that can be taken of the screen at anytime. A screenshot is simply a "what you see is what you get" picture file. Most of the pictures in this book, including the ones in this section, were created using this feature. The File Type option allows users to choose between two different picture formats. The options are either a bitmap (bmp) or Joint

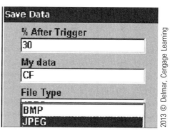

Photographic Experts Group (jpeg or jpg) format. Both file types are very PC friendly and can be viewed and edited using basic PC software. They differ mainly in their ability to compress the image size to save memory. Jpeg is a newer file type with higher compression and is more easily integrated into newer PC programs but either file type should work fine for the average technician wanting to post something online or e-mail the screen shot to a fellow technician. Refer to the Data Management section for more information.

6. **Printer:** The printer dialogue box configures the type of printer and port that the MODIS will use to print documents. The first box allows the user to choose from different printer drivers. The correct printer driver is needed for the MODIS to communicate with different printers. Supported printers include most Hewlett-Packard (HP) printers that use the PCL 3 or higher driver standard, as well as Epson ESC/P2 and Stylus printer drivers that work on most Epson inkjet and Stylus printers. These drivers may allow the MODIS to print on other printers, but it works best on the brands mentioned. The last option box configures the MODIS to print from either the USB or IRDA ports. The USB connection only requires a USB cable and printer, while the IRDA port will require a third party infrared printer adapter along with a USB cable and printer. The adapter is sold by ACTiSYS and is part number ACT-IR100MU. After pressing No to exit, you'll be asked if you want to print a test page. For this demonstration, press No to exit. The USB connection is an easier and more reliable way of printing.

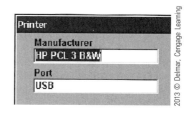

7. **Date:** The date option allows the user to enter the correct date into the MODIS. This is helpful when saving files, as the default file name includes a date and time stamp. This is

also helpful because in the upcoming options the user will have the ability to print sheets with business information as a print header and this will also include the date and time. This is discussed in greater detail under the Shop Info portion coming later in this Utilities section. Use the drop-down menus to enter the date and the format style or arrangement of month, day, and year.

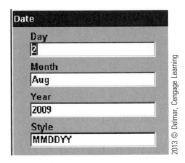

8. **Time:** The time option is very similar to the date regarding set up and benefits. It is part of the default file names as you save data and can also be part of the Shop Info header that can be printed on all sheets. Use the drop down menus to select the correct time, and then choose between the 12 or 24 hour clock option.

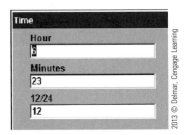

9. **Brightness/Contrast or S-button:** This is where the user can assign different functions to the Brightness/Contrast button on the MODIS. Understand that the MODIS does not have a dedicated S-button like the Solus Pro or Vantage Pro but instead uses software to allow the underused Brightness/Contrast button to become a handier feature for the user. Therefore, it is programmed and used exactly like the S-button on the other pieces of equipment. Whatever option the user selects in this screen will then be activated from this point forward every time the Brightness/Contrast or S-button is pressed. When used efficiently, this can be a big time saver for the technician. Here is an outline of what the Brightness/Contrast or S-button can be programmed to do.

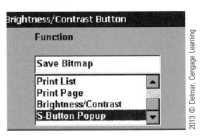

Save Image: This will make the S-button function like a camera, as a picture of the screen will be taken each time the button is pushed. The file will either be saved as a bitmap or jpeg depending on the user's selection under the Save Data options we discussed earlier. It is important to remember that these are just image files and cannot be manipulated like Movies or Snapshots. What you see is what you get.

Freeze/Run: When viewing any screen that has the ability to be frozen or paused, this setting will allow this action to be done without navigating to that icon on the screen. Simply toggle the S-button to freeze or run the data stream.

Save Frame: This option saves all of the frames seen on the graphing data screen, depending on the zoom level, for all of the PIDs being collected. Example: Assume that you are viewing graphing data at a 1x zoom level which allows you to see 500 frames of data across the screen. Saving the frame will then save these 500 frames of data for *all* the PIDs. This means that while reviewing the Frame you will not be able to scroll left and right to see more data but you will be able to scroll up and down to view all of the PIDs.

Print List: This is a shortcut to the print command when viewing either the PID list or Text view in the scanner. The full text list of all the PIDs will print when the S-button is pushed when viewing either screen.

```
2001 CHEVROLET TAHOE 4WD Demo 5.3L V8 VORTEC SFI Copyright 2008 Snap-on Inc.
04/11/2009   12:47PM

RPM                          760      TPS(%)                        16
TPS(V)                      0.57      O2 B1-S1(mV)                 768
O2 B2-S1(mV)                 456      O2 B1-S2(mV)                 716
O2 B2-S2(mV)                 686      INJ PW B1(mS)                5.2
INJ PW B2(mS)                5.1      ST TRIM-1(%)                  -3
LT TRIM-1(%)                  12      ST TRIM-2(%)                  -3
LT TRIM-2(%)                  12      ST TRM AVG1 %                  0
LT TRM AVG1 %                  9      ST TRM AVG2 %                  0
LT TRM AVG2 %                  9      FT LEARN                     YES
FT CELL                        4      MAF(Hz)                     3370
MAF(gm/Sec)                10.89      MAP("Hg)                    14.2
MAP(V)                      1.99      BARO("Hg)                   28.7
BARO(V)                     4.59      COOLANT( F)                  199
INTAKE AIR( F)               120      START CLNT( F)               196
IAC POSITION                 148      DES IAC g/sec              21.62
VTD FUEL                     OFF      DESIRED IAC                  148
SPARK ADV( )                22.5      KNOCK RET( )                   0
EGR CLSD(V)                 0.86      DES EGR(%)                     0
DESIRED IDLE                 550      EGR POS(V)                  0.86
OPEN/CLSD LOOP              CLSD      MIL                          OFF
IGNITION 1(V)               13.7      VEH SPEED(MPH)                 2
BRAKE SW                    OPEN      ENGINE LOAD(%)                 4
COOLANT( F)(2)               199
```

Print List Sample

Brightness/Contrast: This is the default factory setting on the MODIS and can be set back to allow the user to adjust the brightness and contrast each time the button is pushed.

Print Page: This option will print out exactly what is displayed on the screen. It is a printed screen shot.

S-button Popup: This feature allows a quick way to perform any of the available tasks from this menu. By pressing the S-button the drop down menu appears but has a slightly different function. At this time one highlights the required action and then presses Yes, at which time the action is performed. The next time you press the S-button, the drop down menu will again appear allowing you to choose either the same action or a different one by selecting it from the list and then pressing Yes. The benefit of this is the user does not have to navigate around the screen as much to accomplish these actions as they are instantly brought up in a drop down menu with a push of the S-button.

10. **Color Theme:** This option allows the user to choose between different backgrounds. The Blue and White backgrounds both have a more graphical approach to the main menu/screen. One can choose to revert back to the Classic Gold menu that has both graphic icons and a label identifying each of the main menu functions.

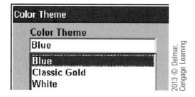

Chapter 4 MODIS

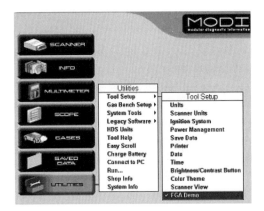 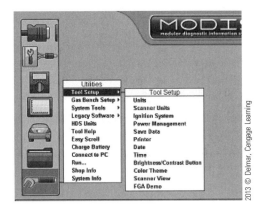

11. **Scanner View:** The default view when entering the Scanner is Text view. This is the view most commonly recognized by past users of the MT-2500 (red brick) and resembles two basic columns of PIDs with their current values. Newer software now has this information in a single column format. Newer scan tools have other options for viewing data such as the PID List and Graphs. This option allows the user to change the default view as the scanner is entered. If you like to see the Text first then leave this option alone, but if you would rather go into the PID List or Graph view first then scroll and select to change your view of choice. There is also an option to have the last scanner view you used before exiting the scanner to become the default view. This gives multiple users of a piece of equipment the option to have the tool automatically change the default view based on what view is being used most by the current technician

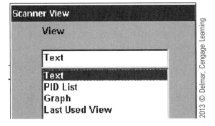

12. **FGA Demo:** This is a toggle on and off command. When selected a check mark will appear to the left and at that point if a user then opens the Flexible Gas Analyzer Drawer, it will open in a Demo mode and allow the user to use the Gas Analyzer and its functions. It is a good way to test drive this accessory and see its capabilities or just become more familiar with exhaust gas analysis.

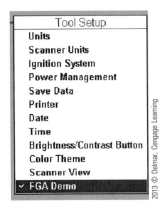

Gas Bench Setup Options

When done completing the necessary Tool Setup options it is then time to scroll back over to the left and continue with the rest of the Utilities options by scrolling down to Gas Bench Setup. The Flexible Gas Analyzer is an optional accessory for the MODIS unit and this sub menu will allow the user to configure that piece of equipment.

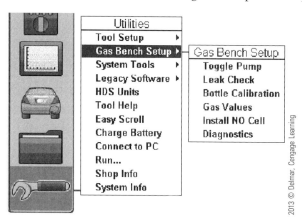

System Tools

The options found under the System Tools menu will most likely be used under the direction of a Customer Care representative or your account manager.

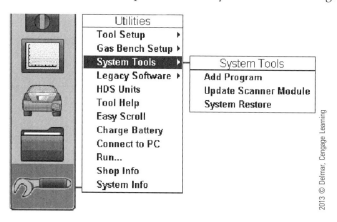

Add Program: This allows the user to add optional software such as European or Heavy Duty Truck applications.

Update Scanner Module: If files need to be added or updated to the scanner in the future this is where a Customer Care or Help Desk person will instruct you to go.

System Restore: This is similar to the System Restore feature you may have used on your home or business PC. It allows the user to reinstall their original software files to reboot the machine after a major software failure.

Legacy Software

Legacy Software will only be used under the direction of a Customer Care representative. It is essentially an older version of software that is installed on the machine to help Snap-On engineers diagnosis problems with their newer software. Assume you have a no communication problem with a vehicle and are using version X.4 software. While talking to Customer Care they may have you connect to the vehicle using Legacy Software to check for communication, if you then have communication with that vehicle, they now know that it is not a hardware issue or a systemic software issue but a glitch in the release you are currently using. This is a very simple example but the main point is to understand that this is only used as a diagnostic procedure for the tool and software. If a technician uses this on a routine basis they are missing out on all the new and updated information and functionality they received, and paid for, with their latest software update.

HDS Units (Heavy Duty Truck): With the optional heavy duty truck adapter and software the user can gather data and diagnosis from medium and heavy duty trucks. This option allows the user to switch between English and Metric units.

Tool Help: All of the printed user manuals that come on the MODIS CD/DVD are also stored in electronic format on the MODIS itself. These manuals are able to be printed out if need be and can be quickly browsed to find a needed answer without the inconvenience of having to dig out the actual user manual. To scroll through all of the information, one will need to highlight the information box and activate it by pressing Yes. You will know the information box is activated by the dark blue border around it.

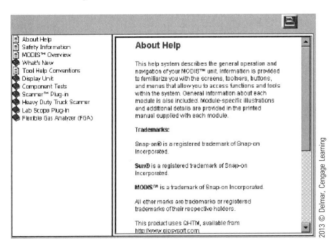

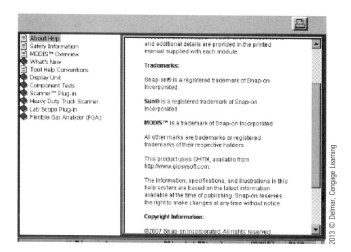

Easy Scroll: This feature is toggled on and off by pressing the Yes button while highlighted in the Utilities menu. When Easy Scroll is enabled a check mark will appear next to it. The Easy Scroll feature makes it quicker for the user to navigate the menus, simplifying the selection process by incorporating the Yes and No buttons with the thumb pad. While enabled, the up and down arrows will function as normal but the right arrow can be used as the Yes button and the left arrow can be used as the No button. The user will not have to move his or her thumb off of the directional pad to navigate and make any selections. Easy Scroll also allows you to more quickly adjust the lab scope as all of the settings can be changed by scrolling to the feature button and pressing the Up and Down arrows to change the values. This is an intelligent feature that works differently depending on what module or screen the user is in. It will automatically be disabled on any screen in which the right and left arrows are needed to move or navigate around the screen. This does not disable the Easy Scroll feature in the Utilities menu but limits its use to certain screens. Once the user gets accustomed to this feature, it will save navigation time and allow for more efficient use of the tool.

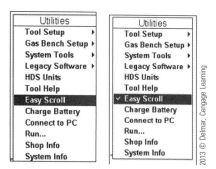

Charge Battery: This is another feature that is toggled on and off by pressing the Yes button while highlighted. When toggled on, a checkmark will appear next to Charge Battery and will override the internal software that controls the automatic battery charger and thus force a charge upon the battery. When you turn on the battery charger you will hear the internal fan start up, to cool the charger and battery. This should only be done if you need to top off the battery but should be left off to allow the automatic charger to function under most conditions. When plugged into an external power source the icon in the lower right corner of the screen will be green, which indicates that the MODIS is either charging or being powered by an external source. Again, with the MODIS on and external power connected, charging will be turned on or off automatically by the software unless the user forces a charge by enabling Charge Battery. If the icon in the lower right is a blue battery then the MODIS is operating from a fully charged battery. The red bar that will appear at the top of the blue battery icon indicates a decreasing battery level. A completely red battery icon indicates a dead battery and a low battery warning message will appear on the screen. It is important to plug the MODIS into an external power source immediately or risk having the unit shut down.

Connect to PC: This command allows the MODIS to communicate with a PC. First the user will have to physically connect the tool to their PC by using the mini USB port on top of the tool and a standard USB port on their PC.

You will need a typical Mini-B USB cable, which probably came with your diagnostic tool, but is also available from any office or electronics supply store to accomplish this. This is the same type of cable used to connect digital cameras and some external hard drives to PC's. To perform this task, from the Utilities menu select Connect to PC. At this time a dialogue box will appear and indicate that the system will have to restart. After pressing Yes, the tool will reboot in Connect to PC mode and the screen will have a different background. If the MODIS does not automatically reboot, you may have to manually power off and reboot the machine yourself. It should then default to restarting in Connect to PC Mode. At this point connect the mini USB end to the tool and the other end to your PC's USB port and then follow the on screen instructions. Any data storage device you have connected to the tool such as the internal storage or the top CF card will now be recognized as eternal drives on your PC. These can be viewed under the My Computer option when using Windows. From here you can cut, copy, paste, or drag and drop files between your PC and diagnostic tool. To make the transfer of files even easier Snap-On has developed a program that can be loaded on your PC. This program is called ShopStream Connect and it is a *free* download available on the Snap-On Diagnostics website at http://www1.snapon.com/diagnostics/us

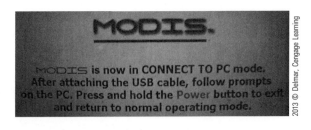

and then search "Shopstream Connect." It can also be found under the diagnostic software page located at: http://www1.snapon.com/diagnostics/us/Software.

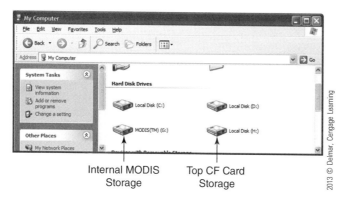

Internal MODIS Storage   Top CF Card Storage

Run...: This feature is very similar to the Run feature on your home PC. It is used to run special application from CF cards. This is something you will be doing under the instruction of your sales representative or a Customer Care person.

Shop Info: Selecting Shop Info will bring up a dialogue box that enables the user to enter their business information using a USB keyboard. After you are finished typing in your information follow the direction at the bottom of the screen that indicates to either press Esc on the keyboard or No if the keyboard is not connected. That will bring up another dialogue box that gives the option of turning on or off the print header. If you press Yes, every page you print will include the information you just entered as a header on the sheet. By pressing No, you *do save* the information you just entered but it will not be printed as a header. The information, once entered, is saved either way, but the option is whether you want it to be printed or not.

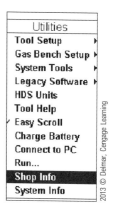

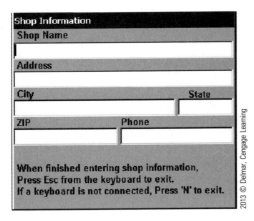

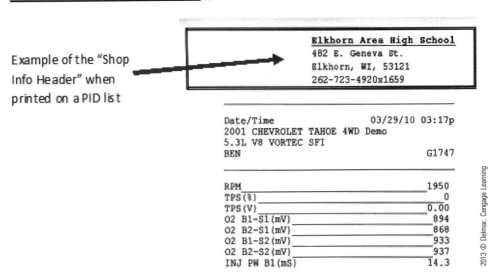

Example of the "Shop Info Header" when printed on a PID list

System Info: This option allows you to see the configuration information of your diagnostic equipment. This is normally used while working with the Technical Help line. Remember to scroll to the right (double arrow in the lower right corner version 9.2 and earlier) or scroll down (version 9.4 and later) to view all of the information as it may be necessary for you to find this information to help Snap-On help you. If the tool is ever stolen, you can report the stolen serial number to customer care, and they will flag the number so if it ever shows up for service or they get a call for some assistance, they can track the tool and possibly locate it for you.

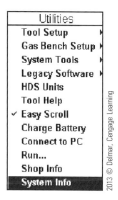

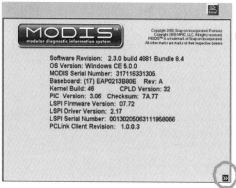

For versions 9.2 and earlier

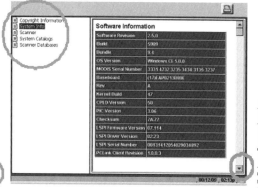

For versions 9.4 and later

MODIS Summary

The MODIS is an all-in-one diagnostic tool that combines the scanner and lab scope in one compact unit. MODIS serves as a multifunction scan tool capable of displaying 8 PID graphs at one time, pulling diagnostic trouble codes, gathering Global OBDII data, performing Functional Tests, and providing access to the Fast-Track Troubleshooter software while also having lab scope and meter capabilities of displaying four channels simultaneously. With optional accessories MODIS can perform pressure, amps, ignition testing, and exhaust gas analysis while providing access to the Component Test Meter (CTM), which brings together all of the required VIN specific information a technician will need to diagnosis a particular component including step-by-step instructions, component locations, wiring diagrams, connector views, and a library of known good waveforms. Since the MODIS is run from a Windows based operating system we can get PC like functions from the tool including the ability to use an USB keyboard to input data, an USB storage device (jump drive) to expand storage space, and the ability to connect it directly to a PC to cut, copy, paste, or drag and drop files between the PC and the diagnostic tool. Time saving features that fit specific needs and personal preferences such as the S-button, Easy Scroll, and Utilities menu options are simple ways to improve diagnostic productivity and allow more focus to be placed on the vehicle and less on the tool.

Review Questions

1. Technician A states that the MODIS is run on Windows CE. Technician B states that the MODIS has bi-directional or Functional Test capabilities. Who is correct?
    a. Tech A Only
    b. Tech B Only
    c. Both Tech A and Tech B
    d. Neither Tech A nor Tech B

2. Technician A states that the MODIS has two CF card slots on the top of the unit. Technician B states that you can access the Fast-Track Troubleshooter database but not the Component Test Meter through the MODIS. Who is correct?
    a. Tech A Only
    b. Tech B Only
    c. Both Tech A and Tech B
    d. Neither Tech A nor Tech B

3. Technician A states that the mini USB port is for connecting to printers. Technician B states that the large USB port is for connecting keyboards and data devices (jump drives). Who is correct?
    a. Tech A Only
    b. Tech B Only
    c. Both Tech A and Tech B
    d. Neither Tech A nor Tech B

4. Technician A states that it is alright to "hot swap" the top CF card while the tool is on. Technician B states that when powered by the AC/DC adapter the MODIS is capable of charging the internal battery. Who is correct?
    a. Tech A Only
    b. Tech B Only
    c. Both Tech A and Tech B
    d. Neither Tech A nor Tech B

5. Technician A states that there is a 5 amp fuse protecting the MODIS and it is easily accessible from the back of the tool. Technician B states that the RPM inductive pick-up is commonly connected to the AUX port. Who is correct?
    a. Tech A Only
    b. Tech B Only
    c. Both Tech A and Tech B
    d. Neither Tech A nor Tech B

Review Questions 79

6. Technician A states that the VGA port can be used to connect the MODIS to a computer monitor or LCD projector. Technician B states that the mini-USB port is located on top of the tool next to the large USB port. Who is correct?

   a. Tech A Only
   b. Tech B Only
   c. Both Tech A and Tech B
   d. Neither Tech A nor Tech B

7. Two technicians are discussing the MODIS below. Technician A states that the connector indicated by the circle is the connector for the vehicle's scanner data cable. Technician B states that all of the lab scope channels sample at the same maximum rate. Who is correct?

   a. Tech A Only
   b. Tech B Only
   c. Both Tech A and Tech B
   d. Neither Tech A nor Tech B

8. Two technicians are discussing the MODIS below. Technician A states that the connector indicated by the circle is the connector for the exhaust gas analyzer. Technician B states that the MODIS has infrared wireless printing capabilities. Who is correct?

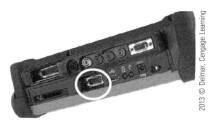

   a. Tech A Only
   b. Tech B Only
   c. Both Tech A and Tech B
   d. Neither Tech A nor Tech B

9. Two technicians are discussing the screen shot below. Technician A states that pressing Yes now will allow you to change lab scope and gas analyzer units of measure. Technician B states that the FGA Demo is currently active and can be used. Who is correct?

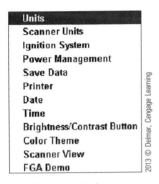

   a. Tech A Only
   b. Tech B Only
   c. Both Tech A and Tech B
   d. Neither Tech A nor Tech B

10. Two technicians are discussing the screen shot below. Technician A states that this screen is accessed through the System Tools menu. Technician B states trickle charging the battery will increase its long term service life. Who is correct?

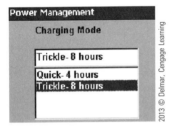

   a. Tech A Only
   b. Tech B Only
   c. Both Tech A and Tech B
   d. Neither Tech A nor Tech B

11. Two technicians are discussing the screen shot below. Technician A states that pressing Yes now will save the current settings and exit out of the Save Data box. Technician B states that any file saved while using the tool will be stored on the top CF card. Who is correct?

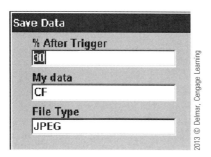

   a. Tech A Only
   b. Tech B Only
   c. Both Tech A and Tech B
   d. Neither Tech A nor Tech B

12. Technician A states that the date comes set from the factory so all that can be changed is the date display format, which is found in the Utilities menu. Technician B states that the time zone can be changed through the Utilities menu. Who is correct?

   a. Tech A Only
   b. Tech B Only
   c. Both Tech A and Tech B
   d. Neither Tech A nor Tech B

13. Two technicians are discussing the screen shot below. Technician A states that pressing Yes now will allow you to adjust the brightness and contrast of the screen. Technician B states that this menu option is under System Tools. Who is correct?

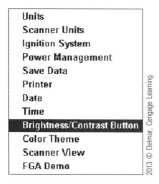

   a. Tech A Only
   b. Tech B Only
   c. Both Tech A and Tech B
   d. Neither Tech A nor Tech B

14. Technician A states that to perform a System Restore you would have to navigate to the System Tools menu. Technician B states that Legacy Software is designed for working on older vehicles. Who is correct?

    a. Tech A Only
    b. Tech B Only
    c. Both Tech A and Tech B
    d. Neither Tech A nor Tech B

15. Technician A states that Tool Help is an electronic collection of all the user manuals that comes with your MODIS. Technician B states that no information can be printed from the Tool Help feature as all the material is copyrighted. Who is correct?

    a. Tech A Only
    b. Tech B Only
    c. Both Tech A and Tech B
    d. Neither Tech A nor Tech B

16. Two technicians are discussing the screen shot at the right. Technician A states that Easy Scroll is enabled. Technician B states that when navigating a screen that requires movement to the left or right Easy Scroll automatically turns off temporally while navigating that screen. Who is correct?

    a. Tech A Only
    b. Tech B Only
    c. Both Tech A and Tech B
    d. Neither Tech A nor Tech B

17. Two technicians are discussing the screen shot below. Technician A states that pressing Yes will override the automatic battery charging software and force a charge on the battery in order to "top it off." Technician B states that the MODIS must be plugged in to the AC/DC adapter and be powered up (on) in order for the battery to charge. Who is correct?

    a. Tech A Only
    b. Tech B Only
    c. Both Tech A and Tech B
    d. Neither Tech A nor Tech B

18. Technician A states that you must use the "Run…" command in order for the MODIS to connect to a PC. Technician B states that the "Run…" command is located under the System Tools menu. Who is correct?

    a. Tech A Only
    b. Tech B Only
    c. Both Tech A and Tech B
    d. Neither Tech A nor Tech B

19. Technician A states that filling in the Shop Info requires the use of an USB keyboard. Technician B states that pressing No to turn off the print header also erases the information from the Shop Info fields. Who is correct?

    a. Tech A Only
    b. Tech B Only
    c. Both Tech A and Tech B
    d. Neither Tech A nor Tech B

20. Two technicians are discussing the screen shot below. Technician A states that pressing Yes right now will allow you to view vehicle system information. Technician B states that System Info is sometimes used by Customer Care to help diagnosis problems. Who is correct?

a. Tech A Only
b. Tech B Only
c. Both Tech A and Tech B
d. Neither Tech A nor Tech B

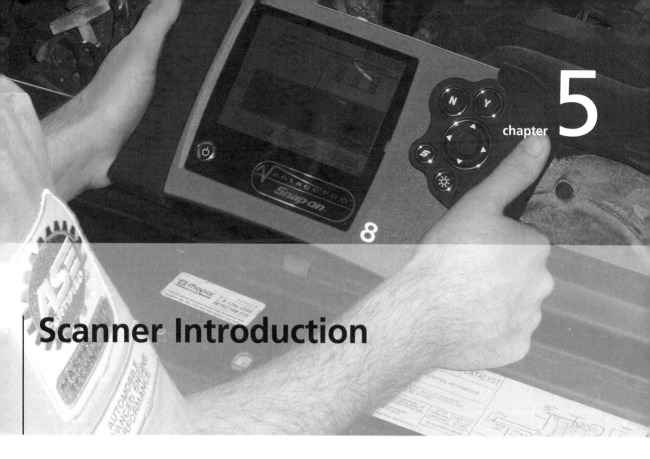

chapter 5

Scanner Introduction

Upon completion of the Scanner Introduction module, you will be able to:

- Explain the advantages and limitations of using a Scanner
- Clarify the difference between front-door and back-door data
- Explain when to use a Scanner versus a different piece of diagnostic equipment
- Identify vehicles through VIN, Global OBDII, and the Previous Vehicles menu options
- Demonstrate how to properly connect a Scanner to both an OBD-I and an OBD-II vehicle

REMINDER: This section is written so the reader can follow along using the Solus Pro or MODIS scan tools, if available. The Scanner software is identical on both of these tools, so all of the procedures will be the same. Most of the screen shots come from a MODIS, but the Solus Pro will be the same except for the screen size, so while the MODIS may show eight graphs, the Solus Pro will only show four. Everything else will function exactly the same. It is strongly encouraged that you follow along using the diagnostic tools, as this provides the hands-on knowledge necessary for passing the Certification Exams. The only required equipment is the Scanner unit itself and the AC power adapter cord. The demonstrations built into the Scanner software will provide the rest of the required material.

Scanner Overview

A scan tool is commonly used to diagnose a vehicle that comes into the shop with a Malfunction Indicator Light (MIL) on. This light is also referred to as the "Check Engine Light," "Service Vehicle Soon (SVS) Light," or "Service Engine Soon Light." **See Figure 5-1.** A Scanner will be able to gather data and trouble codes from the various vehicle control modules. Today's vehicles require most technicians, not just the drivability specialist, to access scan tool data. Brakes, transmission, or collision specialists will need to access scan tool data, perform functional tests, and read trouble codes in their normal work. Scan tools are becoming a necessary tool for working on any vehicle system.

Figure 5-1 MIL Light Configurations

This information is accessed through a Data Link Connector (DLC). All vehicles manufactured after 1996 are OBDII compliant, which means they use a standard 16 pin DLC. **See Figure 5-2.** Of the 16 pins, some are standard between all manufacturers and the remaining pins are left open for use at the manufacturer's discretion.

Figure 5-2a 16 Pins 1996 and Newer Vehicles

PIN	ASSIGNMENTS
1.	Manufacturer's Discretion
2.	Bus positive Line of SAE-J1850
3.	Manufacturer's Discretion
4.	Chassis ground
5.	Signal ground
6.	CAN high (ISO 15765-4 and SAE-J2234)
7.	K line of ISO 9141-2 and ISO 14230-4
8.	Manufacturer's Discretion
9.	Manufacturer's Discretion
10.	Bus negative Line of SAE-J1850
11.	Manufacturer's Discretion
12.	Manufacturer's Discretion
13.	Manufacturer's Discretion
14.	CAN low (ISO 15765-4 and SAE-J2234)
15.	L line of ISO 9141-2 and ISO 14230-4
16.	Battery voltage

Figure 5-2b OBDII Data Link Connector (DLC)

It is important to understand the limitations of a Scanner and what it is intended to do for the technician. A Scanner should be thought of like a compass and not a GPS unit. A vehicle comes into the shop with the MIL on. This indicates a problem somewhere between the front and rear bumper. Using the Scanner as a compass, a technician can begin to diagnose the problem and start to focus on more specific areas or systems for further pinpoint testing. The Scanner, just like an actual compass, only points you in the right direction and rarely gets you to a specific diagnostic destination.

This limitation of the Scanner is because all the information gathered comes directly from the vehicle's computer, which is only going to show you the *effect* of the problem and *not* the *cause*. When connected to the vehicle through the DLC, we are looking at *back-door data* that has been processed by the vehicle's computer. This will show us what the vehicle computer *sees* and the effects the inputs and outputs have on the vehicle's performance. This is why Scanner data is helpful. We can see why the vehicle computer makes the decisions it does, but analyzing these effects is only the first step in the diagnostic procedure. The cause of the problem is what we actually want to find and repair. This cause is found using *front-door data,* gathered by using a digital multimeter or lab scope that reports actual electrical signals and inputs to the vehicle's computer. See **Figure 5-3**. Multimeters and lab scopes are covered in separate sections of this book.

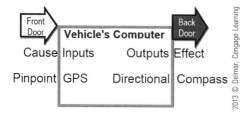

Figure 5-3 Understanding each tool's capabilities and intended purpose is essential to a fast and accurate diagnosis.

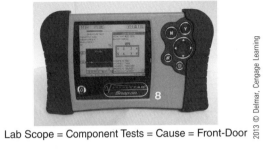

Lab Scope = Component Tests = Cause = Front-Door

Scanner = DLC = Effects = Back-Door

Scanner Scenario

We can use the effects gathered by the Scanner to help narrow down the cause, but the effect will never pinpoint this cause. Additional testing will be needed. Let's look at what this means in a more realistic situation. A customer comes with a MIL light on.

1. To narrow down the problem area, it is quickest and easiest to connect our scan tool to the DLC and gather back-door data so we know what the vehicle computer is "seeing" and reacting to.

2. After connecting our scan tool to the DLC, identifying the vehicle, and retrieving stored trouble codes, we find that a TPS fault is present.

3. Next, after building a custom PID view, we can begin to interpret the effects of the problem as perceived by the vehicle's computer and focusing on the possible problem area indicated by the trouble code.

4. What do you see? At this point, there is a spike in RPM, Mass Air Flow (MAF), and Engine Load (%), but there is no response (0 volts) from the throttle position (TPS) sensor. Because an increase in RPM (someone pressing the gas pedal) should affect the other collected values, including the TPS sensor, we should see this effect on the scan

tool. The vehicle's computer *sees* no effect in this situation so the TPS sensor would be a good place to start looking for our problem.

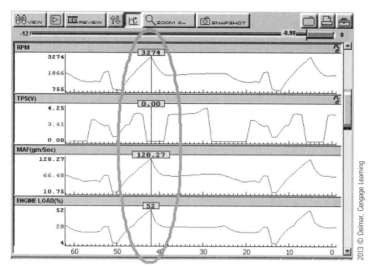

5. Since there is no reading from the TPS sensor, is the sensor the cause of the problem? At this point we don't have enough information to be sure. Remember, a Scanner can not pinpoint a specific cause; it can only show us the effect. Additional testing is needed, so next we would use a lab scope such as the Vantage Pro, MODIS, or Verus to gather *front-door data* directly from the sensor itself. We would need to check for a five-volt reference signal coming to the sensor from the PCM, as well as do a "Sweep" test on the signal from the TPS sensor. After this additional testing, we would be able to determine if the sensor itself was at fault and needed replacement or if the cause was a broken or defective wire or wiring harness that needed repair. The cost difference between these two repairs can be large, so we need to be sure of our diagnosis. This is done using front-door data, but testing every possible component right from the start would take too long, so we always start with back-door data to narrow down the problem quickly and easily.

Systematic Scanner Procedure

As vehicles become more complex and problems harder and more expensive to diagnose, it is important for technicians to use proven procedures in making repairs, and this includes diagnostics. Following a systematic, or routine, procedure each and every time a vehicle is diagnosed helps eliminate oversights and assumptions that can lead to lost time and more expensive repairs. This does not mean that the systematic procedure is the most efficient method each and every time, but rather it means that if implemented and followed over a longer time period, both time and money will be saved. One must trust in the procedure to be right and efficient more often then it is wrong and inefficient. Although some time may have been "lost" at the end of a specific day, the total amount of time saved by finding problems more efficiently by using a systematic procedure will add up by the end of the week, month, and year.

Here is an example of a systematic procedure that can be used every time a Scanner is used. While parts of it may not be intuitive, the rational for each will become clear as you read the rest of the Scanner section. The details of each step will be discussed further in the upcoming parts of the Scanner chapter.

Systematic Scanner Procedure

Always start with a Global OBDII scan first, as there may be trouble codes stored here that are not listed under the VIN specific side of the PCM. This is raw data, not manipulated by the manufacturer, so no substituted values will be present. The federal Environmental Protection Agency (EPA) governs the Global OBDII side of the PCM and whether or not the MIL is illuminated.

Identify and Connect the Vehicle Using Global OBDII Software Database

Step 1: Service $03 Display Trouble Codes

Scan for trouble codes and record the number and description of each. There may be codes here that do not show up in the VIN specific databases.

Step 2: Service $02 Display Freeze Frame Data

Using the trouble codes gathered in the previous step, look at a snapshot of the data stream taken at the time the codes were set. This is very similar to "black box" data on airliners.

Step 3: Service $07 Pending DTC's Detected During Last Drive Cycle

If it is an intermittent problem, this is a good place to start looking for clues. Check here to see if there are any other minor problems that are about to become major problems.

Step 4: Service $01 Display Current Data

This data is important to check, as it is different than the VIN specific data. Substituted values will be discussed in more detail in the Global OBDII section. Don't be fooled by substituted values! Catch them early, viewing current data under Global OBDII.

Identify and Connect the Vehicle using the VIN Specific Software Databases

Step 5: Gather Current and History VIN Specific Trouble Codes

If no codes are present, then look in Troubleshooter for known symptom problems, TSBs, or view current data and start to compare suspected bad PID values against the known good PID values in Fast-Track Data Scan (Normal Values).

Step 6: Use Troubleshooter to Look Up TSBs and Diagnostic Procedures for Codes, Symptoms, or Known Good PID Value Specifications

Step 7: Follow Diagnostic Procedures Outlined in Troubleshooter

This will involve viewing VIN specific data and possibly performing Functional Tests with the Scanner. This may also require the use of a lab scope, multimeter, and other test equipment to verify the problem cause.

Step 8: Make Required Repairs

Step 9: Clear Codes Using the VIN Specific Clear Codes Command or Global OBDII Service $04

Step 10: Verify Repairs Using DTC Status or Global OBDII Service $06

After a test drive of the vehicle, use DTC status under the VIN specific Codes Menu to dial in one the previous codes that should have been repaired. The following information will help you determine whether the repair was successful. Service $06 under Global OBDII can also be useful in verifying repairs if the technician has the necessary test identification information. Each of these procedures is discussed in more detail in their respective sections.

Starting with a Global OBDII scan will save you time in the long run, quickly spotting problems that are masked when viewing the VIN specific data information. Snap-On has made this easier, as some of the Global OBDII information is now available through the VIN specific database. The menu option Generic Functions will allow you access to some Global OBDII data while in the VIN specific database. Now OBDII data can be accessed without having to back all the way out of the VIN specific database and then reidentifying the vehicle under Global OBDII. Remember, Generic Functions equals Global OBDII. The details of each step may be unclear at this point, but as you read the remainder of this section, as well as the other sections in this book, you will better understand the reasons for a systematic procedure like this one.

Vehicle Connection

Connecting a Scanner to a vehicle is a fairly simple process, which is one of the advantages of using a Scanner. The connection process is outlined in the following flowchart. See **Figure 5-4.** Each step of the process will be broken down and explained.

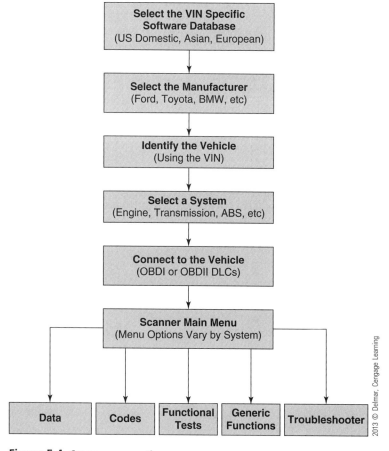

Figure 5-4 Scanner connection process

1. Select the "Vehicle Comm" icon from the Main Screen.

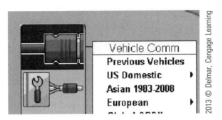

2. Scroll over to the right and highlight "Previous Vehicles" and press Yes. If the vehicle you are connecting to has already been identified previously, this is a quicker way to access the vehicle's information. Previous Vehicles stores the last 25 vehicles that have been identified.

3. Scroll down to the vehicle you are working on and press Yes to automatically identify that vehicle based on previously entered information.

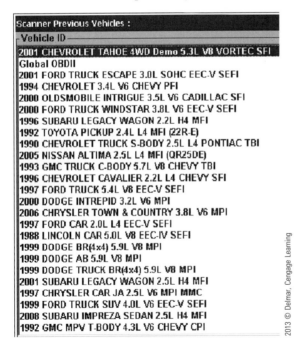

4. **Select the VIN Specific Software Database.** This assumes that we are using a VIN specific software database. However, connecting using Global OBDII is very similar, even quicker and easier, in procedure. The standard software packages come with full access to the U.S. Domestic and Asian databases, while the European database is an extra feature that must be purchased separately. When connecting to an actual vehicle, simply scroll down to the appropriate choice and press Yes. For demonstration purposes, we have a different option, which will be explained shortly. So, for now, do not select a database but continue reading the following steps.

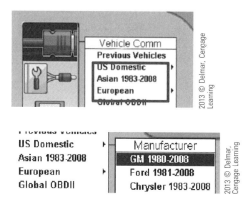

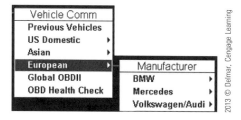

5. **Select the Manufacturer.** After choosing the correct data base the specific manufacturer will have to be chosen. For the U.S. Domestic and European databases, simply scroll to the right (following the arrow) and then scroll up and down to highlight your choice.

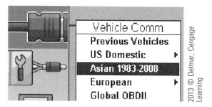

6. **Select the Manufacturer.** The Asian manufacturer menu is a bit different, due to the larger number of manufacturers to choose from. To select an Asian manufacturer, scroll and highlight Asian and press Yes.
7. A new screen will open with a list of all the available Asian manufacturers.

8. Scroll down to the appropriate manfacturer and press Yes to enter that database, and proceed to identify the vehice.

9. **DEMONSTRATION PURPOSES ONLY.** Scroll down and highlight Scanner Demo, and press Yes.

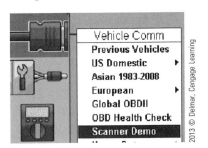

10. **Identify the Vehicle.** When identifying an actual vehicle, use the VIN number to input the requested VIN characters. Change the VIN character by using the Up and Down arrow buttons and select the correct VIN character by pressing Yes. Since this is a demonstration, the VIN characters can not be changed, so continue to press Yes to input the requested VIN characters. Press Yes to select the 10th VIN character.

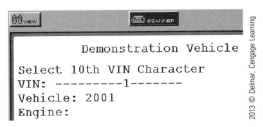

11. Use the Up and Down arrow buttons to select the appropriate Make (Chevrolet) and then press Yes.

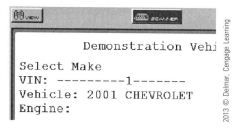

12. Select and confirm the 5th VIN character (K) by pressing Yes.

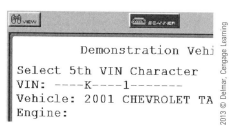

13. Select and confirm the 8th VIN character (T) by pressing Yes.

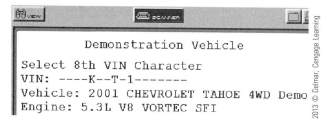

14. Press Yes to confirm the vehicle identification and continue.

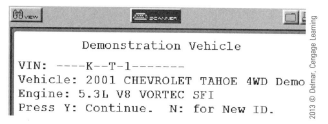

15. Since this is a demonstration, a screen will confirm this and tell you to not connect to a vehicle. Press Yes to continue.

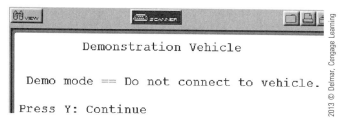

16. **Select a System.** Remember you are still *not* physically connected to the vehicle at this point. The cables, connectors, and keys come after this step. Depending on the year, make, and model of the vehicle selected, there will be a variety of systems that can be chosen. The name of each basic system is fairly self-explanatory regarding the information you are going to retrieve from it. Here are the basic systems that almost all vehicles will have available. To select a system, scroll the double-arrow marker to the desired system name and press Yes.

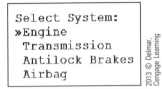

17. Many of today's vehicles will have other systems that can be communicated with. The name of the systems and the information found there is dependent on the vehicle manufacturer. The following is a listing of the common modules found in these systems based on manufacturer.

General Motors

BCM (Body Control Module)

IPC (Instrument Panel Cluster)

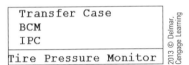

Ford

Body Systems: Similar to the BCM, but it is broken down into many more specific body control systems. Press Yes to view the other Body Systems.

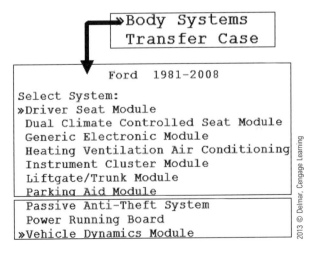

Chrysler

Similar to the BCM, but it is broken down into a couple more specific body control systems. Press Yes to view the other Body systems.

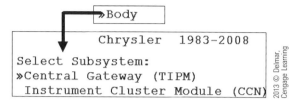

18. **Connect to the Vehicle.** This is the step in which the scan tool is physically connected to the vehicle by a data cable and vehicle adapter. Physically connecting to the vehicle is divided into two categories. These categories are OBDI, which was the standard up through 1995, and OBDII, which is found on 1996 and newer vehicles. The male end of the data cable plugs into the scan tool, and the female end plugs into the vehicle adapter.

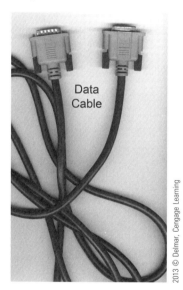

OBDI

OBDI connections are a bit more difficult and time consuming than OBDII. First, each manufacturer created their own vehicle data link connector, so a large array of different adapters is needed to service the various vehicle makes. Second, there was no standard place on the vehicle to position the data link connector, so just finding it can be a bit frustrating and time consuming. Third, rarely was there vehicle power found at the data link connector, so extra cables are needed to provide power to the scan tool while connected to and diagnosing an OBDI vehicle. Snap-On's scan tools help make OBDI connections less frustrating by telling the user which adapter is needed, where the data link connector is located on the vehicle, as well as any other required information required to make a proper connection.

OBDI Connector Examples:

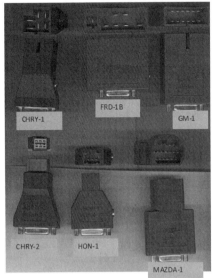

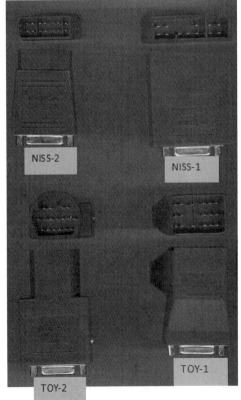

Vehicle Connection

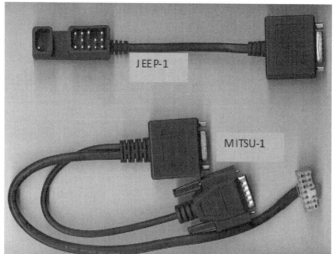

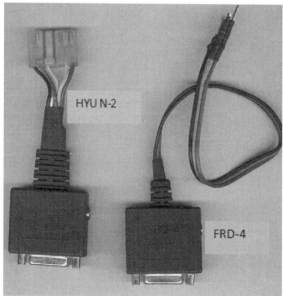

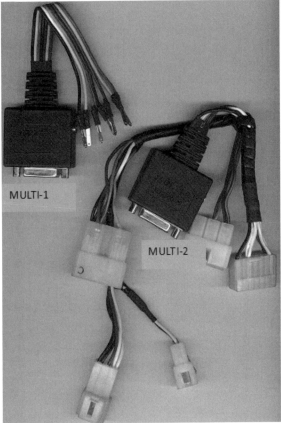

OBDI Connection Steps

Step 1: After identifying the vehicle, read the on-screen instructions to find which adapter is required and where the data link connector is located on the vehicle.

```
              Asian 1983-2008

Connect: TOY-1 Adapter.
Location: Box marked diagnosis under
hood on either fender well or intake
manifold. Press Y: Continue
```

```
              Asian 1983-2008

Connect: HON-1 Adapter.
Location: Behind right kickpanel.
```

```
              GM   1980-2008

Connect: GM-1 Adaptor.
Location: 12 pin connector under left
side of dash.
Except: conversions & special builds.
```

```
              Ford  1981-2008
Connect: FORD-1A or 1B Adapter.
Location: Left side of engine bay.
```

```
              Ford  1981-2008
Connect: Multi-1 Adapter, Red to
Green/Black, Blue to Green/Red, Black to
Black.
Location: Under driver's seat.
```

```
            Chrysler  1983-2008
Connect: CHRY-1 Adapter.
Location: Left side of engine
compartment behind battery, near ECU.
```

```
            Chrysler  1983-2008
Connect: CHRY-2 Adapter.
Location: Blue 6 or 8 pin connector in
or near fuse panel.
```

Step 2: Connect the adapter to the vehicle's data link connector. Most of the adapters will have a port on the side to accept an external power source. Also attach the external power cable to the adapter now. Failing to attach external power will drain the scan tool's battery and can also lead to communication issues with the vehicle.

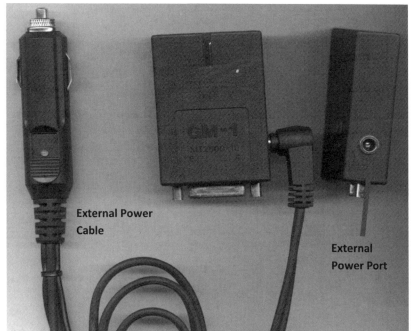

External Power Cable

External Power Port

OBDII

OBDII systems are a bit easier to connect to. First, you only need one adapter. Second, the data link connector (DLC) is located is the same general area on all vehicles. Most 1996 and newer vehicles equipped with OBDII have the DLC located under the dash of the driver's side of the vehicle. Third, power is included at the OBDII DLC, so no external power cables are needed in order to finalize this connection.

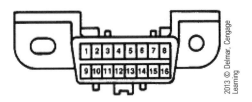

PIN	ASSIGNMENTS
1.	Manufacturer's Discretion
2.	Bus positive Line of SAE-J1850
3.	Manufacturer's Discretion
4.	Chassis ground
5.	Signal ground
6.	CAN high (ISO 15765-4 and SAE-J2234)
7.	K line of ISO 9141-2 and ISO 14230-4
8.	Manufacturer's Discretion
9.	Manufacturer's Discretion
10.	Bus negative Line of SAE-J1850
11.	Manufacturer's Discretion
12.	Manufacturer's Discretion
13.	Manufacturer's Discretion
14.	CAN low (ISO 15765-4 and SAE-J2234)
15.	L line of ISO 9141-2 and ISO 14230-4
16.	Battery voltage

OBDII Connection Steps

Step 1: After identifying the vehicle, read the on-screen instructions. Locate the OBDII adapter, and then check to see a more specific description of where the OBDII DLC is located on the vehicle and what personality key is needed. Remember, most vehicles locate the DLC under the dash on the driver's side of the vehicle. Personality keys are plugged into the OBDII adapter and allow communication with many different years, makes, models, and vehicle systems.

Vehicle Connection

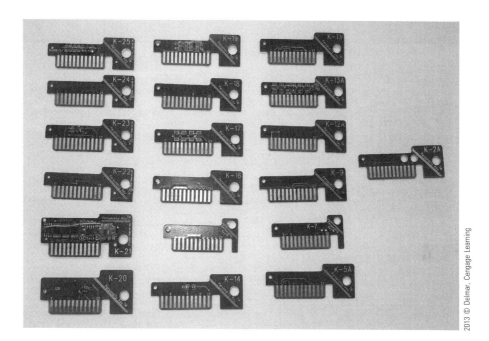

```
          GM  1980-2008
Connect: OBD-II Adapter with K-17 Key.
Location: Under left side of dash.
```

```
          GM  1980-2008
Connect: OBD-II Adapter, no key
required.
Location: 16 pin connector under left
```

```
         Ford  1981-2008
Connect: OBD-II Adapter with K-2A or
K-20 Key.
Location: Under left side of dash.
```

```
         Ford  1981-2008
Connect: OBD-II Adapter with K-19 Key.
Location: Under left side of dash.
```

```
            Chrysler  1983-2008
Connect: OBD-II Adapter with K-25 key.
Location: Under dash to left of steering
column.
```

```
            Chrysler  1983-2008
Connect: OBD-II Adapter with K-13 key.
Location: Under dash to left of steering
column.
```

```
            Asian 1983-2008
Connect: OBD-II Adapter with K-24 key.
Location: Under dash, near left
kickpanel.
```

```
            Asian 1983-2008
Connect: OBD-II Adapter with K-2A or
K-20 Key.
Location: Under drivers side dash.
```

Step 2: Connect the OBDII connector to the data link cable and insert the required personality key. Most of the time, personality keys are stored together on a plastic key chain, and the proper key is inserted with the remainder hanging off to the side. All of the extra personality keys, along with the OBDII connector itself, and the data cable connector add up to bulky package that is fairly heavy. This can make it difficult to connect the OBDII adapter to the vehicle and keep it from falling out. Also, the weight could damage the DLC itself or the area surrounding it. To solve this problem, you can use an OBDII extension cable.

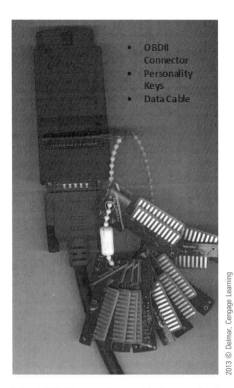

The OBDII extension cable is attached to the output of the OBDII adapter, and then the other end of the extension cable is attached to the vehicle's DLC. This way, the weight of the OBDII connector, personality keys, and data cable connector are kept away from the DLC. This makes it easier to connect to the vehicle and less likely to damage the vehicle's DLC.

Step 3: Connect the OBDII adapter to the vehicle's DLC.

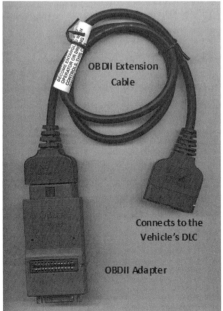

19. **Scanner Main Menu.** After connecting the proper adapter to the vehicle's OBDII data link connector, press Yes to continue. This brings you to the Scanner Main Menu for the specific system you selected earlier (Engine, Transmission, ABS, etc.). Each Main Menu will be slightly different depending on the make of the vehicle and the specific system selected.

```
Main Menu (Engine)
»Data Display
 Codes Menu
 Functional Tests
 Generic Functions
 Troubleshooter
```

```
Main Menu (ABS)
»Data Display
 Trouble Codes
 Clear Codes
 Automated Bleed
 Troubleshooter
```

```
Main Menu (Trans)
»Data Display
 Codes Menu
 Functional Tests
 Troubleshooter
```

```
Main Menu (Airbag)
»Data Display
 Trouble Codes
 Clear Codes
```

```
Main Menu (NVG-246 Auto T-Case)
»Data Display
 Codes Menu
 Functional Tests
 Troubleshooter
```

```
Main Menu (BCM)
»Data Display
 Trouble Codes
 Clear Codes
 Functional Tests
```

```
Main Menu (IPC)
»Data Display
 Trouble Codes
 Clear Codes
 Functional Tests
```

20. **Data, Codes, Functional Tests, Generic Functions, and Troubleshooter.** These are the common options available from the Scanner Main Menu for the various vehicle systems. Some systems will have all of these options, while others may only have a few, as this depends on the make of the vehicle and the system selected. All of these options will be discussed in detail in their respective section of this book. This section only covers the Scanner procedure up through physical connection and final navigation to the Scanner Main Menu. Please see the other sections of this book to continue the diagnostic procedure using the Scanner.

Summary

Scan tools are an invaluable and necessary tool when diagnosing and performing routine repairs on today's vehicles. Something as simple as bleeding the brakes may require a scan tool to perform the job correctly. Scan tools are valuable in the diagnostic process because they allow us to see the effect that the incoming data has on the vehicle's computer. Scan tools are a window into the operation of the vehicle's computer that allows us to view and gather data, read and clear trouble codes, as well as command the vehicle's computer to perform functional tests. Because the information has already been processed by the vehicle's computer by the time we see it on the scan tool, we call this information back-door data. Back-door data shows us effects and points us in the direction of a problem. It will not show us the cause of the problem, however. Front-door data is live signals going into the vehicle's computer. These are not captured with a scan tool but require lab scopes and similar diagnostic tool to measure and interpret. Remember, all scan tool information is back-door data. All of this information is gathered by connecting the scan tool to the vehicle through a data link connector (DLC), which is different based on whether the vehicle is OBDI or OBDII. On Board Diagnostics generation one (OBDI) was the standard up through 1995. Under this standard, the manufacturer could develop and use their own DLC and place it at their discretion somewhere on the vehicle. Starting in 1996, On Board Diagnostics generation two (OBDII) established rules to make this system more universal. OBDII requires all manufacturers to use the same 16 pin DLC and place it in the vicinity of the driver's side under-dash area. The Scanner navigation to establish communication with a vehicle is simple and straightforward. First, use the manufacturer of the vehicle to select the correct VIN specific database. Second, select the specific manufacturer of the vehicle. Third, enter the VIN characters requested by the scan tool. Fourth, select the vehicle system (Engine, Transmission, ABS, etc.) that you wish to communicate with. Next, physically connect the Scanner to the vehicle's DLC via the data cable and the correct vehicle adapter. Lastly, select the information you wish to retrieve from the vehicle.

Review Questions

1. Technician A states that a scan tool is only used when diagnosing a drivability issue. Technician B states that a scan tool receives back-door data. Who is correct?
   a. Tech A only
   b. Tech B only
   c. Both Tech A and Tech B
   d. Neither Tech A nor Tech B

2. Technician A states that a scan tool should be thought of like a compass, pointing in the general direction of a problem but not pinpointing causes. Technician B states that scan tools connect to the vehicle at the DLC or data link connector. Who is correct?
   a. Tech A only
   b. Tech B only
   c. Both Tech A and Tech B
   d. Neither Tech A nor Tech B

3. Technician A states that OBDII started in 1996. Technician B states that OBDII data link connectors can be located at the manufacturer's discretion anywhere on the vehicle. Who is correct?
   a. Tech A only
   b. Tech B only
   c. Both Tech A and Tech B
   d. Neither Tech A nor Tech B

4. Technician A states that, when following the Systematic Scanner Procedure, it is best to perform a VIN specific scan first, as this will give you the most data. Technician B states that Global OBDII data is controlled by the manufacturer. Who is correct?
   a. Tech A only
   b. Tech B only
   c. Both Tech A and Tech B
   d. Neither Tech A nor Tech B

5. Technician A states that step one of the Systematic Scanner Procedure is to look at current data and see if the problem present. Technician B states that, if a problem is intermittent, the technician should be sure to look at Pending Codes. Who is correct?
   a. Tech A only
   b. Tech B only
   c. Both Tech A and Tech B
   d. Neither Tech A nor Tech B

6. Technician A states that after a repair and before the vehicle is released back to the customer, a technician should always verify the repair using either Service $06 under Global OBDII or DTC Status under a VIN specific database. Technician B states that using the Generic Functions option under a VIN specific database will allow access to some Global OBDII functions. Who is correct?

   a. Tech A only
   b. Tech B only
   c. Both Tech A and Tech B
   d. Neither Tech A nor Tech B

7. Technician A states that the first step in using a Scanner is physically connecting it to the vehicle. Technician B states that the entire VIN must be entered to identify a vehicle. Who is correct?

   a. Tech A only
   b. Tech B only
   c. Both Tech A and Tech B
   d. Neither Tech A nor Tech B

8. Technician A states that some available vehicle systems to communicate with include Engine, Transmission, and ABS. Technician B states that some manufacturers listed Body as a vehicle system but have many other subsystems listed under that title. Who is correct?

   a. Tech A only
   b. Tech B only
   c. Both Tech A and Tech B
   d. Neither Tech A nor Tech B

9. Technician A states that, when connecting to an OBDI system, the scan tool will explain what adapter is required and where the DLC is located. Technician B states that OBDI systems have power built in so no external power cables will be required. Who is correct?

   a. Tech A only
   b. Tech B only
   c. Both Tech A and Tech B
   d. Neither Tech A nor Tech B

10. Technician A states that OBDII systems usually require personality keys to properly communicate with the vehicle. Technician B states that OBDII systems have no common pins between manufacturers. Who is correct?

    a. Tech A only
    b. Tech B only
    c. Both Tech A and Tech B
    d. Neither Tech A nor Tech B

Viewing and Interpreting Scan Data

Viewing and Interpreting Scan Data Section Outline:

1. Viewing Data
2. Text View
3. PID List View
4. Graphing Data View
5. Custom Data List
6. Advanced Capturing Techniques
7. Summary

Upon completion of the Viewing and Interpreting Data module, you will be able to:

- Analyze the features and appraise the benefits unique to each type of data display
- Compare and contrast digital (number) data verse graphed (plotted) data
- Construct custom data lists by prioritizing relative data and using higher refresh rates
- Construct custom PID views to allow for easy viewing of PID relationships
- Capture, retrieve, and manipulate saved data views
- Construct and modify multiple graphed views

REMINDER: This section is written so the reader can follow along using the Solus Pro or MODIS scan tools if available. The Scanner software is identical on both of these tools, so all of the procedures will be the same. Most of the screen shots come from a MODIS, but the Solus Pro will be the same except for the screen size—so while the MODIS may show eight graphs, the Solus Pro will only show four. Everything else will function exactly the same. You are strongly encouraged to follow along using the diagnostic tools, as this provides the hands-on knowledge necessary to pass the Certification Exams. The only required equipment is the Scanner unit itself and the AC power adapter cord. The demonstrations built into the Scanner software will provide the rest of the required material.

Viewing and Interpreting Data Overview

To fully understand this section, it is important to read and understand the Scanner Introduction section first. The Scanner Introduction section will explain the required steps to get to this point in the navigation of the software. Some of the data navigation and features covered in this section were outlined and briefly discussed in the Global OBDII section, so after completing this section you should be able to transfer some of this more detailed knowledge to Global OBDII when viewing data.

Parameter Identifications (PIDs)

Parameter Identifications or PIDs are the specific data streams collected by the scan tool from the vehicle's computer. RPM, TPS, MAF, are all examples of PIDs. The number of PIDs collected and displayed is different for every vehicle and specific vehicle system. Some vehicles have more PIDS than others. The term PID will be used a lot in this section, so it is important to understand that we are talking about a specific data stream that is being collected by the scan tool.

Viewing Data

There are a number of different ways data can be viewed on the scan tool. Each specific view will be discussed in detail. For simplicity and to allow you to more clearly see the advantages and disadvantages of each view, we'll be using the built in Scanner Demo, selecting the Engine System, and looking at the Engine Data 1 list. The navigation and features discussed using this data list is identical to *all* the other data lists in *all* the other vehicle systems. The knowledge gained here can easily be transferred to the rest of the vehicle systems and any other data list found on the scan tool including Global OBDII Service $01 and $02. No matter what view you are using it is important to remember that ALL of the PIDs are being collected and recorded to the scan tools memory buffer. Even the PIDs not shown on the screen are still being collected and recorded automatically all of the time. No problem can escape. The information you need to diagnose the problem will be captured and recorded by the scan tool. You just have to find it, which will be discussed when creating custom data lists, custom data views, saving Movies, Snapshots, and using PID triggers.

Please review the Scanner Introduction section to begin the Scanner demo, so you can follow along with the rest of this section using the dignostic tool.

1. Here are the systems that this vehicle has available. Scroll to Engine and press Yes.

```
Select System:
»Engine
 Transmission
 Antilock Brakes
 Airbag
 Transfer Case
 BCM
 IPC
```

2. From the Main Menu (Engine) screen scroll and select Data Display and Press Yes.

```
Main Menu (Engine)
»Data Display
 Codes Menu
 Functional Tests
 Generic Functions
 Troubleshooter
```

3. This is the Data Menu. The scan tool automatically arranges several different PIDs together in like groups. Engine Data 1 will show all engine related PIDs. Misfire Data will show the misfire cycles and history for each cylinder. EGR, EVAP, and Accessories will group PIDs of these specific systems together and finally Sensor Data will put all input sensor PIDs in one list. Some PIDs will be found in multiple groups, but when you are looking for a specific PID, be sure to check all of the groupings until you find it. For the demonstration, let's press Yes to view "Engine Data 1."

```
Data Menu
»ENGINE DATA 1
 MISFIRE DATA
 EGR, EVAP, ACC
 SENSOR DATA
```

Text View

Text view is based on the MT2500 Scanner. Commonly and reverently referred to as the Red Brick scanner, this scan tool has been a longstanding favorite of many technicians. Even though the MT2500 is no longer available and the software updates and support have also ended, the legacy of this tool is living on in other Snap-On scan tools such as the Solus Pro and MODIS. The text view replicates the MT2500 scanner in many ways and until recently functioned just the same. Except for the first line of data, which contains three PIDs, the remainder of the PIDs are in a two-column view with the sampled data immediately following each PID. Also notice the four simulated LED lights across the bottom of the screen. The actual MT2500 had real LED lights built into the scan tool. The newer scan tools still have this LED indicator functionality, but it is now built into the on-screen features and options rather than physical hardware built into the tool itself. The 9.4 software update bundle brought some big changes to the Text view and its functionality. To accommodate the most users, regardless of their software version, there will be two different Text View sections. If you are using version 9.2 and earlier, please continue with the following steps. If you are using version 9.4 and newer, please turn to the next section.

Version 9.2 and *earlier* Text View Functions

1. Follow the *Scanner Introduction Review* (outlined just previously in this section) to navigate to the Data Menu using the Scanner Demo and selecting the Engine System.

114 Chapter 6 Viewing and Interpreting Scan Data

2. From this screen Press Yes to view Engine Data 1.
3. The default setting on the scan tools will bring you to Text view. If, when using the utilities menu and Tool Setup options, the Scanner View options were changed, then a different view might be the default. This is a good time to remind you that after reading this section and finding your favorite view, it may be useful to change this option so that each time you enter the scanner your favorite view will be shown. If your screen does not look like the example screenshot please scroll over to View, press Yes, scroll down to Text, and press Yes.

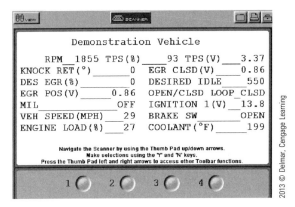

4. With the Scanner button (top center of the screen) highlighted, use the Up and Down arrow buttons to scroll and view all the available PIDs.

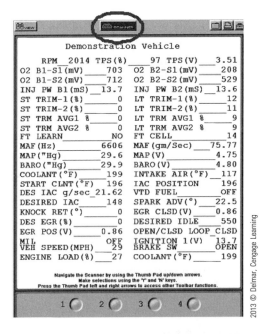

5. To Pause or Freeze the Text data, make sure the Scanner Button (top center of the screen) is highlighted and press Yes. Just to the left of RPM in the upper left corner of the screen the letters "HLD" will appear and the data numbers will freeze.

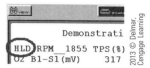

6. With the data frozen, scroll through the PIDs and look for problematic values.
7. Press Yes to remove the hold on the screen and begin gathering data once again. As long as the Scanner button is highlighted, pressing Yes will start and stop the Text data display.
8. Scroll over to the right and highlight the File Folder icon and press Yes. Notice that when the Scanner button is no longer highlighted, the data screen is grayed out and cannot be scrolled through.

   Save: The options available include Save Movie, which will save all 2000 frames of data, Save Frame, which will save all the PIDs in one frame of data, or Save Image, which will take a picture of the current screen. It is important to remember that when a Movie is saved using Text view, it will only be played back in Graphing view. This is the same for PID List view. All saved Movies, no matter what view they were saved in, are played back in graphing view. Please refer back to the Data Management section for more detailed information about these options and how to use them.

9. Scroll over to the right and highlight the Printer icon and press Yes.

   Print: "Full Screen" will perform its normal function of printing the screen just as you see it. This will include the visible PIDs on the screen at the time of the printing. "Full PID List" will print a text version of all the PID values, both on and off the screen, at the time of the printing. An example of what the Full PID List looks like is included. "Full Codes List" will not function under the Text view. This option only works when viewing trouble codes under the Codes Menu.

```
2001 CHEVROLET TAHOE 4WD Demo 5.3L V8 VORTEC SFI      Copyright 2008 Snap-on Inc.
04/11/2009  12:47PM

RPM                          760     TPS(%)                        16
TPS(V)                      0.57     O2 B1-S1(mV)                 768
O2 B2-S1(mV)                 456     O2 B1-S2(mV)                 716
O2 B2-S2(mV)                 686     INJ PW B1(mS)                5.2
INJ PW B2(mS)                5.1     ST TRIM-1(%)                  -3
LT TRIM-1(%)                  12     ST TRIM-2(%)                  -3
LT TRIM-2(%)                  12     ST TRM AVG1 %                  0
LT TRM AVG1 %                  9     ST TRM AVG2 %                  0
LT TRM AVG2 %                  9     FT LEARN                     YES
FT CELL                        4     MAF(Hz)                     3370
MAF(gm/Sec)                10.89     MAP(*Hg)                    14.2
MAP(V)                      1.99     BARO(*Hg)                   28.7
BARO(V)                     4.59     COOLANT( F)                  199
INTAKE AIR( F)               120     START CLNT( F)               196
IAC POSITION                 148     DES IAC g/sec              21.62
VTD FUEL                     OFF     DESIRED IAC                  148
SPARK ADV( )                22.5     KNOCK RET( )                   0
EGR CLSD(V)                 0.86     DES EGR(%)                     0
DESIRED IDLE                 550     EGR POS(V)                  0.86
OPEN/CLSD LOOP              CLSD     MIL
IGNITION 1(V)               13.7     VEH SPEED(MPH)               OFF
BRAKE SW                    OPEN     ENGINE LOAD(%)                 2
COOLANT( F)(2)               199                                   4
```

Print Full PID List Sample

116 Chapter 6 Viewing and Interpreting Scan Data

10. Scroll over to the right and highlight the Tool Box icon and press Yes.

11. Press Yes to enter Custom Setup

    Custom Setup: This is a shortcut back to the Scanner Units option under the main Utilities menu of the scan tool. Navigate the dialogue box to change between standard and metric units. Refer back to the Tool Setup and Utilities options explained under the specific section for the diagnostic tool you are using for more information.

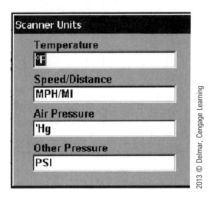

12. Press No to return to the Tool Box icon.
13. Press Yes again and scroll down to Save Data and press Yes.

    Save Data: This is a shortcut back to the Scanner Units option under the main Utilities menu of the scan tool. Navigate the dialogue box to change the percent after trigger option, where data is stored, and what file type all of the screen images will be when saved. Refer back to the Tool Setup and Utilities options explained under the specific section for the diagnostic tool you are using or the Data Management section for more information.

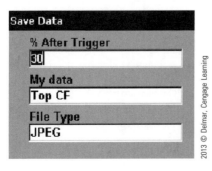

14. Press No to accept any changes and exit the Save Data menu box.
15. Scroll back over to the left and highlight the Scanner button.

16. Press No once. This brings us to an intermediate menu with multiple options. This menu is where we can customize the Text view and also access other options used in other scan tool views. This menu is based on the MT2500 platform. Many of the options are not commonly used by many technicians today, but are still available for the technicians that prefer and are more comfortable with the MT2500 type navigation and options. The *only* way to access this menu is from Text view. From there, highlight the Scanner button, and press No once.

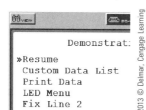

Resume: Pressing Yes will bring you right back to the normal Text view.

Custom Data List: Pressing Yes will allow you to create a Custom Data List that will greatly improve the refresh rate of the scan tool, allowing for easier diagnosis of vehicle problems. It helps produce excellent graphs of the data streams. This is still commonly used today and is really the only option not obsolete in this menu. It is now also accessible through the Toolbox icon in the upper right corner of the screen. This will be discussed in more detail at the end of this section.

Print Data: This option is really no longer used as it is much easier to use the Printer icon near the upper right hand corner of the screen. Pressing Yes will bring you to another screen that will give you the choice between Print Screen or Print Frame. This is the same as the options under the Printer icon. Print Screen shows identical wording, while "Print Frame" in this menu is equal to "Full PID List" under the Printer icon options.

LED Menu: This option is especially designed for the MT2500 technicians who used the LEDs built into that scan tool to help gather information about the vehicle. Pressing Yes will bring you to a set up screen where the LEDs can be assigned to display certain values. Obviously since the LEDs can only be on or off, they can only display on/off, rich/lean, yes/no, and other binary information.

Fix Line 2: This is how a custom data view is built in Text view. This allows the PIDs to be rearranged and organized so it is easier to view relationships between their values as the vehicle is running. Once "Fix Line 2" is activated this option will change to Fix Line 3. The functionality is the same.

17. Scroll down to LED Menu and press Yes.

    This is where tasks for the LEDs are assigned. The LEDs are numbered across the bottom of the screen (1-4).

    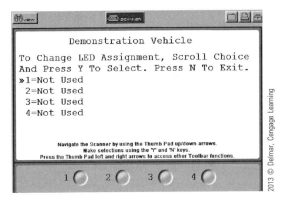

18. Follow the on screen instructions to assign the LEDs a task available in the list. Using the Up and Down arrow button, scroll to the desired task and press Yes to move to the next LED. Continue until all the LEDs have been assigned to your liking.

    After this point when viewing Text data the LEDs will function according to the tasks assigned in this menu and will flash red to display data.

    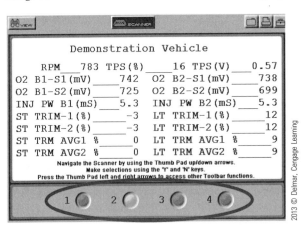

19. Press No until you return to the intermediate menu, then scroll up to Resume, and then press Yes. Text view is the desired screen.

20. For demonstration's sake, let's say you needed to view the relationship between the data found in line 1 (RPM, TPS (%), and TPS (V)) and MAF, Injector Pulse Width, and Engine Load %. How can we get all of these PIDs on the screen at the same time?

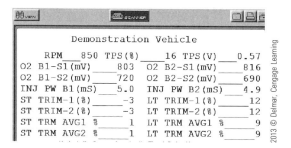

21. Scroll down until MAF (Hz) and MAF (gm/Sec) is in the line 2 position as shown in the screen shot.

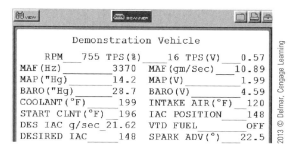

22. Press No once to go to the intermediate menu.
23. Scroll down to "Fix Line 2" and press Yes.

24. You will automatically be brought back to the Text view, but Line 2 will no longer move when you scroll up and down through the data.
25. Now scroll until INJ PW B1 (mS) and INJ PW B2 (mS) are in the Line 3 position. When you scroll up or down, Line 2 will not move from its position.

120 Chapter 6 Viewing and Interpreting Scan Data

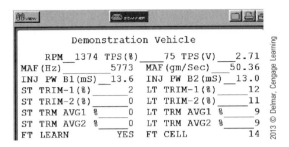

26. Press No once to go to the intermediate menu.
27. Scroll down to Fix Line 3 and press Yes. Notice that the option of Releasing Line 2 is also available.

28. You will be automatically brought back to the Text view, but now Lines 2 and 3 will be fixed into position.
29. There is *no* Fix Line 4 option available, so the final PID values are scrolled to the Line 4 position and left there while you view the data and look for relationships or problems between those PIDS.
30. Scroll down until Engine Load (%) and Coolant (°F) occupy the Line 4 position. This would be the final screen with all of the desired PIDs available to view together on the same screen. This is how to build a Custom Data View in the Text view.

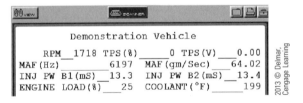

31. Scroll over to the View button and Press Yes.
32. Be sure PID List is highlighted and press Yes.

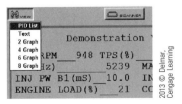

Please skip the next section that deals with versions 9.4 and later, and page ahead to the **PID List View** section.

Version 9.4 and *later* Text View Functions

1. Follow the *Scanner Introduction Review* (outlined just previously in this section) to navigate to the Data Menu using the Scanner Demo and selecting the Engine System.

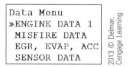

2. From this screen Press Yes to view Engine Data 1.
3. The default setting on the scan tools will bring you to Text view. If, when using the utilities menu and Tool Setup options the Scanner View options were changed, a different view might be the default. This is a good time to remind you that after reading this section and finding your favorite view it may be useful to change this option so that each time you enter the scanner your favorite view will be shown. If your screen does not look like the example screen shot please scroll over to View, press Yes, scroll down to Text, and press Yes.

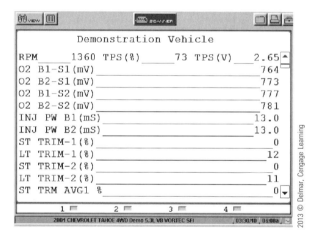

4. With the Scanner button (top center of the screen) highlighted, use the Up and Down arrow buttons to scroll and view all the available PIDs. The only way to scroll the data is to have the Scanner button highlighted.

122 Chapter 6 Viewing and Interpreting Scan Data

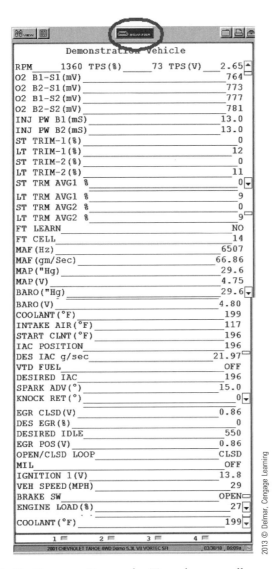

5. To Pause or Freeze the Text data, scroll over to the left and highlight the Freeze/Run button and press Yes.

The button will toggle from Freeze to run each time Yes is pressed with it highlighted.

6. With the data frozen, scroll through the PIDs and look for problematic values by scrolling back over to the right to highlight the Scanner button and then using the Up and Down arrows to scroll through the data.

7. Scroll over to the right and highlight the File Folder icon and press Yes. Notice that when the Scanner button is no longer highlighted the data screen is grayed out and cannot be scrolled through.

   Save: The options available include Save Movie, which will save all 2000 frames of data, Save Frame, which will save all the PIDs in one frame of data, or Save Image, which will take a picture of the current screen. It is important to remember that when a Movie is saved using Text view, it will only be played back in Graphing view. This is the same for PID List view. All saved Movies, no matter what view they were saved in, are played back in graphing view. Please refer back to the Data Management section for more detailed information about these options and how to use them.

8. Scroll over to the right, highlight the Printer icon and press Yes.

   Print: "Full Screen" will perform its normal function of printing the screen just as you see it. This will include the visible PIDs on the screen at the time of the printing. "Full PID List" will print a text version of all the PID values, both on and off the screen, at the time of the printing. An example of what the Full PID List looks like is included. "Full Codes List" will not function under the Text view. This option only works when viewing trouble codes under the Codes Menu.

```
2001 CHEVROLET TAHOE 4WD Demo 5.3L V8 VORTEC SFI      Copyright 2008 Snap-on Inc.
04/11/2009  12:47PM

RPM                       760      TPS(%)                     16
TPS(V)                   0.57      O2 B1-S1(mV)              768
O2 B2-S1(mV)              456      O2 B1-S2(mV)              716
O2 B2-S2(mV)              686      INJ PW B1(mS)             5.2
INJ PW B2(mS)             5.1      ST TRIM-1(%)               -3
LT TRIM-1(%)               12      ST TRIM-2(%)               -3
LT TRIM-2(%)               12      ST TRM AVG1 %               0
LT TRM AVG1 %               9      ST TRM AVG2 %               0
LT TRM AVG2 %               9      FT LEARN                  YES
FT CELL                     4      MAF(Hz)                  3370
MAF(gm/Sec)             10.89      MAP("Hg)                 14.2
MAP(V)                   1.99      BARO("Hg)                28.7
BARO(V)                  4.59      COOLANT( F)               199
INTAKE AIR( F)            120      START CLNT( F)            196
IAC POSITION              148      DES IAC g/sec           21.62
VTD FUEL                  OFF      DESIRED IAC               148
SPARK ADV( )             22.5      KNOCK RET( )                0
EGR CLSD(V)              0.86      DES EGR(%)                  0
DESIRED IDLE              550      EGR POS(V)               0.86
OPEN/CLSD LOOP           CLSD      MIL                       OFF
IGNITION 1(V)            13.7      VEH SPEED(MPH)              2
BRAKE SW                 OPEN      ENGINE LOAD(%)              4
COOLANT( F)(2)            199
```

Print Full PID List Sample

124 Chapter 6 Viewing and Interpreting Scan Data

9. Scroll over to the right and highlight the Tool Box icon and press Yes.

10. Press Yes to enter Custom Setup

    Custom Setup: This is a shortcut back to the Scanner Units option under the main Utilities menu of the scan tool. Navigate the dialogue box to change between standard and metric units. Refer back to the Tool Setup and Utilities options explained under the specific section for the diagnostic tool you are using for more information.

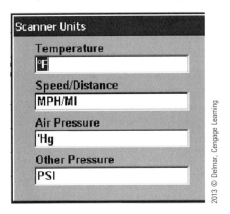

11. Press No to return to the Tool Box icon.
12. Press Yes again, then scroll down to Save Data and press Yes.

Save Data: This is a shortcut back to the Scanner Units option under the main Utilities menu of the scan tool. Navigate the dialogue box to change the percent after trigger option, where data is stored, and what file type all of the screen images will be when saved. Refer back to the Tool Setup and Utilities options explained under the specific section for the diagnostic tool you are using or the Data Management section for more information.

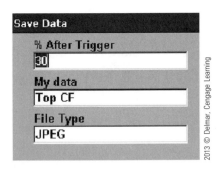

13. Press No to accept any changes and exit the Save Data menu box.
14. Press Yes again and scroll down to Custom Data List, but do *not* press Yes at this time. This option will only work if the data stream is live or running and not paused or frozen.

Custom Data List: Pressing Yes will allow you to create a Custom Data List that will greatly improve the refresh rate of the scan tool, allowing for easier diagnosis of vehicle problems. It helps produce excellent graphs of the data streams. This will be discussed in more detail at the end of this section.

15. Scroll down to LED Setup and press Yes. This option will only work if the data stream is live or running and not paused or frozen.

LED Menu: This option is especially designed for the MT2500 technicians who used the LEDs built into that scan tool to help gather information about the vehicle. Pressing Yes will bring you to a setup screen where the LEDs can be assigned to display certain values. Obviously, since the LEDs can only be on or off, they can only display on/off, rich/lean, yes/no, and other binary information.

16. Scroll down to LED Menu and press Yes.

This is where tasks for the LEDs are assigned. The LEDs are numbered across the bottom of the screen (1-4).

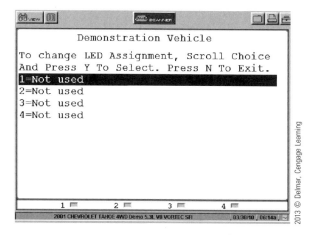

126 Chapter 6 Viewing and Interpreting Scan Data

17. Follow the on-screen instructions to assign the LEDs a task available in the list. Using the Up and Down arrow button, scroll to the desired task and press Yes to move to the next LED number. Continue until all the LEDs have been assigned to your liking.

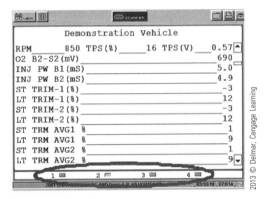

18. After completing your selections, press No to back out to the Text View

    After this point, when you are viewing Text data, the LEDs will function according to the tasks assigned in this menu and will flash red to display data.

19. Scroll over to the View button and Press Yes.
20. Be sure PID List is highlighted and press Yes.

PID List View

PID List view is more organized and easier to search than Text view. In this view, all of the PIDs are listed on the left side of the screen, while all of the values are at the right side of the screen with everything organized in a boxed matrix system. Scroll down to see all of the available PIDs. This view has many other options and features that can be used to help find and diagnose problems. All of these options will be detailed in this section. Pay attention to how the buttons across the top of the screen change as other buttons are activated. This can sometimes be confusing if you're not looking for it. When you need to get back to the top tool bar, press No.

PID	Value
RPM	1374
TPS(%)	75
TPS(V)	2.71
MAF(Hz)	5773
MAF(gm/Sec)	50.36
INJ PW B1(mS)	13.6
INJ PW B2(mS)	13.0
O2 B1-S1(mV)	712
O2 B2-S1(mV)	794
O2 B1-S2(mV)	781
O2 B2-S2(mV)	777
ST TRIM-1(%)	2
LT TRIM-1(%)	12
ST TRIM-2(%)	0
LT TRIM-2(%)	11
ST TRM AVG1 %	0
LT TRM AVG1 %	9
ST TRM AVG2 %	0
LT TRM AVG2 %	9
FT LEARN	YES
FT CELL	4
MAP("Hg)	13.9
MAP(V)	1.94
BARO("Hg)	28.7
BARO(V)	4.59
COOLANT(°F)	199
INTAKE AIR(°F)	120
START CLNT(°F)	196
IAC POSITION	153
DES IAC g/sec	22.72
VTD FUEL	OFF
DESIRED IAC	196
SPARK ADV(°)	13.5
KNOCK RET(°)	0
EGR CLSD(V)	0.86
DES EGR(%)	0
DESIRED IDLE	550
EGR POS(V)	0.86
OPEN/CLSD LOOP	CLSD
MIL	OFF
IGNITION 1(V)	13.8
VEH SPEED(MPH)	30
BRAKE SW	OPEN
ENGINE LOAD(%)	26
COOLANT(°F)	199

Freeze/Run

1. Scroll over to the right and highlight the Freeze/Run Button and press Yes. Notice how it changes when activated.

Freeze/Run: By default when entering the data views the diagnostic tool is automatically collecting and storing data to the memory buffer. The Freeze/Run or play/pause button allows the user to temporarily stop this recording and review the data while frozen.

2. When the data collection is frozen or paused you can view each frame of data stored in the data memory buffer one at a time to look for problems with the values. The memory buffer is indicated by the blue line across the top of the screen. In this example, you can see 2000 frames of data stored in the memory buffer. All of this data is temporary data and is not saved anywhere on the tool. If the tool is turned off, all of this data will be lost. Each frame of data is a snapshot or picture of the PID values at that moment in time. The scan tool takes approximately 10 shots per second of the incoming data stream and displays the results. The 10 shots per second is a rough estimate with many variables. It is dependent on the number of PIDs being collected and the vehicle year, make, and model. Some vehicle computers communicate faster than others. This is also where the Custom Data List mentioned earlier comes into play as we can force the tool to sample faster. Again this will be discussed in more detail later in this section. The maximum amount of frames able to be stored in the buffer can change depending on what version of software you are using. The latest versions have the capability to store at least 2000 frames of data. When the 2001st frame enters the memory buffer, the first frame is erased and dumped from the memory buffer. This way the scan tool always has the most recent 2000 frames of data for review. When the memory buffer is full, a small black check mark moves across the blue memory buffer bar and this indicates the new frames that are being added and the old frames that are being dumped while data is being collected and recorded. When the scan tool is paused and not collecting or recording data, the black check mark indicates the frame that is being viewed. In the example, it is frame 600.

Review

3. Scroll over to the right, highlight the Review button, and press Yes. The button will change color to indicate that it is activated. Remember that the Review button is only visible when the data stream is frozen or paused.

4. Use the Left and Right arrow buttons to view each frame of data. Notice how the frame number on the memory buffer moves and changes as you review the frames.

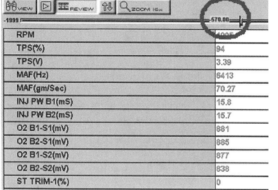

5. To scroll down the PID list, you will need to press No to deactivate the Review button and then press Down to scroll the PID list. When the PIDs you want are on the screen, press No to return to the top tool bar and then press Yes to activate the Review button again to view the frames of data.
6. If you haven't done so already, press No to deactivate Review.
7. Scroll to the left and highlight the Freeze/Run button and press Yes to start recording data to the memory buffer. The PID values should now be changing.

Clear

1. Scroll to the Right and highlight the Clear button and press Yes. Notice that the memory buffer (blue bar) is full in the example screenshot.

2. Press Yes to confirm that you want to Clear the Scanner data and start over. This simply clears all the data frames stored in the memory buffer and starts a fresh recording.

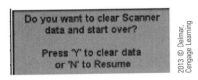

3. The memory buffer is now empty, indicated by the absence of the blue bar.

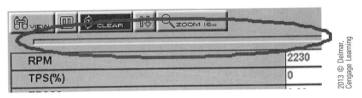

PID Sort

1. Scroll over to the left and highlight the Sort icon (opposite double vertical arrows) and press Yes.

2. The default PID order is Factory Sort, but this can make it more difficult to find a specific PID.

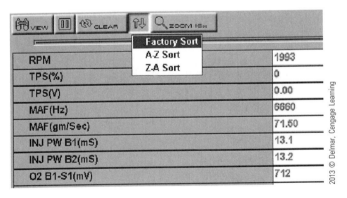

3. To more quickly find a PID that starts with a letter near the beginning of the alphabet, scroll down to highlight A-Z Sort and press Yes to alphabetize the PID list from A-Z.

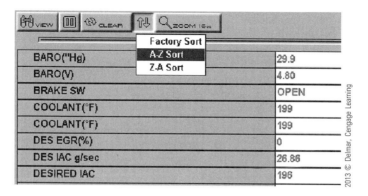

4. If the PID you are looking for starts with a letter that is near the end of the alphabet, it is quicker to sort the PID alphabetically from Z-A. To accomplish this scroll down and highlight Z-A Sort and press Yes.

5. The demonstration will leave the PIDs in the Factory Sort arrangement. It will be easier to follow along if you return yours to the Factory Sort arrangement also.

PID List View Zoom

1. Scroll over to the right, highlight the Zoom button, and press Yes. You are going to see a Zoom button on many different screens within the Scanner and Lab Scope. Zoom is different depending on which screen you are viewing. In the PID List view, Zoom changes the number of PIDs shown on the screen. This can be helpful in many ways. First, it enlarges the font size so the screen is easier to read, and second, it focuses attention on a smaller number of PIDs, which is useful when looking for a problem. Rarely does it take the viewing of 16 PIDs at one time to diagnose a problem, so zooming allows us to focus our diagnostic attention to a select number of PIDs. Later in this section, we'll look at creating a custom PID view where the order of the PIDs are rearranged so only the exact PIDs you want to see are on the screen.

132 Chapter 6 Viewing and Interpreting Scan Data

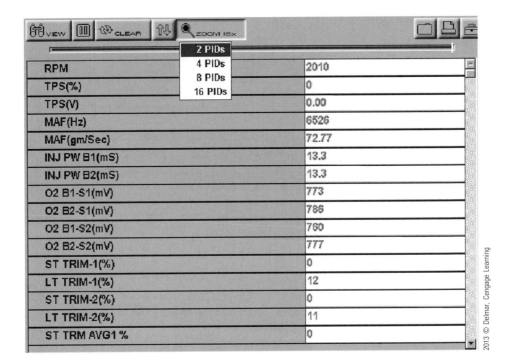

2. Be sure that 2 PIDs is highlighted and press Yes. You can see that this makes it easier to read and clearly focuses your attention on these two PIDs. Remember that all of the other options we have learned and will learn about can be used when zoomed on any number of PIDs. We can still scroll up and down through all the PIDs, Freeze the data recording, Review the frames, Clear the memory buffer, and Sort the PIDs by name. We lost no functionality in the tool. It is also important to remember that, even zoomed, we are still recording *all* PIDs all of the time. Nothing is being lost just because it is not on the screen.

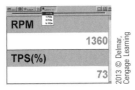

3. Highlight the Zoom button, press Yes, scroll to 4 PIDs and press Yes.

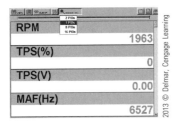

4. Highlight the Zoom button, press Yes, scroll to 8 PIDs and press Yes.

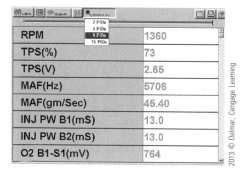

PID List Custom Data View

1. While still in the 8 PID zoomed view, press the down arrow button and scroll if needed to highlight RPM. Notice that the PID display area becomes active, and this is indicated by the thin blue border around the area.

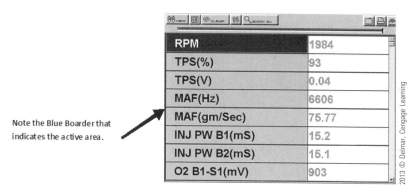

Note the Blue Boarder that indicates the active area.

2. Press Yes with RPM highlighted to open a small dialogue box.
3. Press Yes again to Lock RPM in place. Notice the small padlock that appears next to RPM, indicating that it is locked. This is how we build a Custom Data View by rearranging the PIDs to the order needed to help us diagnose the problem.

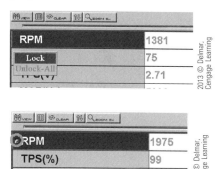

4. Let's build a Custom Data View with the PIDs RPM, MAF (gm/Sec), MAP (V), Intake Air (°F), TPS (%), TPS (V), and MAF (Hz) and place them in this exact order.

5. Scroll down until MAF (gm/Sec) moves up the list and occupies the position under RPM.

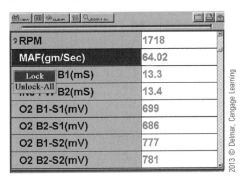

6. Scroll back up to highlight MAF (gm/Sec), press Yes to open the dialogue box, and then press Yes again to lock the PID in place.

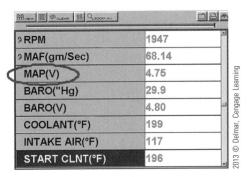

7. Scroll down until MAP (V) moves up the list and occupies the position under MAF (gm/Sec).

8. Scroll back up to highlight MAP (V) press Yes to open the dialogue box, and then press Yes again to lock the PID in place.

9. Scroll down until Intake Air (°F) moves up the list and occupies the position under MAP (V).

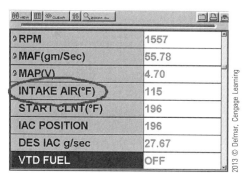

10. Scroll back up to highlight Intake Air (°F), press Yes to open the dialogue box, and then press Yes again to lock the PID in place.

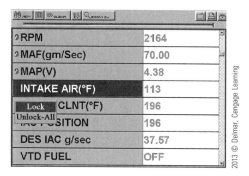

11. The next PID we want to lock in place is TPS (%). Remember that this PID was at the top of the original list right under RPM.

12. Scroll up until TPS (%) occupies the position under Intake Air (°F). Does this work? Can you get to the TPS (%) PID? The answer is no, for right now. There is a way to correct this.

136 Chapter 6 Viewing and Interpreting Scan Data

13. To refresh the PID list and bring back down the "lost" PIDs, press No to get back to the top Tool Bar and highlight the View button.

14. With the View button highlighted, press Yes, highlight PID List, and press Yes again.

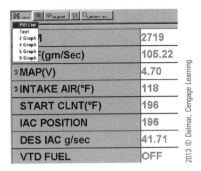

15. This will refresh the screen back to the maximum number of viewable PIDS (8 for the Solus Pro and 16 for the MODIS) but also bring back the "lost" PIDS that were locked out when we first started to build our Custom Data List.

16. If using the MODIS, scroll over to Zoom and change the setting back to 8 PIDs so we can continue to build our Custom Data View. Do this by highlighting 8 PIDS and pressing Yes.

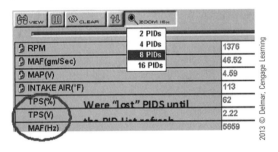

17. Now scroll until TPS (%) occupies the position under Intake Air (°F).

18. Scroll back and highlight TPS (%), press Yes to open the dialogue box, and then press Yes again to lock the PID in place.

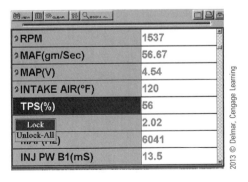

19. Scroll until TPS (V) occupies the position under TPS (%).

20. Scroll back and highlight TPS (V), press Yes to open the dialogue box, and then press Yes again to lock the PID in place.

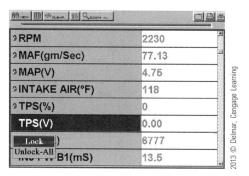

21. Scroll until MAF (Hz) occupies the position under TPS (V).
22. Scroll back and highlight MAF (Hz), press Yes to open the dialogue box, and then press Yes again to lock the PID in place.

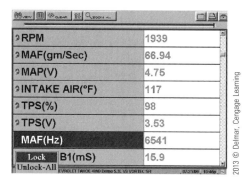

23. With this we have seven of the eight PIDs locked in place. Scroll down to the eighth PID and press Yes. Can you lock this eighth and final PID?

    The answer is no. The scan tool will *not* allow you to fully lock the screen. The only option from here is to Unlock-All as the Lock command is grayed out (ghosted) and inactive.

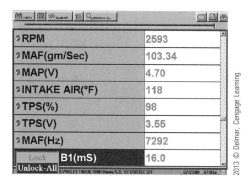

24. Press No and return to the top Tool Bar.

25. Scroll over to highlight the Zoom button and press Yes.
26. Scroll to highlight 4 PIDs and press Yes. What happens when you try to Zoom to a smaller number of PIDs than you currently have locked on the screen? This would be similar to what we tried to do in the previous step, and again the scan tool will not allow you to lock all of the PIDS on the screen. At least one has to be left unlocked so that the other PIDS can scroll through.

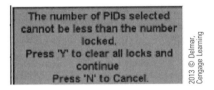

27. Press No to cancel.

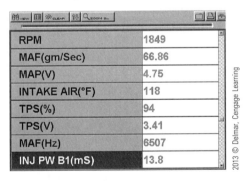

28. Notice that you can scroll through all of the PIDs and position whatever PID you desire into the last (8th) position for viewing purposes, but you cannot lock it.
29. Scroll to any of the PIDs on the screen and press Yes.

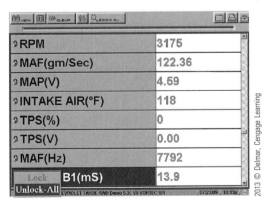

30. Highlight the Unlock-All and press Yes. Notice that all of the padlocks disappear from the PIDs but the PIDs themselves remain in place.

After you build a Custom Data View, all of the functionality of the scan tool remains. You can still Freeze the data recording, Review the data frames, Clear the memory buffer, or Save a Movie. All of the PIDs are still being collected and recorded; even the PIDs not on the screen.

Save, Print, and Utilities in PID List View

1. Scroll over to the right and highlight the File Folder icon and press Yes.

   Save: The options available include Save Movie, which will save all 2000 frames of data, Save Frame, which will save all the PIDs in one frame of data, or Save Image, which will take a picture of the current screen. It is important to remember that when a Movie is saved using PID List view it will only be played back in Graphing view. All saved Movies, no matter what view they were saved in, are played back in graphing view. Please refer back to the Data Management section for more detailed information about these options and how to use them.

2. Scroll over to the right and highlight the Printer icon and press Yes.

   Print: "Full Screen" will perform its normal function of printing the screen just as you see it. This will include the visible PIDs on the screen at the time of the printing. "Full PID List" will print a text version of all the PID values, both on and off the screen, at the time of the printing. An example of what the Full PID List looks like is included.

   ```
   2001 CHEVROLET TAHOE 4WD Demo 5.3L V8 VORTEC SFI      Copyright 2008 Snap-on Inc.
   04/11/2009  12:47PM

   RPM                              760      TPS(%)                    16
   TPS(V)                          0.57      O2 B1-S1(mV)             768
   O2 B2-S1(mV)                     456      O2 B1-S2(mV)             716
   O2 B2-S2(mV)                     686      INJ PW B1(mS)            5.2
   INJ PW B2(mS)                    5.1      ST TRIM-1(%)              -3
   LT TRIM-1(%)                      12      ST TRIM-2(%)              -3
   LT TRIM-2(%)                      12      ST TRM AVG1 %              0
   LT TRM AVG1 %                      9      ST TRM AVG2 %              0
   LT TRM AVG2 %                      9      FT LEARN                 YES
   FT CELL                            4      MAF(Hz)                 3370
   MAF(gm/Sec)                    10.89      MAP("Hg)                14.2
   MAP(V)                          1.99      BARO("Hg)               28.7
   BARO(V)                         4.59      COOLANT( F)              199
   INTAKE AIR( F)                   120      START CLNT( F)           196
   IAC POSITION                     148      DES IAC g/sec          21.62
   VTD FUEL                         OFF      DESIRED IAC              148
   SPARK ADV( )                    22.5      KNOCK RET( )               0
   EGR CLSD(V)                     0.86      DES EGR(%)                 0
   DESIRED IDLE                     550      EGR POS(V)              0.86
   OPEN/CLSD LOOP                  CLSD      MIL                      OFF
   IGNITION 1(V)                   13.7      VEH SPEED(MPH)             2
   BRAKE SW                        OPEN      ENGINE LOAD(%)             4
   COOLANT( F)(2)                   199
   ```

   Print Full PID List Sample

3. Scroll over to the right and highlight the Tool Box icon and press Yes.

4. Press Yes to enter Custom Setup.

Custom Setup: This is a shortcut back to the Scanner Units option under the main Utilities menu of the scan tool. Navigate the dialogue box to change between standard and metric units. For more information, refer back to the Tool Setup and Utilities options explained under the specific section for the diagnostic tool you are using.

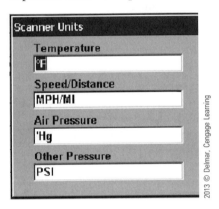

5. Press No to return to the Tool Box icon.
6. Press Yes again, scroll down to Save Data, and press Yes.

Save Data: This is a shortcut back to the Scanner Units option under the main Utilities menu of the scan tool. Navigate the dialogue box to change the percent after trigger option, where data is stored, and what file type all of the screen images will be when saved. For more information, refer back to the Tool Setup and Utilities options explained under the specific section for the diagnostic tool you are using or the Data Management section.

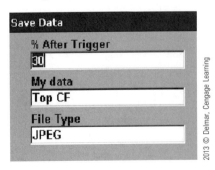

7. Press No to accept any changes and exit the Save Data menu box.
8. Scroll Down to Custom Data List and press Yes.

Custom Data List: This option allows you to select the PIDs that will be collected by the scan tool, thus increasing the sampling rate of the tool. There are two ways to access this option. On newer software this is found through the Toolbox icon in the upper right of the scanner screen. On older software you need to navigate to Text view and press No once to access the intermediate menu. This option is discussed in more detail later in this section.

9. Press No to exit.

Graphing Data View

We have previously discussed in detail how to gather data using the PID List and Text views where all of the values are displayed as numbers. This can be useful, but as vehicle communication and tool refresh rates get faster it can become difficult to "see" what is happening as the numbers are moving way too fast. When comparing relationships between different PIDs, it is important to see how the numbers are reacting to each other. This is very difficult for the human eye and brain to achieve when the only input is fast-moving numeric values. Remember that at any time the scan tool has the previous 2000 screen shots of data stored in its memory buffer. We could pause and look at each frame of numeric data. This will establish relationships between PIDs. But let's take it one step further. Assume we pick the RPM PID and plot the RPM value for each of the 2000 frames on a piece of graph paper and then connect the 2000 dots with a pen. Would this not be easier for our eyes and brain to interpret? It will be very apparent when the RPM is falling, raising, or steady, and its relationship to other PID values can also be more easily interpreted. The saying "A picture is worth a thousand words" is a very good summary of why graphing data is so valuable to technicians today. When viewing graphed data, it is important to remember that it is truly plotted PID List numeric values, not actual measured values. If you have some lab scope experience you may see data plots that resemble waveforms. But lab scope waveforms are front-door data, whereas we are currently still connected to the DLC and getting processed information from the PCM, so this graphed data is clearly back-door data. Graphed data is *not* front-door data (*not a waveform*) so it still only shows the effect and *not* the cause.

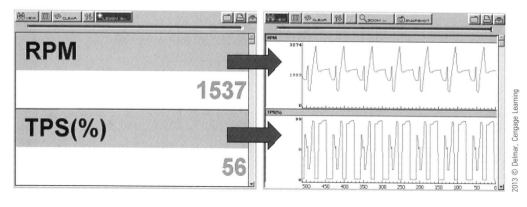

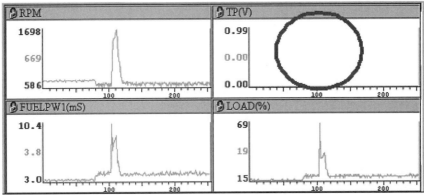

It is much easier to "see" the problem in this Graphing Data view. When the throttle is depressed we would expect to see a reaction (effect) from all four PID values. Clearly TPS (V) did not respond. This Back Door data quickly lead us to a problem area, now further Front Door testing will be needed to pinpoint the actual cause.

1. Depending on whether you are using a Solus Pro or a MODIS, the number of graphs able to be viewed at one time is slightly different. Here is a quick run down of the different graph views.

SOLUS PRO 1 Graph

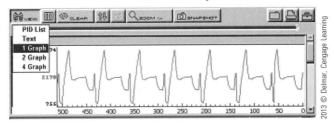

SOLUS PRO 2 Graph

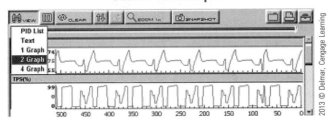

Graphing Data View **143**

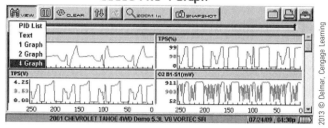

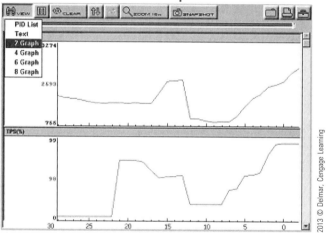

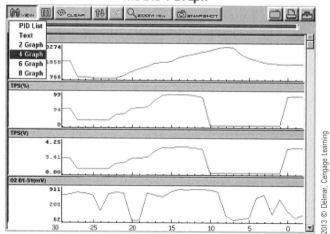

144 Chapter 6 Viewing and Interpreting Scan Data

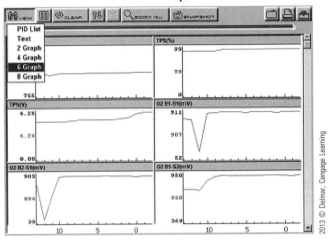

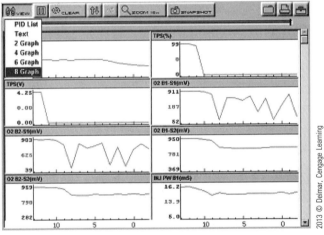

2. Scroll over to the View button and press Yes.
3. Scroll down to "2 Graph" and press Yes.

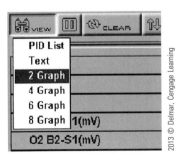

4. The basic graph displays the minimum, maximum, and current numeric values to the left of the graph.

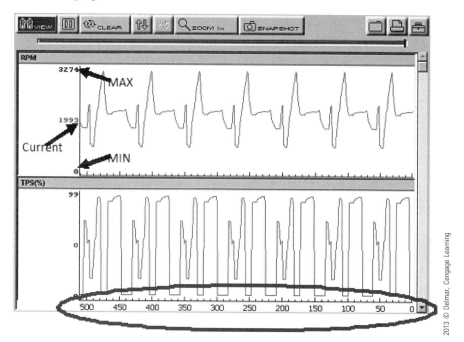

5. It is important to note that the horizontal scale does *not* use a time measurement. The horizontal scale is actually the data Frame number. In the example, we have 500 Frames of data visible. To reiterate what was said earlier, graphed data is just the PID List numeric values plotted for each Frame of data. The dots are then connected by a line to more clearly and visually show the changes in the values.

Freeze/Run

6. Scroll over to the right and highlight the Freeze/Run Button and press Yes. Notice how it changes when activated.

Freeze/Run: By default, when entering the data view, the diagnostic tool is automatically collecting and storing data to the memory buffer. The Freeze/Run or play/pause button allows the user to temporarily stop this recording and review the data while frozen.

7. When the data collection is frozen or paused, you can view each frame of data stored in the data memory buffer one at a time to look for problems with the values. The memory buffer is indicated by the blue line across the top of the screen. In this example, you can see 2000 frames of data stored in the memory buffer. All of this data is temporary data and is not saved anywhere on the tool. If the tool is turned off, all of this data will be lost. Each frame of data is a snapshot or picture of the PID values at that

moment in time. The scan tool takes approximately 10 shots per second of the incoming data stream and displays the results. The 10 shots per second is a rough estimate with many variables. It is dependent on the number of PIDs being collected and the vehicle year, make, and model. Some vehicle computers communicate faster than others. This is also where the Custom Data List mentioned earlier comes into play, as we can force the tool to sample faster. Again, this will be discussed in more detail later in this section. The maximum amount of frames able to be stored in the buffer can change depending on what version of software you are using. The latest versions have the capability to store at least 2000 frames of data. When the 2001st frame enters the memory buffer, the first frame is erased and dumped from the memory buffer. This way the scan tool always has the most recent 2000 frames of data for review. When the memory buffer is full, a small black check mark moves across the blue memory buffer bar, and this indicates the new frames that are being added and the old frames that are being dumped while data is being collected and recorded. When the scan tool is paused and not collecting or recording data, the black check mark indicates the frame that is being viewed. Just as with the PID List, the memory buffer will capture and hold the immediately previous 2000 frames of data.

8. There are 2000 frames of data stored in this screenshot, but the number by the memory buffer shows four. To explain this, you need to look at the bottom of the screen and see how many frames of data are visible to you at any one time. In this example, 500 frames are visible. In Graphing view, the number next to the memory buffer is the number of screens that can be reviewed not the number of frames as was the case in the PID List view. Simple math makes sense of this. Viewing 500 frames in a single screen will require four screens to show all 2000 captured data frames. This will be more clearly shown next when explaining the Review function.

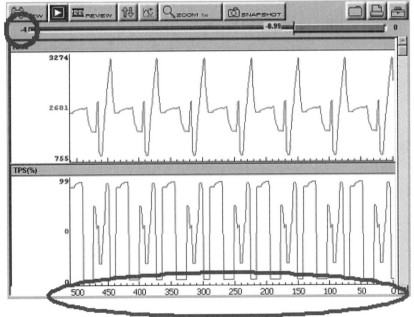

Reviewing Graphed Data and the Cursor Function

1. Scroll over to the right and highlight the Review button.
2. Press Yes to activate the review button. Note the change in color to indicate that it is activated. Remember that the Review button is only visible when the data stream is frozen or paused.
3. Use the Left and Right arrow buttons to review each screen of data. In this instance, there are only four screens to Review as each screen shows 500 frames of data, and there are 2000 frames of data total. Note the frame numbers along the bottom of the screens.

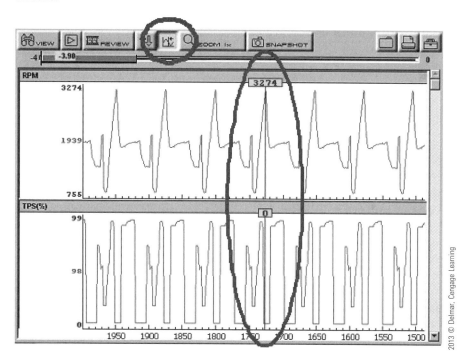

4. To scroll down the graphed PIDs you will need to press No to deactivate the Review button and then press Down to scroll the graphed PIDs. When the graphs you want are on the screen press No to return to the top tool bar and then press Yes to activate the Review button again to view the frames of data.

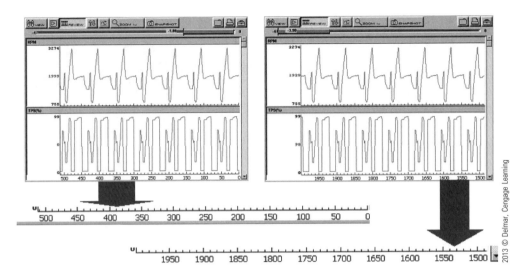

5. If you haven't done so already, press No to deactivate Review button.
6. Another function that is only available when the data stream is frozen or paused is the Cursor function. Scroll over to the right to highlight the Cursor button and then press Yes to activate it.
7. The Cursor button changes color just like the Review button to indicate it is active. Use the Left and Right arrow buttons to move the cursor along the graph. The numeric PID value for that Frame where the Cursor hits the graph is displayed in a box at the top of the Cursor. This can be described as overlaying the PID List view onto the Graph view. Now you can gather numeric data while easily viewing individual PID data over time and relationships between different PIDs. Displaying the numbers over the graphed data can make problems easier to detect. In the sample screenshot, you may miss the fact that the TP (%) dropped to zero while the RPM was peaking by just looking at the graph itself, but displaying the numbers using the Cursor function makes this problem stand out.
8. Press No to deactivate the Cursor Function. A dialogue box will appear, asking you if you would like the cursors to remain visible for reference or to completely turn them off. When left on for reference, the Cursor can no longer be moved. It will automatically disappear if the data stream is turned back on and starts recording new data. Press No to turn the Cursor off.

> Press 'Y' to leave cursors on for reference, 'N' or down-arrow to turn cursors off.

9. Scroll to the left and highlight the Freeze/Run button and press Yes to start recording data to the memory buffer. The PID values should now start to be graphed.

Clearing the Data Memory Buffer

1. When you restart the data recording, you will notice a vertical black line on the graph. This indicates that the data recording has been interrupted, frozen, and is now restarted. It does not indicate the amount of time elapsed between the interruption and the start of recording new data; it just indicates a break in the graphed data stream.

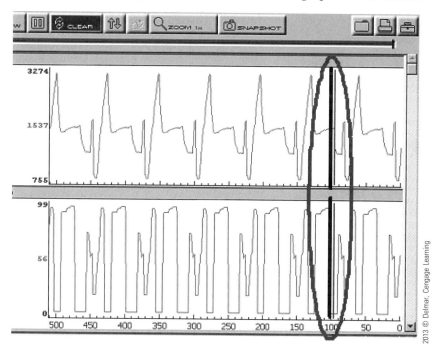

2. If this old graphed data is going to be a distraction to diagnosing the problem and you wish to start a new graph, without the black line, use the Clear function.

3. Scroll over to the right to highlight the Clear button and press Yes.
4. A warning box will appear asking you to confirm that you wish to clear the memory buffer and start a new data recording. Press Yes to confirm this.

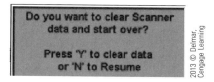

5. Notice that all of the old graphed data is now erased and a completely new graph is started. The blue memory buffer line has also been erased and is starting over. This will reset the captured minimum and maximum values also.

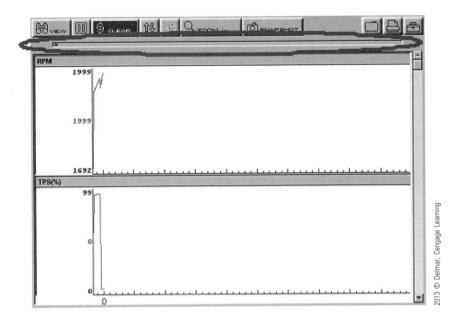

Graphing PID Sort

1. Scroll over to the left and highlight the Sort icon (opposite double vertical arrows) and press Yes.
2. The default PID order is Factory Sort, but this can make it more difficult to find a specific PID.

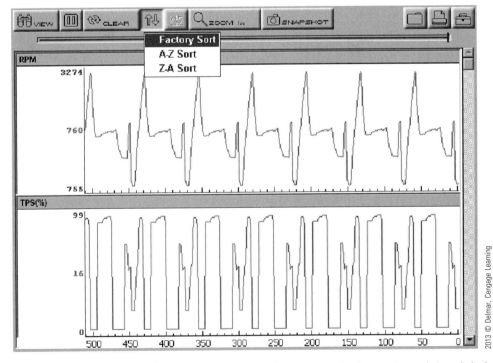

3. To more quickly find a PID that starts with a letter near the beginning of the alphabet, scroll down to highlight A-Z Sort and press Yes to alphabetize the PID list from A-Z.

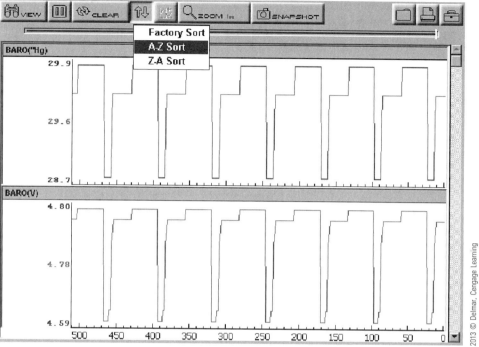

4. If the PID you are looking for starts with a letter that is near the end of the alphabet, it is quicker to sort the PID alphabetically from Z-A. To accomplish this scroll down and highlight Z-A Sort and press Yes.

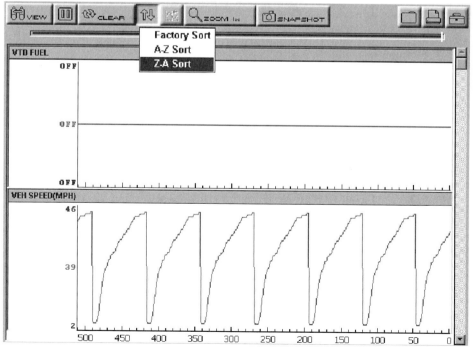

152 Chapter 6 Viewing and Interpreting Scan Data

5. The demonstration will leave the Graphed PIDs in the Factory Sort arrangement. It will be easier to follow along if you return yours to the Factory Sort arrangement also.

Graphing View Zoom

As stated in the PID List section the Zoom feature is different depending on what view is currently being used. The Zoom for the Graphing view is different from the Zoom in the PID List view. The Zoom in the PID List view would change the number of PIDs that were viewed on the screen. If we want to change the number of graphs on the screen, we have to go to the View button, *not* the Zoom button. The Zoom button in Graphing view changes the number of data frames that are being viewed on the screen, not the number of graphs. Keep the different Zoom functions clear so it will not cause confusion when working with the scan tool in the future.

Zooming in a graph will allow you to more easily see the changes in the values, especially if the change is small. It focuses your attention to a more specific portion of the graph, which can make it easier to analyze data and diagnose problems.

1. Scroll over to the right to highlight the Zoom button and press Yes.
2. Notice that at the current 1x Zoom we are viewing 500 frames of data: Frames 0-500.

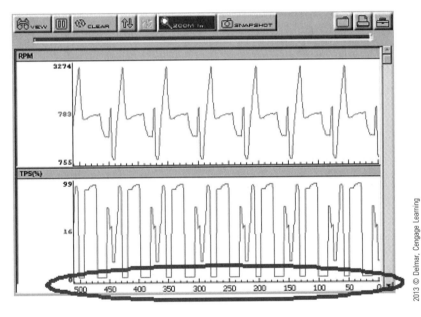

3. Scroll down to highlight 2x Zoom and press Yes. Notice that we are now only viewing 250 frames of data: Frames 0-250.

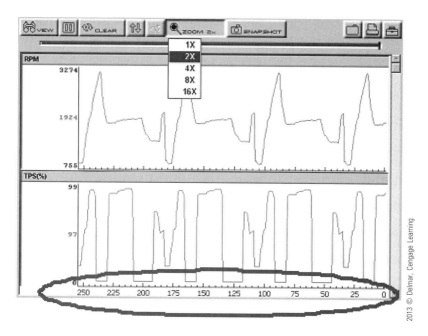

4. Press Yes to open the Zoom options and now scroll down to 4x Zoom and press Yes. Notice that we are now only viewing only about 126 frames of data: Frames 0-126.

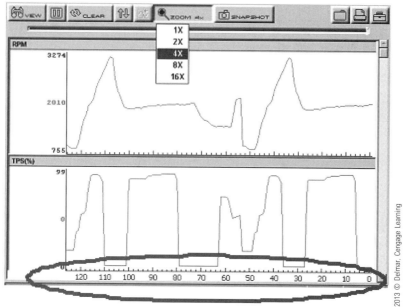

5. Press Yes to open the Zoom options and now scroll down to 8x Zoom and press Yes. Notice that we are now viewing only about 62 frames of data: Frames 0-62.

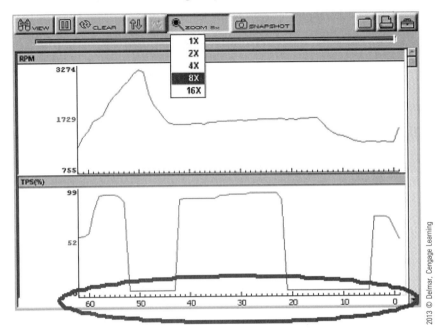

6. Press Yes to open the Zoom options and now scroll down to 16x Zoom and press Yes. Notice that we are now viewing only about 30 frames of data: Frames 0-30.

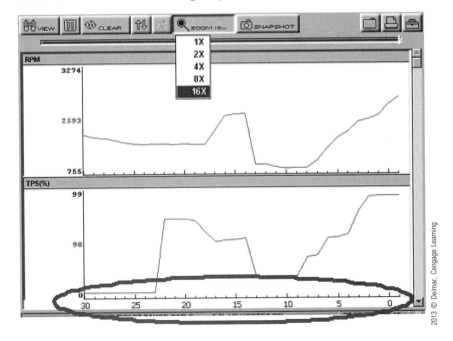

7. There is one more option found under the Zoom button but it is only available when the data stream is paused. Scroll to the Freeze/Run button and press Yes to Freeze the data stream.

8. Scroll over to the Zoom.

9. Press Yes to open the Zoom options and now scroll down to "Zoom Out" and press Yes. Notice that we are now viewing *all* of the 2000 frames of data: Frames 0-2000. This is a complete overview of all the data that was captured.

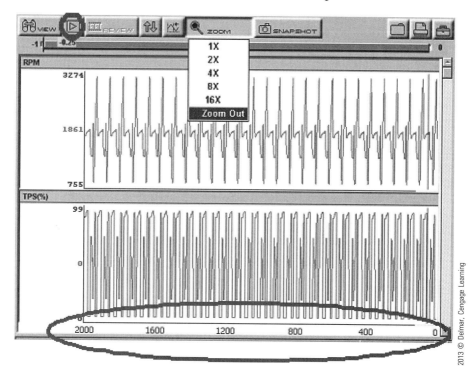

Zoom Usage Example

1. From the "Zoom Out" screen, highlight the Zoom button, press Yes, scroll to highlight 1x, and press Yes again.

2. Scroll and activate the Cursor button as described in the previous section.

3. Move the Cursor until you find the RPM value of 3274 and at the same time the TPS (%) value at 0.

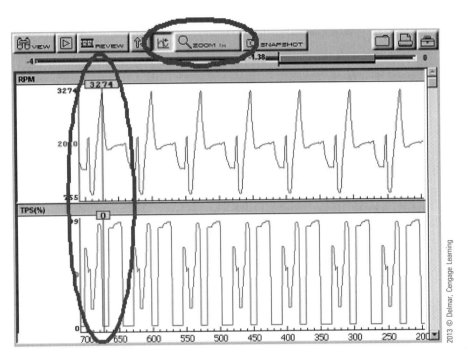

4. As stated earlier, with the graphed data in such close proximity, it may have been difficult to see this problem, but we can use the Zoom feature to make this easier to see.
5. The Zoom feature will always follow the Cursor. The Cursor will be placed near the center of the screen after the Zoom has taken place.
6. Press No to deactivate the cursor, but this time, when the dialogue box appears, press Yes to leave the Cursors on the screen for reference.

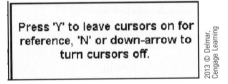

7. Scroll over to the Zoom button and press Yes.
8. Scroll down to 4x Zoom and press Yes. First notice how the Cursor, which was way over to the left, is now centered on the screen. Second, notice how the problem is more obvious with the graph spread out a bit.

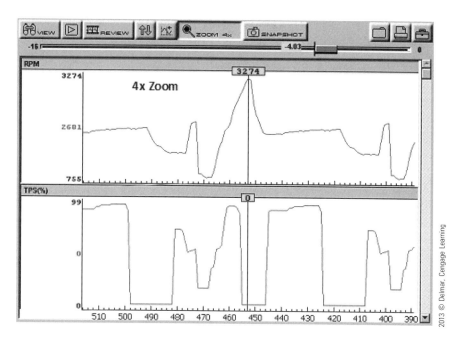

9. Press Yes to open the Zoom options and now scroll down to 16x Zoom and press Yes. The Cursor still remains at the center of the screen and now the problem is obvious. It is easy to see the peaking RPM and the dropping out of the TPS (%).

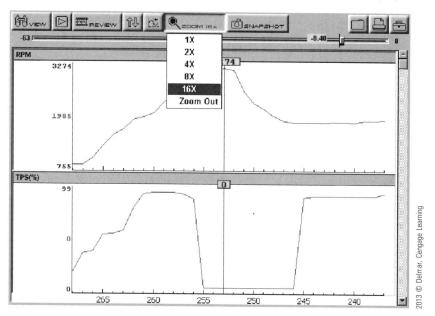

Snapshot

1. Scroll over to highlight the Snapshot button and press Yes.

2. There are two options under the Snapshot function; PID Trigger and Manual. Both of these options will be explained in detail near the end of this section in Advanced Capturing Techniques. Please refer to that section on the procedures on how to use the Snapshot functions.

3. Press No to exit the Snapshot functions.

Save, Print, and Utilities in Graphing View

10. Scroll over to the right and highlight the File Folder icon and press Yes.

    Save: The options available include Save Movie, which will save all 2000 frames of data, Save Frame, which will save all the PIDs in one frame of data, or Save Image, which will take a picture of the current screen. It is important to remember that when a Movie is saved using any view, it will only be played back in Graphing view. All saved Movies or Snapshots, no matter what view they were saved in, are played back in graphing view. Please refer back to the Data Management section for more detailed information about these options and how to use them.

11. Scroll over to the right and highlight the Printer icon and press Yes.

    Print: "Full Screen" will perform its normal function of printing the screen just as you see it. This will include the visible Graphs on the screen at the time of the printing. This is the only way to print the Graphs. To print all the graphs, you would have to print multiple screens scrolling to change the PIDs that are visible each time. "Full PID List" will print a text version of all the PID values, both on and off the screen, at the time of the printing. An example of what the Full PID List looks like is included.

```
2001 CHEVROLET TAHOE 4WD Demo 5.3L V8 VORTEC SFI    Copyright 2008 Snap-on Inc.
04/11/2009  12:47PM

RPM                    760        TPS(%)                  16
TPS(V)                 0.57       O2 B1-S1(mV)           768
O2 B2-S1(mV)           456        O2 B1-S2(mV)           716
O2 B2-S2(mV)           686        INJ PW B1(mS)          5.2
INJ PW B2(mS)          5.1        ST TRIM-1(%)            -3
LT TRIM-1(%)           12         ST TRIM-2(%)            -3
LT TRIM-2(%)           12         ST TRM AVG1 %            0
LT TRM AVG1 %           9         ST TRM AVG2 %            0
LT TRM AVG2 %           9         FT LEARN               YES
FT CELL                 4         MAF(Hz)               3370
MAF(gm/Sec)           10.89       MAP("Hg)              14.2
MAP(V)                 1.99       BARO("Hg)             28.7
BARO(V)                4.59       COOLANT( F)            199
INTAKE AIR( F)         120        START CLNT( F)         196
IAC POSITION           148        DES IAC g/sec         21.62
VTD FUEL               OFF        DESIRED IAC            148
SPARK ADV( )          22.5        KNOCK RET( )             0
EGR CLSD(V)           0.86        DES EGR(%)               0
DESIRED IDLE           550        EGR POS(V)            0.86
OPEN/CLSD LOOP         CLSD       MIL                    OFF
IGNITION 1(V)         13.7        VEH SPEED(MPH)           2
BRAKE SW               OPEN       ENGINE LOAD(%)           4
COOLANT( F)(2)         199
```

Print Full PID List Sample

12. Scroll over to the right and highlight the Tool Box icon and press Yes.
13. Press Yes to enter Custom Setup.

    Custom Setup: This is a shortcut back to the Scanner Units option under the main Utilities menu of the scan tool. Navigate the dialogue box to change between standard and metric units. For more information, refer back to the Tool Setup and Utilities options explained under the specific section for the diagnostic tool you are using.

    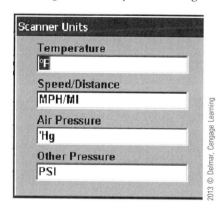

14. Press No to return to the Tool Box icon.
15. Press Yes again and scroll down to Save Data and press Yes.

    Save Data: This is a shortcut back to the Scanner Units option under the main Utilities menu of the scan tool. Navigate the dialogue box to change the percent after trigger option, where data is stored, and what file type all of the screen images will be when saved. For more information, refer back to the Tool Setup and Utilities options explained under the specific section for the diagnostic tool you are using or to the Data Management section.

    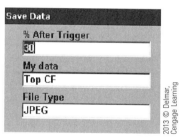

16. Press No to accept any changes and exit the Save Data menu box.
17. Scroll Down to Custom Data List and press Yes.

    Custom Data List: This option allows you to select the PIDs that will be collected by the scan tool, thus increasing the sampling rate of the tool. There are two ways of accessing this option. On newer software this is found through the Toolbox icon in the upper right of the scanner screen. On older software you need to navigate to Text view and press No once to access the intermediate menu. This option is discussed in more detail later in this section

18. Press No to exit.

Custom Data View with Graphs

When you are creating a Custom Data View in graphing mode, it is important to remember that each *column* of graphs needs to have one unlocked PID spot to allow the other PIDs to be scrolled through. Column is emphasized because this will be different for different tools. The most a Solus Pro can graph is four graphs at one time. They are displayed in two columns of two. Since each column has to have one unlocked spot, the maximum number of lockable graphs is two. The MODIS can graph eight graphs at one time, and this equates to two columns of four. The maximum number of lockable graphs here would be six. The confusion to avoid between the tools is that the MODIS will also display four graphs but in one column of four, so, in this case, three graphs can be locked in total, whereas four graphs on the Solus Pro, arranged in two columns, can only lock two. It is not the number of graphs on the screen that determines the maximum number that can be locked, but rather the number in each column. The general rule is that, at maximum, you can lock one less than the total number of graphs in that column. If your column has four graphs, then three can be locked. If there are two columns of four, then the total for the entire screen would be six. Remember that Custom Data View is when the PIDs/Graphs are just rearranged so we can view the relationships we need to diagnose the problem. Later we'll discuss the Custom Data List that actually limits the number of PIDs being collected from the vehicle. While these two features are similar in name, they are quite different.

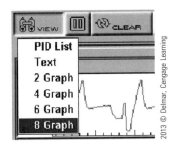

1. Scroll to the View button and press Yes.
2. Scroll down to 8 graphs (Solus users choose 4 graphs) and press Yes.
3. Press the Down arrow to activate the lower PID/graph display area. RPM should be highlighted.
4. Lets build a Custom Data View with the PIDs RPM, MAF (gm/Sec), MAP (V), Intake Air (°F), TPS (%), MAF (Hz), BARO ("Hg), and Coolant (°F) and place them in this order. It needs to be noted that when creating a Custom Data View when the graphs are in two columns, there are limitations to what PIDs can be placed where. The PIDs are automatically aligned in the two columns and cannot be moved from that column. This means if there are five PIDs in the first column you want to view together it would be impossible to do that with the eight graph view, as only three of them could be locked and four viewed due to the fact they were all in the first column. This can sometimes be manipulated by using the Sort option. When the PIDs are arranged alphabetically, they may no longer be in the same column, so now they could be rearranged and locked to provide the desired view. This is not a common occurrence, but it is possible.

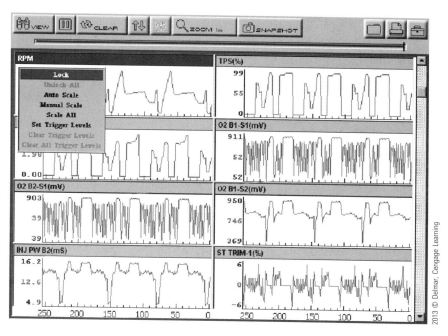

5. With RPM highlighted, press Yes, and then with Lock highlighted, press Yes. Notice the small lock icon that appears in the upper right corner of the graph. Just as before, in the PID List view RPM will be locked and no longer scroll with the rest of the PID streams.

6. Just as with the PID List, we will need to scroll down until our desired PID is positioned under RPM and then scroll back up to that desired PID and lock it in place.

7. Note the lock icon at the top right of the RPM PID and that RPM will not scroll with the rest of the data PIDs. Scroll down until MAF (gm/Sec) is positioned under RPM.

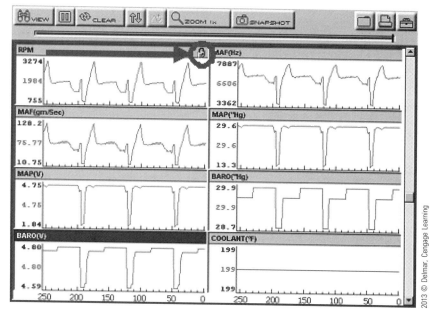

162 Chapter 6 Viewing and Interpreting Scan Data

8. Now scroll back up until MAF (gm/Sec) is highlighted, press Yes, highlight Lock, and then press Yes to lock it into place.

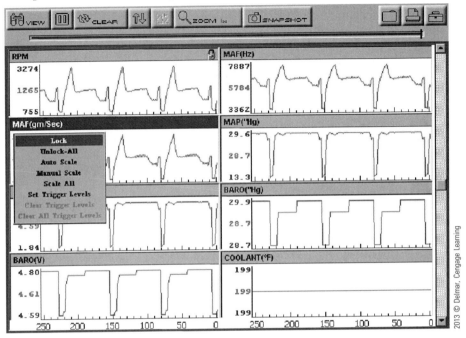

9. Scroll down until MAP (V) is positioned under MAF (gm/Sec), and then scroll back up until MAP (V) is highlighted, press Yes, highlight Lock, and then press Yes again to lock it into place.

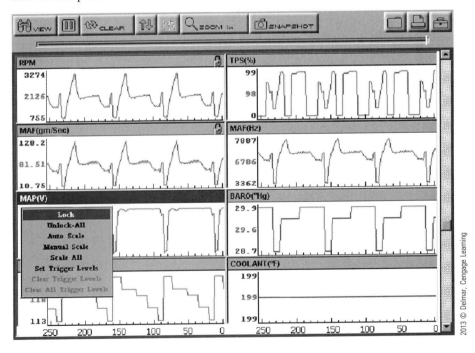

Graphing Data View **163**

10. Scroll over to the right and then all the way back up to the top of column two until TPS (%) is highlighted.
11. With TPS (%) highlighted, press Yes, and then with Lock highlighted press Yes.

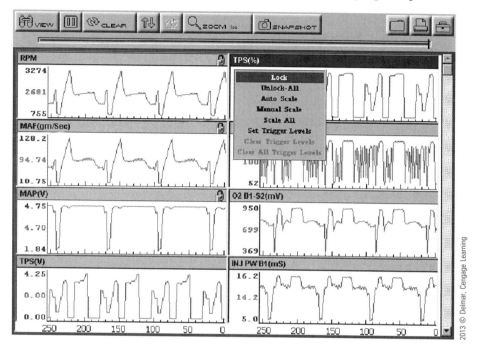

12. Scroll down until MAF (Hz) is positioned under TPS (%), and then scroll back up until MAF (Hz) is highlighted, press Yes, highlight Lock, and then press Yes again to lock it into place.

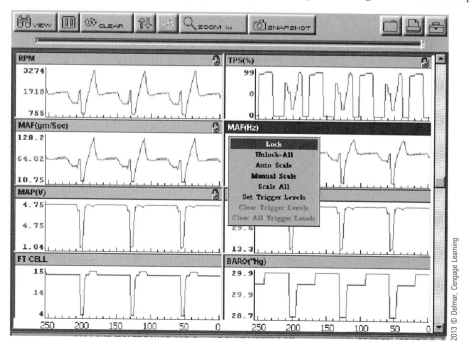

13. Scroll down until BARO ("Hg) is positioned under MAF (Hz), and then scroll back up until BARO ("Hg) is highlighted, press Yes, highlight Lock, and then press Yes again to lock it into place.

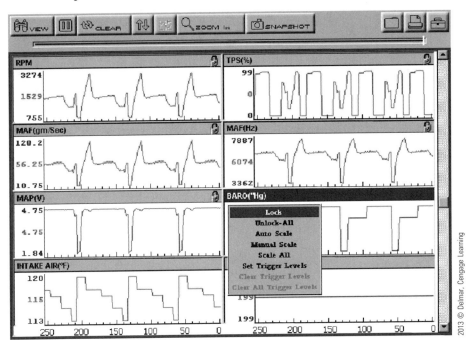

14. Scroll down until Intake Air (°F) and Coolant (°F) occupy the bottom row.

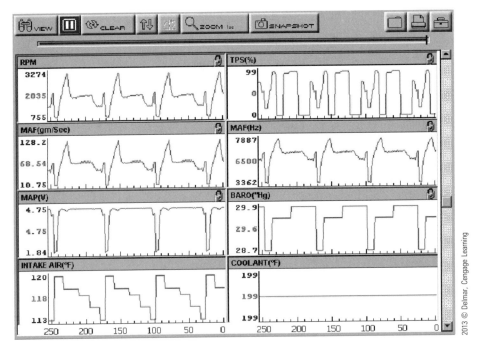

15. At this point, if you attempt to lock either of the bottom PIDs your will not be able to do so. The dialogue box will open with the Lock command ghosted (grayed) out and only the Unlock-All command as an option.

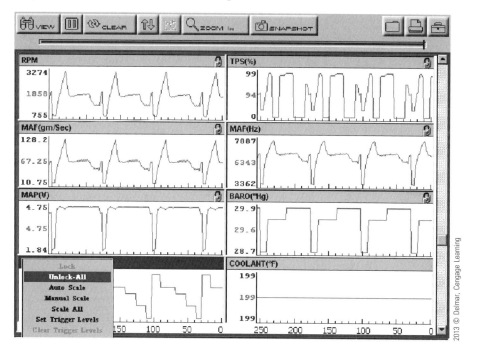

16. At this point, you can start to analyze this Back Door data and look for relationships between the desired graphs thus beginning to narrow down problem areas which will require further testing using front-door data equipment.
17. Let's put a number of the functions we have discussed together and see how they can help us diagnose problems.
18. Scroll to the Freeze/Run button and press Yes to Freeze the data stream.

19. If necessary, we can use the Review button to analyze all of the captured data.

20. Scroll and activate the Cursor button as described in the previous section.

21. Move the Cursor until you find the RPM value of 3274 and at the same time the TPS (%) value at 0.

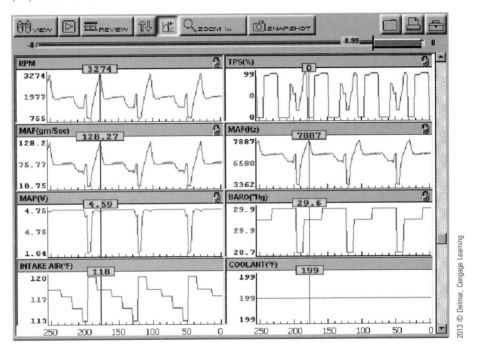

22. As stated earlier, with the graphed data in such close proximity it may have been difficult to see this problem but we can use the Zoom feature to make this easier to see.

23. Press No to deactivate the cursor. But this time, when the dialogue box appears, press Yes to leave the Cursors on the screen for reference.

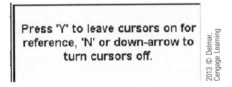

24. Scroll over to the Zoom button and press Yes.

25. Scroll down to 4x Zoom and press Yes for a closer analysis of the data.

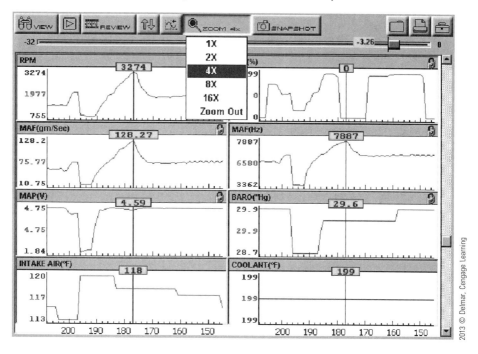

26. Press Yes to open the Zoom options and now scroll down to 16x Zoom and press Yes. This gives the closest view possible when analyzing the data.

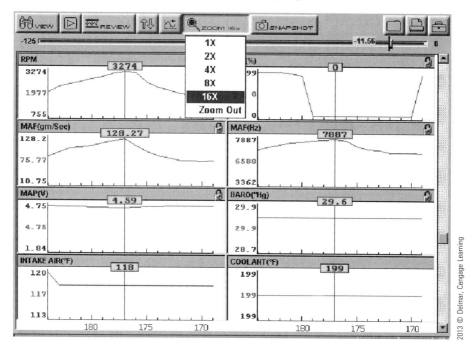

27. When finished viewing this data, and as a small review, return the screen to a more neutral setting. The following directions will not be as detailed as earlier ones. Please refer back to the proper section if needed.
28. Zoom to 1x.
29. Unlock All PIDs.
30. Remove the Cursors.
31. Run/Play the data recording.
32. View 2 Graphs.
33. When you have finished, the screen should look like this.

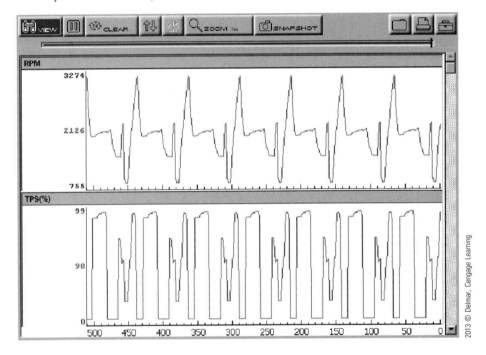

Graphing PID Scaling Options

1. Scroll down to highlight the RPM PID and press Yes.
2. Scroll down to Manual Scale and press Yes.

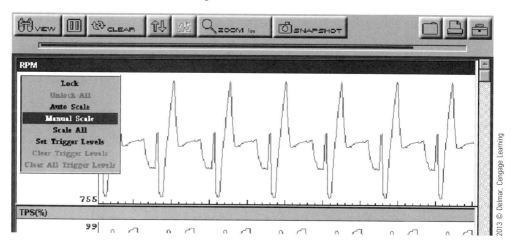

3. A horizontal line will appear across the graph with the maximum value displayed in a box near the center of the screen. This is the maximum value scaling line.

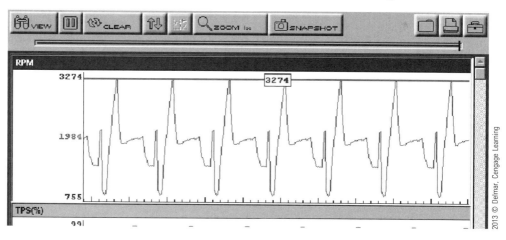

4. Press the Down arrow button to lower the line and the value. Notice what happens to the graph as you do this. Bring the value down close to 2500. Notice how the graph begins to stretch to show more detail. This cuts off the top part of the graph.

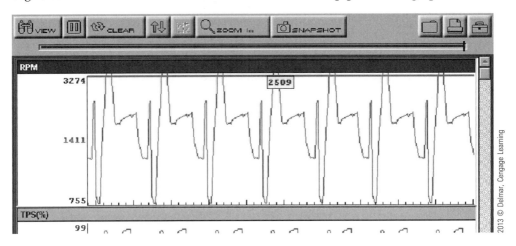

5. Press Yes to activate the minimum value scaling line.

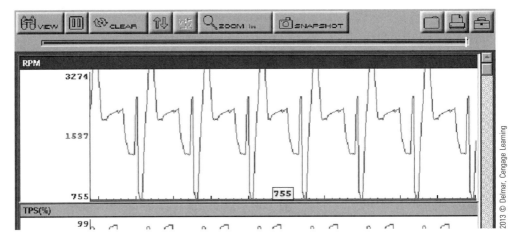

6. Press the Up arrow button to set the minimum value just as you did the maximum. Bring this value close to 1500. This cuts off the bottom part of the graph while further stretching the graph to provide more detail.

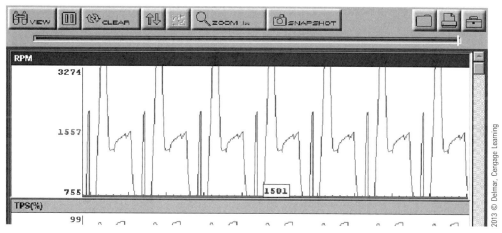

7. Press No to exit the scaling option. Notice that the graph remained scaled to the upper and lower values you determined. It is important to remember that the minimum and maximum values on the left side of the graph do not change; only the graph view changes, as indicated by the missing peaks and valleys of the graph. This can be done to any number of individual PIDs to allow you the best view possible of the data you need to see to make a high-quality diagnostic decision.

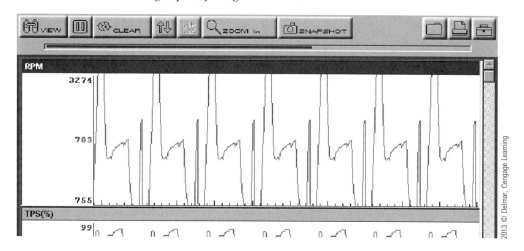

Before Manual Scale

Look at the center section of the graph and notice the lack of detail.

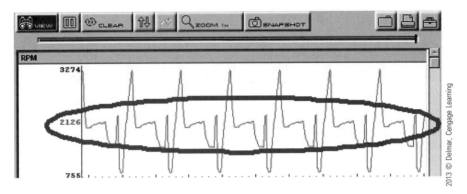

After Manual Scale

Look at the center section of the graph and notice the more detailed view and how little glitches become more obvious.

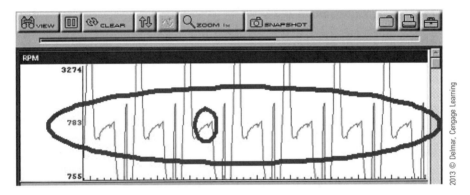

8. The next two scaling options, "Auto Scale" and "Scale All," are very similar, with one important exception. They both reset the Manual Scale option and return the graph to the current minimum and maximum values. The difference is that Auto Scale will do this only for the PID that is currently highlighted, while Scale All resets all the PIDs being collected by the scan tool.

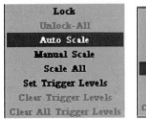 

9. With RPM highlighted press Yes to open the options dialogue box.

Custom Data List **173**

10. Scroll down to select Auto Scale and press Yes. Notice how the graph automatically captured the current minimum and maximum values and displays the graph accordingly. The example screen shot also has the TPS (%) PID Manually Scaled. Note the missing peaks and valleys, but watch how it remains scaled while RPM is returned to a normal un-scaled view when using Auto Scale.

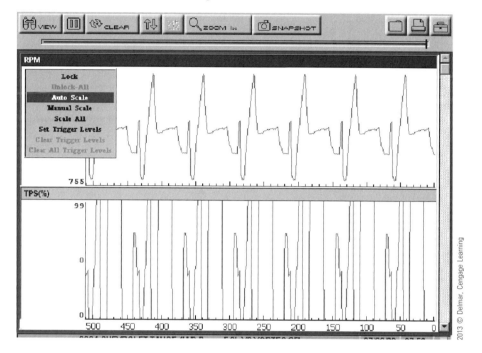

11. When using Scale All, all PIDs are returned to an un-scaled view.

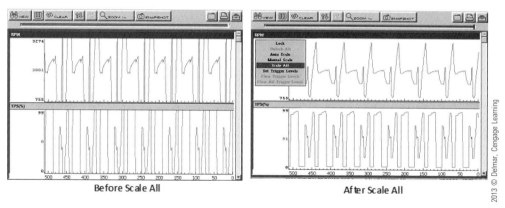

Before Scale All   After Scale All

Custom Data List

When a Custom Data List is created the number of PIDs being collected by the scan tool is limited, but the advantage is having a higher display or refresh rate for the remaining PIDs.

174 Chapter 6 Viewing and Interpreting Scan Data

Do not confuse a Custom Data List with the previously discussed Custom Data View. Again, the Custom Data List limits the number of PIDs being collected by the tool, while the Custom Data View simply rearranges the order of the PIDs that have been collected. The reason the display rate increases when using a Custom Data List can be described using an engine analogy. Assume the scan tool is a 100 horsepower engine and always runs at this peak horsepower. If there are 50 PIDs of information to collect from a vehicle, the scan tool will use all 100 hp to collect these PIDs allocating about two horsepower per PID. Now assume we decided that we only need to look at 10 PIDs to determine the cause of the problem, so we use the Custom Data List to collect *only* those 10 PIDs. Now the scan tool will still use all 100 hp but will allocate 10 hp per PID instead of two, as before. This means that these 10 PIDs will be collected, analyzed, and displayed at a much higher rate. Most problems do not require a lot of PIDs to make a diagnostic decision, but many vehicle data streams may collect 50 plus PIDs. It is beneficial to the technician to focus on the data that is relative to the problem and have that data be of the highest quality possible. The Custom Data List option makes this happen.

1. Be sure the data stream is *not* Frozen and that data is being recorded into the memory buffer.
2. If not in Text view, scroll to the View button and select Text view.
3. Look at the numbers as they appear on the screen. Remember the speed at which they are changing.
4. With the Scanner button highlighted, press No once.

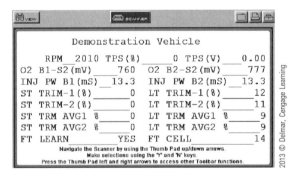

5. This will take us to that intermediate menu we discussed earlier. Now we are going to look at the Custom Data List option.
6. Scroll down to Custom Data List (version 9.2 and earlier) and press Yes.

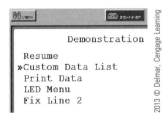

OR

Custom Data List 175

7. Scroll over to the far right and highlight the Toolbox icon and press Yes. Scroll down to Custom Data List and again press Yes (version 9.4 and later).

8. Notice that all of the PIDs listed have a small asterisk next to the PID name. This indicates that the PIDs are currently being collected. By default, *all* PIDs are collected.

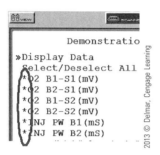

9. Scroll down to Select/Deselect All. The quickest way to build a Custom Data List is to deselect all of the PIDs and then go and pick the PIDs you wish to view. Press Yes to Deselect All and remove the asterisk from each PID.

When no PIDs are selected, there are still three default PIDs that will be collected. RPM, TPS (%), and TPS (V) are always collected and can *not* be turned off on the current software version as of this writing. This may change in the future and the software is upgraded. Any additional PIDs you select will always be added to these three.

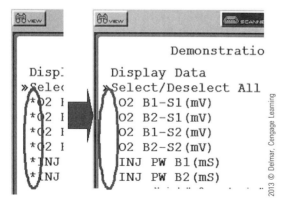

For the demonstration, scroll down and select O2 B1-S1 (mv) and press Yes. Notice an asterisk appears next to the PID name. This indicates it is going to be collected.

176 Chapter 6 Viewing and Interpreting Scan Data

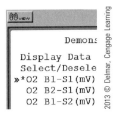

10. Scroll down to select MAF (gm/Sec) and press Yes.

11. Scroll down to select MAP ("Hg) and press Yes.

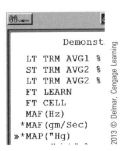

12. Now hold down the Up arrow button until you return to the top of the page and select Display Data.

13. Press Yes to Display Data. This will display six PIDs of data; the default three and the additional three we chose.

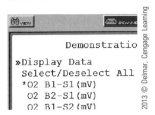

14. Do you notice a difference between the speed at which the PID numbers are changing? It should be much faster. This is more easily seen using the graphing view.

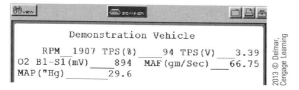

15. Scroll over to the View button and select 6 Graph.
16. Here is a before and after Custom Data List screen shot of this 6 Graph view. Notice that each graph is showing 250 Frames of data but the "After" graph displays many more cycles of the data because the tool was able to display the data faster.

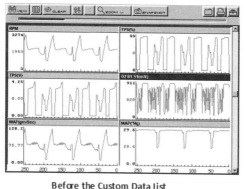
Before the Custom Data List

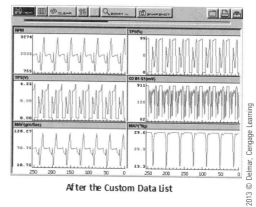
After the Custom Data List

17. Remember that *all* of the options and features we have already discussed and will discuss in this section can be used when using a Custom Data List. Choose any number of Graphs you wish to view and then Lock to rearrange PID positions, Scale, Sort, and Zoom PIDS while using the Cursor function to gather numeric data from the graph after the data stream is frozen.
18. As a review, and to make the rest of the demonstration easier, return the data list to its original status by selecting Text view, press No once to access the intermediate menu, select Custom Data List, choose Select/Deselect All, and finally select Display Data to collect view all the PIDs.

Custom Data List Summary

The Custom PID List is a great way to increase the data-capturing capabilities of your scan tool. It increases the display rate of the scan tool because fewer PIDs are being collected. There are two ways of accessing this option, depending on the version of software you have. One way is through the Toolbox icon in the upper right of the scanner screen. The other way is to navigate to Text view and press No once to access the intermediate menu. If your diagnostic data is better and more accurate, then your final diagnostic results will be better and more accurate as well. This will save time and money in the long run.

Advanced Capturing Techniques

Snapshot: Snapshots are covered in detail in the Data Management Section. Here is a quick summary. A Snapshot is slightly different from a Movie; it can be made up of both past data stored in the buffer and new data that is continually being fed into the buffer. The moment the user activates or takes the Snapshot is called the trigger. The % after trigger option, part of the Save Data dialogue box, allows the user to change how much old information is taken from the buffer and how much new information is collected after the trigger. Save Data options can be quickly accessed using the Toolbox icon in the upper right corner of the screen. The default setting is typically 30% after trigger. Assuming a buffer of 2000 frames of data, this means that once the user triggers a Snapshot, the tool takes 1400 frames (70%) of the data from the buffer that was automatically recorded before the trigger and then continues to capture 600 frames (30%) of data after the trigger. These 2000 frames are bundled together and stored as a Snapshot file, with the triggering event located between the beginning and the end of the entire Snapshot file and at the user specified percentage. If the user wants to see more of the effect of the triggering event, then the percent after trigger should be increased. If the user wants to see more of the cause of the triggering event, then the percent after trigger should be decreased. When viewing this file, unlike a Screen Shot, data can be moved and manipulated just like in a Movie, but now the user can also view the data that was recorded after the triggering event.

Manual Snapshot

Manual Snapshot is when the scan tool is armed to take a Snapshot and then waits until the trigger is activated by the user. When using Manual Snapshot, the trigger is you pressing the Yes button while the Snapshot button is highlighted and blinking on the screen. The procedure for this is discussed next.

1. From a common recording 2 Graph view, Scroll the top tool bar until Snapshot is highlighted.

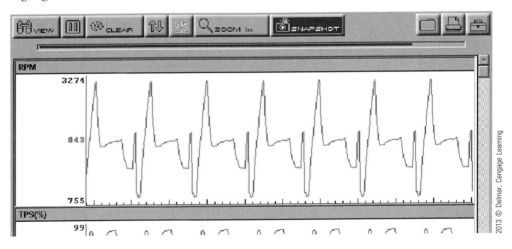

2. Press Yes to open the Snapshot options.

3. Scroll and select Manual.

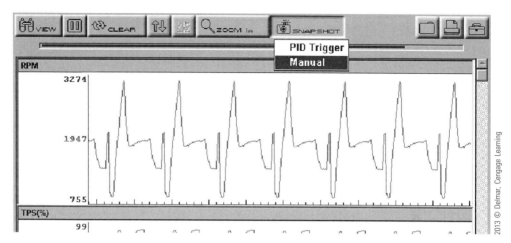

4. Press Yes to arm the Manual Snapshot. You will know it is armed, ready, and waiting for the trigger because the button will be flashing.

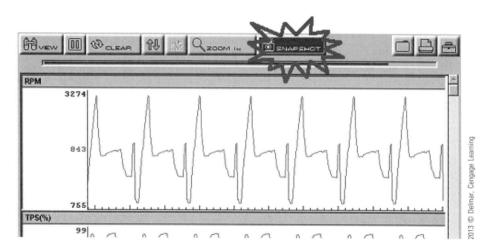

5. When you wish to trigger the Snapshot, just press Yes.

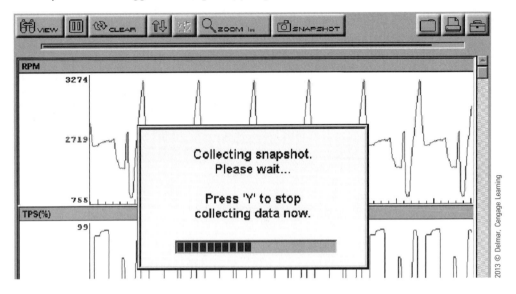

6. Because it is going to continue to collect and record data for the set percent after trigger (default 30%), it may take a minute or more for the Snapshot to finish and completely save the file. A SC(P) Scanner Snapshot will be saved to the data device you selected and can be reviewed by going through the Data Management option from the Main Screen. If you have any questions, refer to the Data Management Section of this book. Remember that *all* of the options and features we have discussed so far in this section can be used when reviewing a Movie or Snapshot file. Choose the number of Graphs you wish to view and then Rearrange, Scale, Sort, and Zoom PIDs while using the Cursor function to gather numeric data from the graph.

PID Trigger

PID Trigger is an auto capturing technique where the scan tool will capture and save data without you having to act as the trigger. We are going to set upper and lower trigger levels on a maximum of three different PIDs. After setting the trigger levels and then arming the PID Trigger, the scan tool will collect data as normal, but as soon as the data stream crosses one of the set trigger levels, the scan tool will automatically save a Snapshot. A generic example of where this may be used could be on a vehicle with an intermittent rough idle. Assume the normal idle of this vehicle is 800 RPM, but intermittently the RPM will flare up to 1200 and then drop down to 400 or less and almost die. You could set the scan tool for Manual Snapshot and then stand there by the vehicle waiting for the intermittent idle problem to happen and capture the Snapshot by pressing Yes. Depending on how intermittent the problem is, a lot of valuable time could be wasted. Instead, you could set the upper RPM trigger level near 1000 RPM and the lower trigger level near 600 RPM (the actual values you would determine based on the diagnostic situation) and arm the PID Trigger. The scan tool would collect data as normal, but as soon as the idle problem begins and either the upper or lower trigger level is reached, the scan tool will automatically capture and save a Snapshot without any additional input from you. When this happens, the scan tool beeps loudly a couple of times. While you are busy performing another job, a more profitable one than sitting and waiting for a problem to occur, you are now alerted to the fact the scan tool just captured the required data needed to begin diagnosing this vehicle.

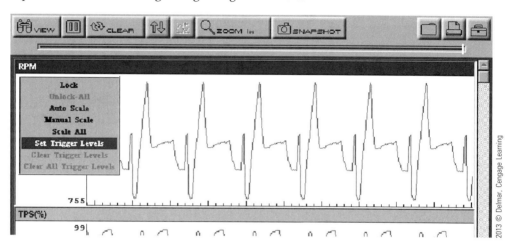

1. From a common recording 2 Graph view scroll to the RPM PID and press Yes.
2. Scroll down to "Set Trigger Levels" and press Yes.
3. This is similar to setting the Manual Scale option discussed earlier. The first horizontal line to appear is the Upper Trigger threshold. For this demonstration, move the trigger to below 3200 RPM and press Yes.

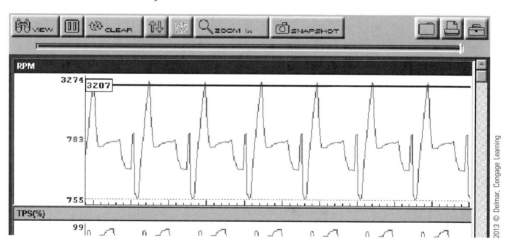

4. Notice that a small dashed line appears, indicating where the upper trigger has been set, and now the Lower Trigger threshold is active and able to be set.

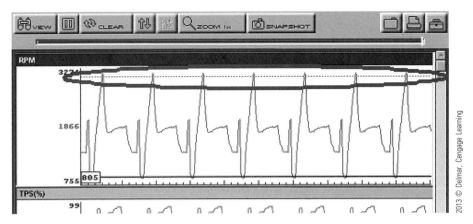

5. For this demonstration, move the lower trigger to above 800.
6. If you press Yes again, you will reactivate the Upper Trigger and be able to change its value. When the upper and lower triggers are properly set to your desired values, press No to exit. There will be a dashed line indicating both the upper and lower trigger levels on the graph.

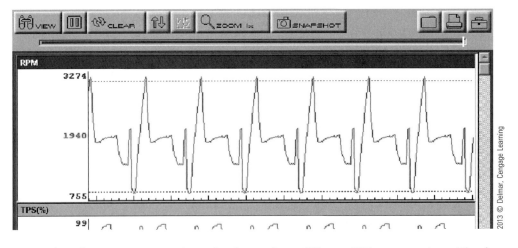

7. Remember that you can set trigger levels on three different PIDs at one time. The first value to hit either the upper or lower trigger level will cause the Snapshot to be saved.
8. Scroll over to the Snapshot button and press Yes.

9. Highlight PID Trigger and press Yes.

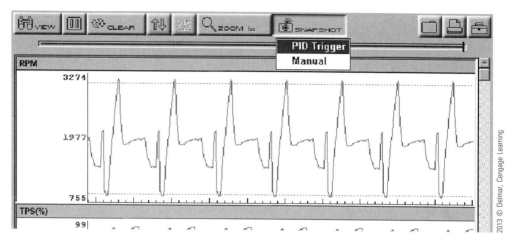

10. Wait. It may take a minute or so for the demonstration data stream to hit either the upper or lower trigger level and cause the Snapshot to be saved. You will hear a beep when it does and see a screen indicating that a Snapshot is in the process of being saved.

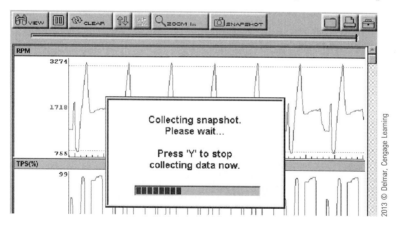

11. Wait a minute for the Snapshot to finish saving.
12. After the Snapshot is saved, notice that the data stream was automatically frozen or paused. This allows you immediate access to the review features described earlier so you can begin to diagnose the problem.

184 Chapter 6 Viewing and Interpreting Scan Data

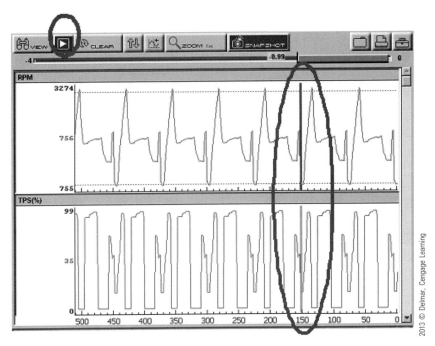

13. Notice that the PID that acted as the trigger (RPM) has a blue line indicating the trigger point, while all of the other PIDs have green trigger-point lines. If we had set multiple triggering PIDs, we would now know which one actually triggered the Snapshot.
14. Scroll to the View button and select 4 Graphs.
15. What was the actual triggering value? Use the cursor function to try to find it. Move the cursor as close to the blue triggering line as possible.

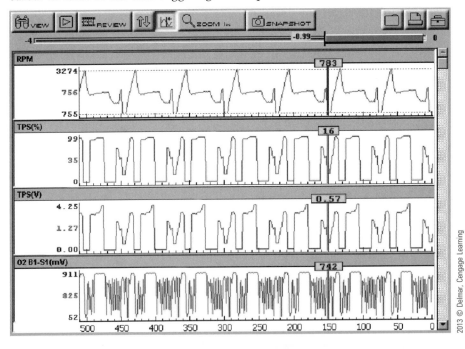

16. You may find it hard to see the actual triggering event. Is there a way we can solve this?
17. Use the Zoom function to take a closer look and easily find the actual triggering value. In this example, a 16x Zoom was used. Was the cursor on the actual triggering point?

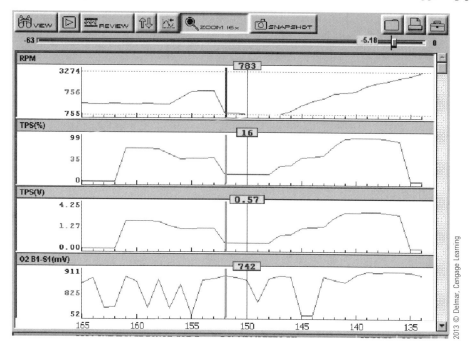

18. Reactivate the Cursor button to scroll over and find the actual trigger value.

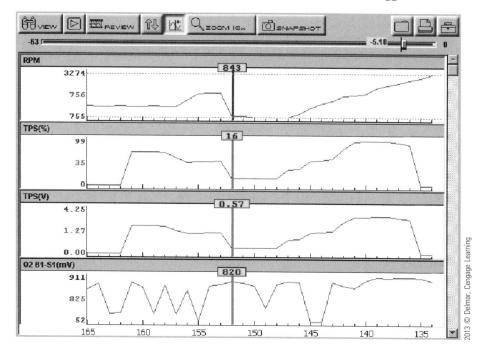

19. Remember that the Snapshot file is saved to the data device of your choosing, set through the Save Data option box available through the Toolbox icon in the upper right corner of the screen. To access this SC(P) or Scanner Snapshot file, you will need to highlight View and press No to return to the Main Screen. From here scroll down to the File Folder icon and select Data Management to be able to Load the Snapshot file. Refer to the Data Management section of this book for more information. A Snapshot file has all the features and options that we have learned about in this section. The only limitation is that you cannot collect any new data as this is a saved file, and the scan tool may no longer even be connected to the vehicle.

20. After you set trigger levels on the maximum number of PIDs, which is three, if you try to set another trigger level on the fourth PID, you will receive a warning message.

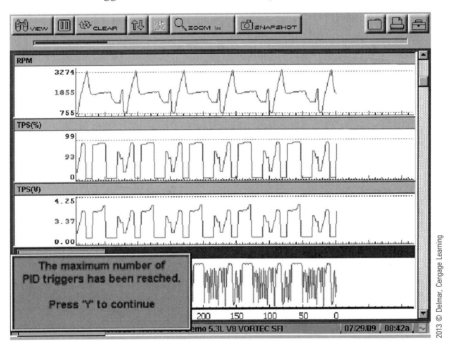

PID Trigger Summary

A PID Trigger can be used to safely trigger a Snapshot while on a test drive. Depending on the vehicle you are connected to, look for a PID value that you can manually operate while driving the vehicle, such as a Brake Switch or A/C Clutch. Set the PID Trigger on this switch. When the problem symptom happens and you wish to capture the data, instead of trying to manipulate the scan tool buttons and possibly getting into an accident, simply tap the brake or turn on the air conditioner and that will trigger the Snapshot. All the PIDs are being recorded, so while the Brake switch or A/C Clutch PIDs will probably not help us in the diagnosis, we just rearrange the PIDs so the data we want is available and ignore the PIDs we used to trigger the Snapshot. One other technique similar to this is to trigger off the MIL PID. If available, have the Snapshot triggered when the MIL light turns on. This is similar to Freeze Frame Data, but you are now using all of the PIDs available through the VIN-specific database. Practice with PID Trigger to become more familiar with it and come up with your own creative ways to capture the data you need.

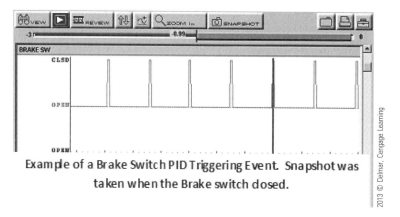

Example of a Brake Switch PID Triggering Event. Snapshot was taken when the Brake switch closed.

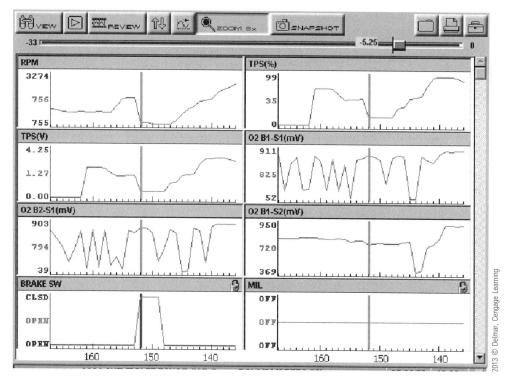

The key concepts to effectively use PID Triggers are:

- Upper and Lower Trigger Levels must be determined by the technician and set for at least one PID.
- A maximum of three PIDs can be used as triggers.
- The Snapshot will be triggered by the first trigger level that is hit. Upper level, Lower level, or PID does not matter. The first trigger level on any triggered PID to be hit will cause the Snapshot to be saved.

- The Blue trigger line indicates the PID that was actual trigger for the Snapshot while the Green trigger lines are used to reference the triggering point on the remaining PIDs.
- PID Triggers can be set to numeric values as with RPM or TPS (V) or can be set to trigger off events such as the Brake switch closing or the MIL turning on. By triggering off the MIL you can basically create your own freeze frame data, but using all of the VIN specific PIDs and information, not just the ones collected under Global OBD-II

Viewing and Interpreting Data Summary

One of greatest advantages of using a scan tool is the ability to view data processed by the vehicle's computer. This window into what the computer thinks is a very important step in the diagnostic process. The vehicle's computer will output multiple data streams called Parameter Identifications or PIDs. There are many viewing options available to choose from when looking at these PIDs. No matter what view is used all of the PIDs are being recorded all of the time, not just the ones on screen. Text view aligns the PIDs into columns. This view is based on the MT2500 or Red Brick Scanner, and it is not commonly used on newer vehicles as the data is sampled so fast the numbers are hard to read. PID List view puts all of the PIDs into a single column matrix, which makes finding a specific PID easier, especially when using the Sort option. The PID List view also allows the technician to review many recorded data frames. This makes it easier to look for changes in the PID values and find any problematic PID values. The easiest way to view data and catch problems is to use the graphing options.

It is important to remember that when viewing graphed data, this is still back-door (effect) data and *not* front-door or lab scope data. Although the graphs may look like waveforms, they actually are not. Graphing data allows you to view changes in a particular PID value over time, no matter how fast the data is coming in, and you can also look for relationships among multiple PIDs. The Cursor function in graph view allows the technician to get numeric data from the graph, so it is like having the PID List and Graph view working at the same time.

There are many options that can help us see the data we need to, easier and more quickly, no matter what view we are in. One option is to boost the effectiveness of the tool by using the Custom Data List feature to increase the sampling rate of the data stream. This is done by handpicking the PIDs you want to collect and view, thus limiting the total number of PIDs being collected. The data you gather will be of higher quality and allow for better diagnostic decisions.

To help view relationships between multiple PIDs, you can rearrange, organize, and lock PIDs into new positions to create a Custom Data View. We can also use the Zoom features to get a better look at the data. The Zoom feature will work differently depending on the selected view. In PID List view, Zoom will change the number of PIDs visible on the screen. This can make the PIDs bigger and therefore easier to read, but it also can be helpful when making a Custom Data list that will focus your attention to a specific number of PIDs and block out other data that may be confusing or distracting. When in Graph view, Zoom changes the number of data frames that are viewed at one time. With the data recording or live you can zoom in and view a small portion of the recorded data. This enlarges the graphs and makes smaller changes in the data easier to see. When the data is Frozen or paused the Zoom Out option is available. This allows you to view all of the recorded data frames at one time. This can be helpful in quickly identifying problem areas in the data streams. Unique to the Graphing view is the ability to Manually Scale the graphs to show details that are lost when the graph is automatically capturing and plotting the minimum and maximum values. These highs and lows can be cut off and bring back the detail lost in the mid part of the graph.

All of the views allow you to save or archive the data for a later time. Printing the screen is a hard copy option, while saving a Movie is the electronic option. All Movies playback in Graphing view no matter what view they were saved in. While in the Graphing view, you can also use advanced capturing techniques, such as the Snapshot function. A Snapshot is similar to a Movie, but it has a triggering event and continues to collect for a set amount of time after the triggering event. A PID Trigger can be used to capture a Snapshot automatically when a PID hits a predetermined value.

The key to becoming a Power User of the scan tool is to combine all of these features to produce the best possible data image on the screen so you can make the right diagnostic decision the first time. Remember that *all* of the options and features we have discussed in this section can be used together, with Graphing view offering the most options and flexibility. Choose the number of Graphs you wish to view and then Lock to rearrange the PIDs, Scale, Sort, and Zoom PIDs to build the best possible view and finally use the Cursor function to gather numeric data from the graph. Viewing scan tool data is an important beginning step in the diagnostic process, so the better the data gathered here, the easier the rest of the diagnosis will be.

Summary Points

- Any scanner connected to the Data Link Connector (DLC) is gathering only back-door data which has been processed by the vehicle's computer.
- The scan tools continuously record incoming data and keep the last 2000 frames in the temporary memory buffer, which is automatically erased after ending communication with the vehicle. For the data to become permanent, the user must select the save Movie option, after which the 2000 frames from the memory buffer are saved as a Movie file to the long-term internal memory, which can only be deleted by the user. Saving a Snapshot is similar to this and will also save 2000 frames of data.
- By default, *all* PIDs are collected and recorded into the memory buffer, and this includes all the PIDs that are not able to be seen on the screen. Everything is collected and recorded all the time, unless you specify otherwise.
- Custom Data Lists can increase the sampling rate of the tool by limiting the number of PIDs being collected while Custom Data Views only rearrange the PIDs so that specific PIDs can be placed next to each other for comparison. Combined, these features allow the technician to make very accurate and efficient diagnostic decisions.
- Graphing data is *not* front-door Data, but only the back-door Data frames plotted and connected. While this makes problems much easier to see, it is still only the effect and not the cause.

Review Questions

1. Technician A states that the features and options available when viewing data are the same for OBDII data as they are for VIN specific data. Technician B states that PID stands for Parameter Identification and refers to the individual data streams such as RPM, TPS (%), and MAF (gm/Sec). Who is correct?

   a. Tech A only
   b. Tech B only
   c. Both Tech A and Tech B
   d. Neither Tech A nor Tech B

2. Technician A states that only the PIDs shown on the screen are being captured and recorded into the memory buffer. Technician B states that the memory buffer automatically captures and holds the most previous 2000 frames of data. Who is correct?

   a. Tech A only
   b. Tech B only
   c. Both Tech A and Tech B
   d. Neither Tech A nor Tech B

3. Two technicians are discussing the screenshot below. Technician A states that pressing Yes now will pause the data stream. Technician B states that pressing No will take you to an Intermediate Menu with additional options to customize this screen. Who is correct?

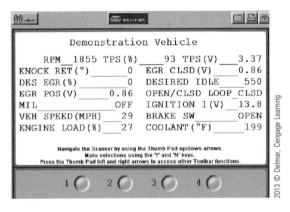

   a. Tech A only
   b. Tech B only
   c. Both Tech A and Tech B
   d. Neither Tech A nor Tech B

4. Two technicians are discussing the screenshot below. Technician A states that pressing Yes now will save all of the data frames stored in the memory buffer along with an additional 30 seconds worth of frames. Technician B states that the Save Frame option only allows you to review the PIDs shown on the screen at the time the frame was saved. Who is correct?

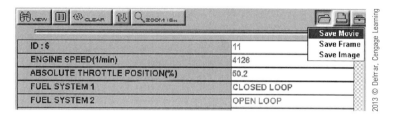

a. Tech A only
b. Tech B only
c. Both Tech A and Tech B
d. Neither Tech A nor Tech B

5. Two technicians are discussing the screenshot below. Technician A states that the data stream is currently frozen or paused. Technician B states that pressing Yes now will allow you to scroll up and down the PID List. Who is correct?

a. Tech A only
b. Tech B only
c. Both Tech A and Tech B
d. Neither Tech A nor Tech B

6. Two technicians are discussing the screenshot below. Technician A states that there are currently two locked PIDs and pressing Yes will add a third. Technician B states that a maximum of six locked PIDs is possible on this screen. Who is correct?

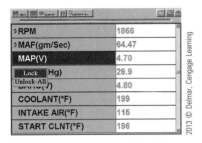

   a. Tech A only
   b. Tech B only
   c. Both Tech A and Tech B
   d. Neither Tech A nor Tech B

7. Two technicians are discussing the screenshot below. Technician A states that pressing No now will take you back to the Text view. Technician B states that the memory buffer is completely full. Who is correct?

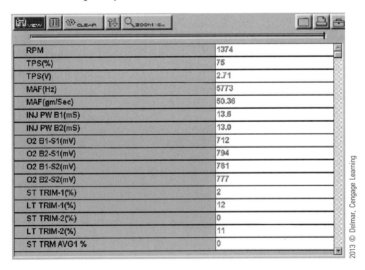

   a. Tech A only
   b. Tech B only
   c. Both Tech A and Tech B
   d. Neither Tech A nor Tech B

8. Two technicians are discussing the screenshot below. Technician A states that when the screen is un-frozen and data starts to be collected again the highlighted button will change to the "Clear" button which can be used to erase the memory buffer. Technician B states that currently data frame number 1999 is being viewed. Who is correct?

a. Tech A only
b. Tech B only
c. Both Tech A and Tech B
d. Neither Tech A nor Tech B

9. Two technicians are discussing the screenshot below. Technician A states that this is back-door graphed data and is *not* a front-door data waveform. Technician B states that the current screen displays about 510 frames of data and pressing Yes now will Freeze the data recording. Who is correct?

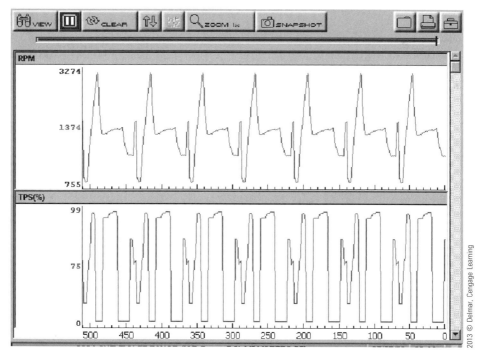

a. Tech A only
b. Tech B only
c. Both Tech A and Tech B
d. Neither Tech A nor Tech B

10. Two technicians are discussing the screenshot below. Technician A states that using the left and right arrow button will move the Cursor across the screen. Technician B states that increasing the Zoom from the current 4x to 8x will increase the number of data frames visible on the screen. Who is correct?

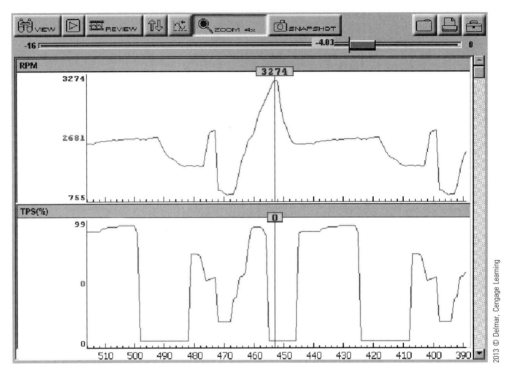

a. Tech A only
b. Tech B only
c. Both Tech A and Tech B
d. Neither Tech A nor Tech B

11. Two technicians are discussing the screen shot below. Technician A states that one more PID is able to be locked on this screen. Technician B states that using the left and right arrows will scroll the data buffer to allow more of the graphed data frames to be viewed. Who is correct?

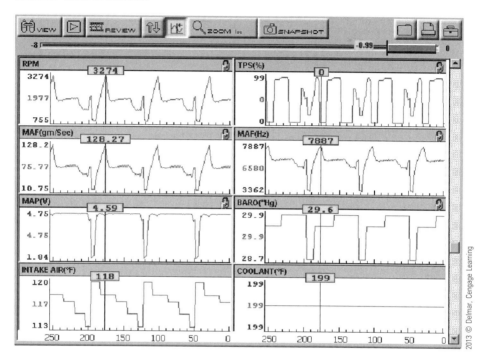

a. Tech A only
b. Tech B only
c. Both Tech A and Tech B
d. Neither Tech A nor Tech B

12. Two technicians are discussing the screen shot below. Technician A states that the TPS (%) PID has been Manually Scaled. Technician B states that pressing Yes now will Auto Scale all PIDs. Who is correct?

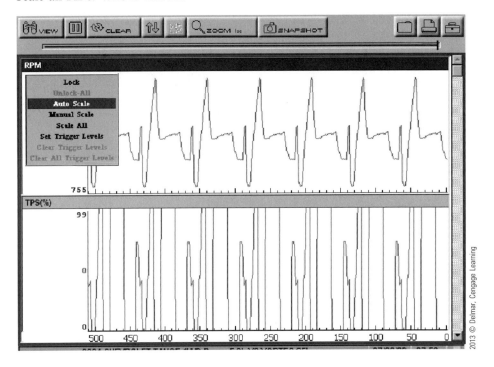

   a. Tech A only
   b. Tech B only
   c. Both Tech A and Tech B
   d. Neither Tech A nor Tech B

13. Two technicians are discussing the screenshot below. Technician A states that pressing Yes will allow you to Create a Custom Data List that increases the sampling rate of the tool. Technician B states that this menu has to be accessed through the Text view. Who is correct?

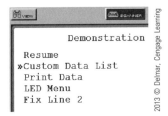

   a. Tech A only
   b. Tech B only
   c. Both Tech A and Tech B
   d. Neither Tech A nor Tech B

14. Two technicians are discussing the screenshot below. Technician A states that trigger levels have been set for both the RPM and TPS (%) PIDs. Technician B states that pressing Yes will arm the PID Trigger, after which a Snapshot will be taken as soon as the data value hits either the upper or lower trigger level. Who is correct?

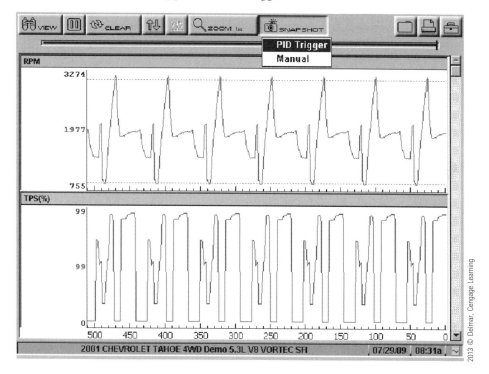

a. Tech A only
b. Tech B only
c. Both Tech A and Tech B
d. Neither Tech A nor Tech B

15. Two technicians are discussing the screenshot below. Technician A states that a maximum of 4 PIDs can be used as triggers. Technician B states that RPM was the PID that caused the PID Trigger to save a Snapshot. Who is correct?

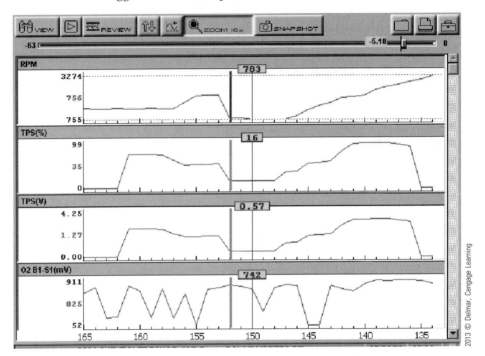

a. Tech A only
b. Tech B only
c. Both Tech A and Tech B
d. Neither Tech A nor Tech B

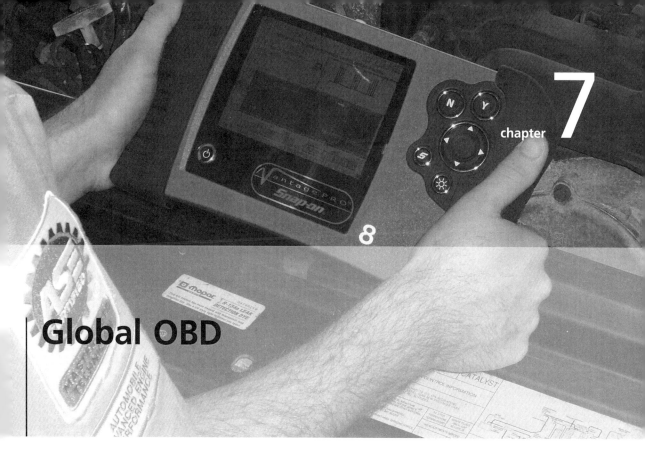

Global OBD

Upon completion of the Global OBD module, you will be able to:

- Compare and contrast Global and VIN specific systems
- List the service modes of Global OBDII and explain the use of each
- Identify when to use Global OBDII data versus VIN specific data while diagnosing a vehicle
- Practice accessing Global OBDII data using the built-in OBD Training Mode
- Efficiently navigate the Global OBDII software and quickly access available data
- Explain the function of the OBD Health Check option

REMINDER: This section is written so that the reader can follow along using the Solus Pro or MODIS scan tools, if available. The Scanner software is identical on both of these tools, so all of the procedures will be the same. Most of the screenshots come from a MODIS but the Solus Pro will be the same except for the screen size, so while the MODIS may show eight graphs, the Solus Pro will only show four. Everything else will function exactly the same. You are strongly encouraged to follow along using the diagnostic tools, as this provides the hands-on knowledge necessary for passing the Certification Exams. The only required equipment is the Scanner unit itself and the AC power adapter cord. The demonstrations built into the Scanner software will provide the rest of the required material.

This section focuses on the Global OBDII part of the scanner software. As the purpose of this textbook is focused on diagnostic tool usage and not vehicle systems, we will not go into great detail about each of the service modes. Instead, we will outline each and show how to access the information available. To fully appreciate and understand everything in this section, it is important for you to get training in the theory and diagnostic procedures using Global OBD principles. Find a certified training center by visiting the NC3.net website and look for schools that offer this type of training and specific certification. The direct link is: www.nc3.net/partners/certification-center-locations/ Under Global OBDII, you can access codes, gather data, and perform other functions. While all of this is outlined in this section, please also refer to the other specific sections of this book that detail each function. If after reading the Global OBDII section you are still a little unclear about using trouble codes, please refer to the Codes Menu section of this book for more information. While that section is written using VIN specific details, the function will be very similar when using Global OBDII. This holds true for gathering data and viewing it in both PID and graphing form. Greater detail on how to do this is located in the Viewing and Interpreting Data section of this book. Use this section as an outline of basic scanner use and then refer to other specific sections in this book for greater detail. All of the same features and options used for gathering and organizing data when identifying the vehicle using the VIN can be applied and used in the Global OBD section. So, remember: Bring this knowledge together to expand the capabilities of both the tool and of yourself.

Global OBDII

When connecting to the vehicle's computer using a scan tool, we have the choice of two different areas for gathering information. All of this information is accessed through the same data link connector (DLC) and comes from the same physical box (vehicle computer). It is easiest to picture the inside of the vehicle's computer, also called a Powertrain Control Module (PCM), with two separate rooms in it; one being the VIN Specific or OEM room and the other being the Global OBDII room. The OEM room is accessed through the VIN specific databases on the scan tool, whereas the Global OBDII is accessed by choosing this option from the Main Screen. Both the VIN and OEM room will give you a lot of information, but some information in the VIN room can be manipulated by the manufacturer. One example of manipulated information is called substituted values.

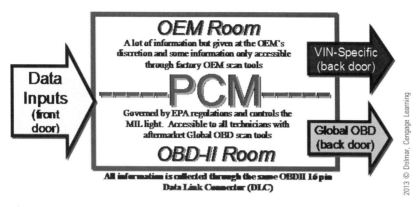

If all parameters on a vehicle are running within specification with the exception of one, the OEM room will then derive a value for the "bad" component based on the inputs of the other "good" components. This calculated good value will then be substituted for the bad value. This will keep the vehicle running fairly well, and most customers will never know the difference. This also keeps the MIL off, and thus keeps customers happy. As problems pile up and finally substituted values can no longer keep the vehicle running within legal limits, the MIL will illuminate and the vehicle will likely be brought in for service. It is because of substituted values that technicians should always look at the Global OBDII information first, as this information cannot be substituted: It is governed by federal law.

An illustration of this point can be seen using a Chrysler intake air temperature (IAT) sensor. Begin by unplugging the IAT sensor; this simulates an open sensor input to the PCM. When you look at this value by identifying the vehicle using the VIN, you see a value of 80 to 85°F. This keeps the vehicle running pretty well, the customer satisfied, and to the technician this may not seem like a bad value and probably will not raise any obvious red flags. Leaving the sensor unplugged, identify the vehicle using the Global OBDII option. Look at the data for this sensor. The reading will be -40°F, as it should be with the sensor disconnected. This will be an obvious red flag to most technicians.

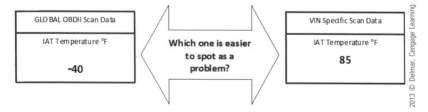

When you looked at the Global OBDII data, this problem was much easier to see and thus diagnose. This is because the Global OBDII side is governed by the Environmental Protection Agency (EPA) and is part of the federal government's clean air laws. When the vehicle malfunctions to the point where excessive emissions are being released into the atmosphere, the MIL light will be turned on by the Global OBDII side of the PCM. The OEM has no control over this information. Always check the Global OBDII side of the PCM for codes and data that may not be located in the VIN room. The advantage you have using Snap-On diagnostic tools is the Generic Functions feature located in the VIN specific databases. After identifying a vehicle using the VIN, choose the Generic Functions option (this will be discussed in greater detail later in this and other chapters) to access some, not all, of Global OBDII data. This can save a lot of time, as it does not require a complete exit out of the VIN specific side or a new identification using the Global OBDII option. Remember, Generic Functions equals Global OBDII.

Global OBDII Navigation

1. Scroll down and highlight "Global OBDII" and press Yes.

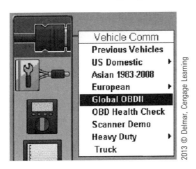

2. Press Yes and select "Generic OBD/EOBD"

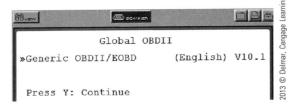

3. Press Yes and select "OBD Training Mode"

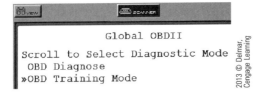

When diagnosing an actual vehicle, you highlight and select OBD Diagnose, the OBD training mode allows you to see what information is available through the Global OBD side of the computer and learn what the different service modes can give you for information in an educational setting. Again, no vehicle has to be connected to the tool for the training mode to work, as it simulates being connected to a live vehicle.

4. Press Yes and select "Start Communication"

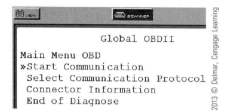

Start Communication will initialize communication with the vehicle's computer so that you can begin your diagnosis. While you are using this option, the communication protocol is automatically detected and the proper one used. If you need to manually select the communication protocol, then scroll down to the second option and select the proper protocol from a list. If you cannot find the OBD connector on the vehicle, scroll and select the third option "Connector Information." At this point, you will need to select the vehicle manufacturer and make and then the tool will provide a description to the location of the connector. Selecting "End of Diagnosis" will bring you back one screen.

5. When you are using the training mode, please ignore this screen, as it is not necessary to connect the tool to the vehicle. Just press Yes.

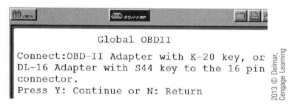

6. This screen identifies the number of computer control modules detected and communication protocol. Press Yes to continue.

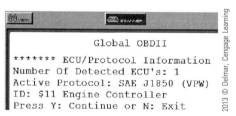

7. This is the Main Global OBDII Menu, where you will gain access to all of the service modes. Each service mode is numbered in hexadecimal format, indicated by the dollar sign before each number.

As the purpose of this textbook is focused on diagnostic tool use and not vehicle systems, we will not go into great detail about each of the service modes. Instead, we will outline each mode and show how to access the information available. To be a power user of any diagnostic tool, you must be comfortable navigating and using the Global OBDII functions within that tool. The OBDII Training Mode is a great way to gain some navigational experience and acquire background knowledge of what types of information are available when using OBDII. Please explore the information found under the different service modes and how the scanner displays this information. You will frequently notice a dollar sign ($)when working in Global OBDII. This is the symbol for hexadecimal, a number format used by computers. For the most part, this is not something you need to be concerned with. You may need to have a deeper understanding of hexadecimal format when working with Service $06, but generally you can ignore the dollar sign and focus on the information.

Readiness Monitors

1. With "Readiness Monitors" selected press Yes.

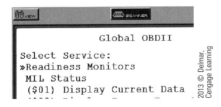

2. With "Monitors Complete Since DTC Cleared" selected press Yes. Scroll down to view the status of all the available readiness monitors. Each monitor is identified by name and then indicates whether the monitor test is not supported, completed, or not complete.

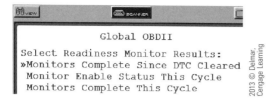

3. When you have finished reviewing the readiness monitors, press No to exit.

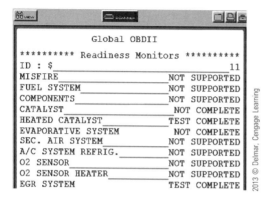

4. Scroll down to "Monitor Enable Status This Cycle" and press Yes.

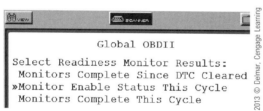

5. Scroll down to view all of the enabled monitors during the current drive cycle. Each monitor is identified by name, and then indicates whether its current status is enabled or disabled.

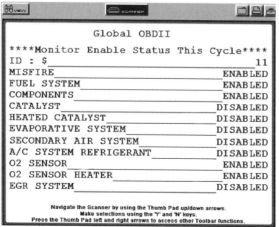

6. When you have finished reviewing the readiness monitors, press No to exit.
7. Scroll down to "Monitors Complete This Cycle" and press Yes.

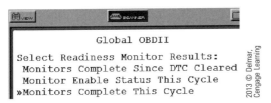

8. Scroll down to view all of the monitors completed during the current drive cycle. Each monitor is identified by name, and then indicates whether its current status is completed or not completed.

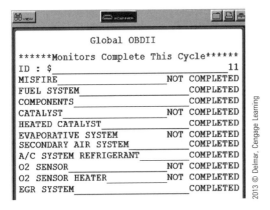

9. When you have finished reviewing the readiness monitors, press No to exit.
10. Press No one more time to return to the Main Global OBDII Menu.

MIL Status

1. Scroll down to select MIL Status and press Yes.

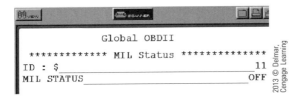

2. This quickly identifies the engine controller identification and whether or not the MIL Status is on or off.
3. When you have finished reviewing the MIL Status, press No to exit.
4. Just as before, once at the intermediate menu press No again to return to the main Global OBDII menu.

($01) Display Current Data

1. Scroll down to "($01) Display Current Data" and press Yes.
2. Please see the "Viewing and Interpreting Scanner Data" section for details on how to navigate, configure, and customize the data menu and displays.

Items covered in detail found in the "Viewing and Interpreting Scanner Data" section include:

- View
  - PID List
  - Text
  - 2 Graph
  - 4 Graph, etc.

- Freeze/Run

- Review

- Clear

- Sorting (PID List)

- Zoom

- File Folder icon
    - Save Movie
    - Save Frame
    - Save Image

- Print Icon
    - Full Screen
    - Full PID List

- Graphing PIDS
    - Review
    - Zoom

- PID Trigger
- Snapshot
- Cursor

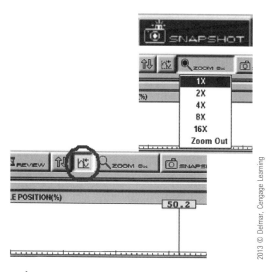

These features are mentioned here to help you become a true power user of the tool and to understand the diagnostic advantages of using OBD. However, you will want to review the "Viewing and Interpreting Scanner Data" section to take advantage of all features and benefits outlined here.

3. If you are in Text view, with the Scanner button at the top center of the screen highlighted press No. Otherwise, press No to return to the OBD Main Menu.

($02) Display Freeze Frame Data

1. Scroll down to "($02) Display Freeze Frame Data" and press Yes.

2. Freeze Frame Data is a snapshot of the data stream at the time the trouble code was set. This can be very useful in diagnosing the problem, especially if it is an intermittent one. Think of this as the black box or flight recorder that is on airliners. This feature functions just as any saved Movie or Snapshot. Please refer back to the Data Management section, parts of the last Service $01 section, and the VIN specific Scanner section for details on how to manipulate the different views to find the information you desire.

3. When you have finished reviewing the Freeze Frame Data, press No to return to the Global OBDII Main menu.

($03) Display Trouble Codes

1. Scroll down to "($03) Display Trouble Codes" and press Yes.

```
              Global OBDII
Select Service:
  Readiness Monitors
  MIL Status
  ($01) Display Current Data
  ($02) Display Freeze Frame Data
»($03) Display Trouble Codes
  ($04) Clear Emissions Related Data
```

2. The current trouble codes will be listed on the screen. Be sure to scroll all the way down to the bottom to view all of the codes. In this example, there are five codes, so be sure the last code you view is identified as 05/05 or the fifth of five codes. See the "Codes Menu" section of this textbook for more details on how to view and print trouble codes.

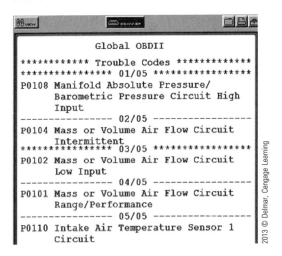

4. Then, with the Scanner button at the top center of the screen highlighted, press No.

($04) Clear Emission Related Data

1. Scroll down to "($04) Clear Emission Related Data" and press Yes.

```
Select Service:
 Readiness Monitors
 MIL Status
  ($01) Display Current Data
  ($02) Display Freeze Frame Data
  ($03) Display Trouble Codes
» ($04) Clear Emissions Related Data
  ($05,06,07) Display Test Param./Results
```

2. This feature simply requires you to read and follow the on-screen directions. Press Yes to continue.

   Remember that all diagnostic information will be erased when this function is performed. This includes all trouble codes and freeze frame data. Plus, all readiness monitors will be reset.

```
          Global OBDII
******** Clear Diagnostic Data ********
This service will clear all diagnostic
data (Codes,Freeze frame,Test results)!
Press Y: Continue or N: Abort
```

3. Read and follow the on-screen directions. Press Yes to continue.

```
          Global OBDII
First try to clear emission related data
with ignition on, engine off. If not
successful, retry with engine running.
Press Y: Continue or N: Abort
```

```
         Global OBDII

DTC Erase Routine (KOEO):
      Erasing DTC's
```

4. The last message will confirm that "Diagnostic data has been cleared!" Press Yes to continue.

```
          Global OBDII

    Diagnostic data has been cleared!

Press Y or N: Continue
```

5. You should automatically be brought back to the Global OBDII Main Menu.

($05) Oxygen Sensor Monitoring

1. Scroll down to "($05, 06, 07) Display Test Param./Results" and press Yes. Note that CAN (Controller Area Network) systems do no support Service $05. In CAN systems, these tests are part of Service $06.

   ```
   ($02) Display Freeze Frame Data
   ($03) Display Trouble Codes
   ($04) Clear Emissions Related Data
   »($05,06,07) Display Test Param./Results
   ```

2. Select ($05) Oxygen Sensor Monitoring and press Yes.

   ```
              Global OBDII
   Select Test Parameters/Results:
   »($05) Oxygen Sensor Monitoring
    ($06) Non-Cont. Monitored Systems
    ($07) DTCs Detected During Last Drive
   ```

3. Listed are the various oxygen sensor tests that can be viewed using the scan tool. We'll look at one as an example, but please take a moment to view each test to become aware of the information available through this service mode.

   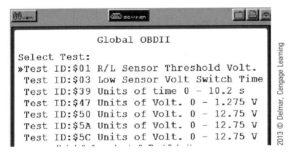

4. Select the first test, "Test ID: $01 R/L Sensor Threshold Volt." and press Yes.

5. The data for this test is given in Text view by default. However, it can be switched to either PID List or Graph View by scrolling over to the View button and selecting a different view, just as we have done while using earlier service modes. See the "Viewing and Interpreting Scanner Data" section for more details.

6. When you have finished viewing the data for this test and want to select another, you will need to first scroll to the View button and return to Text view.

7. Next, while in Text View, make sure the Scanner button at the top center of the screen is highlighted and then press No.

8. When you have finished reviewing all of the available tests under Service $05 press No to return to the Test Parameters/Results menu.

($06) Non-Cont. Monitored Systems

1. Scroll down to "($06) Non-Cont. Monitored Systems" and press Yes.

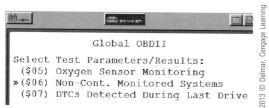

2. This screen lists all of the available tests in Service $06. Be sure to scroll down all the way to the bottom to gain access to all of the tests.

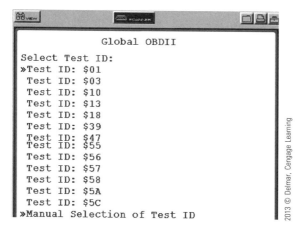

Service $06 allows the technician to request diagnostic test results from the PCM. These results are usually given in a hexadecimal number that will need to be interpreted by the technician in order to make a diagnostic decision. However, as software and technology develop, much of this deciphering is done by the diagnostic tool, and results are displayed in a more user-friendly format. The benefit of this is that the test result is given as a value that can be compared to a minimum and maximum value. These minimum and maximum values are the thresholds at which the PCM will trigger a trouble code. The test value is not just a pass or fail as now—the technician can see how far away from the fault values the current value is. This makes it possible to catch problems before they actually become problems and to easily verify repairs that have been performed to ensure that the customer will not be back with the same problem in a short period of time.

To fully use Service $06 data, the technician needs to have access to manufacturer information labeling the test and component identifications. This type of information is becoming more easily available through automotive reference subscriptions and the internet.

Service $06 Basic Terms

CID: Component Identification. This is the currently monitored sensor or component.

MID: Monitor Identification. Used instead of TID on some newer vehicles.

TID: Test Identification. This is the specific test that is going to monitor the component.

Service $06 Information Websites

Performing an Internet search with Google or a similar search engine will usually result in some Service $06 information. Here are some of the results of an Internet search performed on various manufacturers. The best option is to use a subscription service such as ShopKey5 that has all of the Mode 6 data in one place.

General Motors:	http://service.gm.com/gmspo/mode6/
Ford: (Click on OBDII Theory and Operation)	http://www.motorcraftservice.com/
Honda	https://techinfo.honda.com/rjanisis/RJAAI001_mode.asp

Service $06 Example:

Here is an example of using Service $06 information. The vehicle is a 1998 Mercury Tracer. The first thing to do is research in order to find out the required information about the vehicle you are working on. Luckily, Ford has most of their OBD information online for free at www.motorcraftservice.com/. Navigate to this site and then choose the OBDII Theory and Operation link.

Here is a screenshot of the information provided by Test ID: $21.

As you can see. Test ID $21 and Component $00 both have a minimum and maximum value and are listed as separate tests. The minimum value threshold was breached, and the test failed, whereas the maximum value was not surpassed, so that test passed.

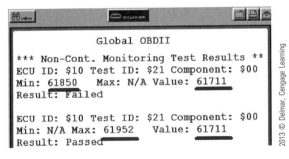

Using the information from Ford, we can find out more about this failure and help decipher the meaning of this information.

J1979 Mode $06 Data			
Test ID	Comp ID	Description	Units
$21	$C0	Initial tank vacuum and minimum limit	in H$_2$0
$21	$C0	Initial tank vacuum and maximum limit Note:	in H$_2$0
$22	$C0	Leak check vacuum bleed-up and threshold	in H$_2$0
$25	$C0	Vapor generation maximum pressure rise	in H$_2$0

Conversion for Test IDs $21 through $25: If value is > 32,767, the value is negative. Take value, subtract 65,535, and then multiply result by 0.00195 to get inches of H$_2$0. If value is <or= 32,767, the value is positive. Multiply by 0.00195 to get inches of H$_2$0.

(Phase 0 - Initial vacuum pulldown):

First, the Canister Vent Solenoid is closed to seal the entire evap system, then the VMV is opened to pull a 7" H2O vacuum. If the initial vacuum could not be achieved, a large system leak is indicated (P0455). This could be caused by a fuel cap that was not installed properly, a large hole, an overfilled fuel tank, disconnected/kinked vapor lines, a Canister Vent Solenoid that is stuck open or a VMV that is stuck closed.

We now know that the initial tank vacuum is not meeting the minimum requirement, but it is pretty close. In fact, it is so close that no trouble code has been set. Although the value does not seem to meet the requirement of a PO455 Large Leak, it is very possible that this could be a code causing problem in the near future.

3. When you have finished viewing the results of the Service $06 Tests, keep pressing No until you return back to the Test Parameters/Results menu.

($07) DTCs Detected During Last Drive Cycle

1. Scroll down to "($07) DTCs Detected During Last Drive Cycle" and press Yes.
2. This lists the Pending trouble codes. These codes will not cause the MIL light to come on, but if the problem is detected again on another drive cycle, these codes could become hard codes, which do trigger the MIL light. If the problem is intermittent, check Service $07 early in the diagnostic procedure, as this can lead to helpful clues. This is also a helpful feature in more quickly verifying repairs. Most problems will have to be detected on two to three drive cycles before a hard code is set and the MIL triggered, but checking here before releasing the vehicle to the customer will let you know if the problem is going to reoccur.

216 Chapter 7 Global OBD

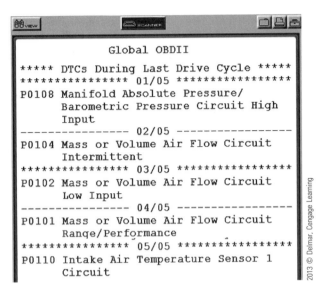

3. When you have finished viewing the pending codes listed in Service $07, keep pressing No until you return to the Test Parameters/Results menu.

4. Then press No one more time to return to the Global OBDII Main Menu.

($08) Request Control On-Board System

1. Scroll down to "($08) Request Control On-Board System" and press Yes.

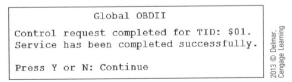

2. Press Yes to initiate Test ID: $01. Follow the on-screen instructions to complete the test. Most systems only have one test programmed into this service mode. This test allows the technician to close the canister vent solenoid so an EVAP system leak (pressure) test can be completed.

```
          Global OBDII
      Evaporative System Leak Test
» Test ID: $01
```

3. Pressing Yes will take you back out to the Global OBDII Main Menu.

```
          Global OBDII
Control request completed for TID: $01.
Service has been completed successfully.

Press Y or N: Continue
```

($09) Read Vehicle Identification

1. Scroll down to "($09) Read Vehicle Identification" and press Yes.

   ```
   ($05,06,07) Display Test Param./Results
   ($08) Request Control On-Board System
   »($09) Read Vehicle Identification
   ($09) In-Use Performance Tracking
   ($0A) Display Permanent Trouble Codes
   ```

2. Select "Vehicle Identification Number (VIN)" and press Yes.

   ```
   Select Service:
   »Vehicle Identification Number (VIN)
    Calibration Identification
    Calibration Verification Number  (CVN)
    ECU's Acronym and Text Name
   ```

3. Press No twice to return to the Service Menu.

   ```
   ****** Vehicle Identification
   ECU ID : $11
   VIN : WF0HXXGAJH4S37338
   Press N: Return
   ```

4. Scroll down to "Calibration Identification" and press Yes. This is the identification of the software installed on the vehicle's ECU.

   ```
   Select Service:
    Vehicle Identification Number (VIN)
   »Calibration Identification
    Calibration Verification Number  (CVN)
    ECU's Acronym and Text Name
   ```

5. Press No twice to return to the Service Menu.

   ```
   ****** Calibration Identification
   ECU ID : $11
   Calib ID01 : F14E203A4VB
   Calib ID02 : G25F314B5WC
   Press N: Return
   ```

6. Scroll down to "Calibration Verification Number (CVN)" and press Yes.

   The CVN number is used to verify the integrity of the vehicle's software. The OEM is responsible for determining the method of calculating the CVN, such as using a checksum or similar formula.

   ```
   Select Service:
    Vehicle Identification Number (VIN)
    Calibration Identification
   »Calibration Verification Number  (CVN)
    ECU's Acronym and Text Name
   ```

7. Press No twice to return to the Service Menu.

```
****** Calibration Verification
ECU ID : $11
CVN 01 : 10A022B1
CVN 02 : 13C244D3
CVN 03 : 25DDAC33
Press N: Return
```

8. Scroll down to "ECU's Acronym and Text Name" and press Yes.

```
Select Service:
  Vehicle Identification Number (VIN)
  Calibration Identification
  Calibration Verification Number   (CVN)
»ECU's Acronym and Text Name
```

9. Press No twice to return to the Service Menu.

```
************** ECU Name ****
ECU ID : $11
ECU Name : ECM1-EngineControl
Press N: Return
```

10. Press No one more time to return to the Global OBDII Main Menu.

```
****** In-Use Performance Tracking *****
ECU ID : $11
OBD Monitoring Conditions          1
Engine Starts                      2
Catalyst Monitor Completion Bank1  3
Catalyst Monitor Conditions Bank1  4
Catalyst Monitor Completion Bank2  5
Catalyst Monitor Conditions Bank2  6
O2 SNS Monitor Completion Bank1    7
O2 SNS Monitor Conditions Bank1    8
O2 SNS Monitor Completion Bank2    9
O2 SNS Monitor Conditions Bank2   10
EGR Monitor Completion            11
EGR Monitor Conditions            12
Secondary Air Monitor Completion  13
Secondary Air Monitor Conditions  14
EVAP Monitor Completion           15
EVAP Monitor Conditions           16
Press N: Return
```

($09) In-Use Performance Tracking

1. Scroll down to "($09) In-Use Performance Tracking" and press Yes.
2. In-Use Performance Tracking requires the on-board computer to keep a count of how often each major monitor has run and therefore could have detected a fault. There is also a counter that tracks the number of times that the vehicle has been used or the number of engine starts.
3. Press No until you return to the Global OBDII Main Menu.

($0A) Display Permanent Trouble Codes

1. Scroll down to "($0A) Display Permanent Trouble Codes" and press Yes.

```
($08) Request Control On-Board System
($09) Read Vehicle Identification
($09) In-Use Performance Tracking
»($0A) Display Permanent Trouble Codes
```

2. This screen displays all of the trouble codes stored in the vehicle's non-volatile memory that cannot be cleared by disconnecting the battery or a scan tool. If a code commanded the MIL one at one time, that code will be placed in the permanent code memory. The display and function is identical to the display of Service $03 and Service $07. The main difference between these features is that permanent trouble codes *cannot* be cleared by any scan tool, generic or enhanced, and disconnecting the battery has no impact on this memory. Only the PCM itself can clear this code. Only the codes stored in Service $03 will illuminate the MIL, so if those codes are cleared using Service $04, then the MIL will turn off but the permanent codes will continue to be stored in Service $0A. This is an excellent place for technicians to see a more complete history of the vehicle and its problems as no one can erase their mistakes or repair attempts under Service $0A.

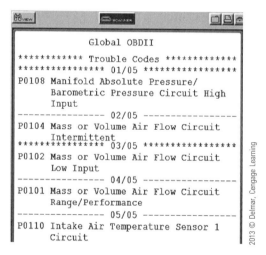

3. Press No until you return to the Global OBDII Main Menu.

OBD Health Check

OBD Health Check is a quick scan of the OBDII system. The scan will indicate whether the MIL is on or off, the number of DTCs present and the status of all readiness monitors for all of the vehicle's ECUs. If a customer needs to know if their readiness monitors are completed before going for an emission test, this can be a convenient way for the technician to gather data and quickly print it out so the customer has something to confirm the reading.

1. From the Main Screen, highlight the Scanner icon and then scroll down to select OBD Health Check and press Yes.

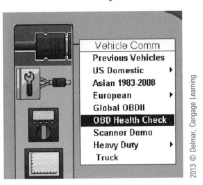

2. Follow the on-screen instruction to connect the scanner to the vehicle and then press Yes.

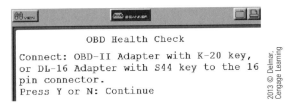

3. The default view is Text, which will display the data in a column format. Scroll down to the bottom to view all of the available information.

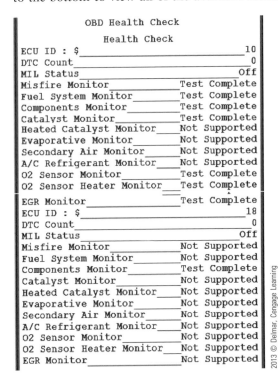

4. This information can also be viewed in PID List and Graph. PID List is a very organized way to quickly view the data. Graph is an option, but it is not very helpful with this type of data. To accomplish this scroll over to the View button, press Yes, and then choose your desired viewing option.

5. Printing the Full PID List is a quick way to ensure the customer that you did, in fact, look at the vehicle and scan it. Print the Full PID List just as we have described in earlier sections. Scroll to the printer icon, press Yes, select Full PID List and press Yes again. An example of the OBD Health Check Full PID List printout is shown. Be sure to turn on the Print Header as described in the Tool Set-Up utilities of the specific diagnostic tool you are using. This way, your shop name and contact information will be included on this printout.

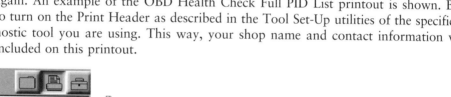

6. To quickly exit back to the Main Screen from here or any feature of the diagnostic tool, scroll over and highlight the View button and press No.

Generic Functions

Generic Functions is another way of accessing Global OBDII information, but this time it is accessed through the VIN specific side of the scan tool. This can save time by allowing the technician to access some Global OBDII data without backing all the way out of the VIN specific database. It is important to remember that only *some* of the Global OBDII service modes will be accessible through the VIN specific database, but all of the service modes can be accessed when initially identifying the vehicle using the Global OBDII option. Remember that Generic Functions is another term for Global OBDII. They are one and the same.

Please review the Scanner Introduction section to begin the Scanner demo, so you can follow along with the rest of this section using the dignostic tool.

1. To select a system, scroll the double arrow marker to the desired system name and press Yes. For the demonstration, pick Engine and press Yes.

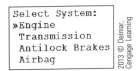

2. Scroll down to Generic Functions and press Yes.

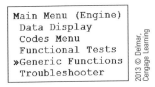

3. To access the available Global OBDII data, select the option you wish with the double arrow cursor and then press Yes. All of the information can be used and navigated exactly the same as was discussed previously in this section. The information and its use is the same; the method of accessing it is different.

First, notice that not all of the Global OBDII Service Modes are available under Generic Functions, but this is a much quicker way to access the available information than backing all of the way out and having to re-identify the vehicle under the Global OBDII option.

Next, notice that we are indeed in a vehicle specific database (GM to be specific), but we still have access to some Global OBDII information. You know this is a VIN specific screen based on the vehicle identification along the bottom.

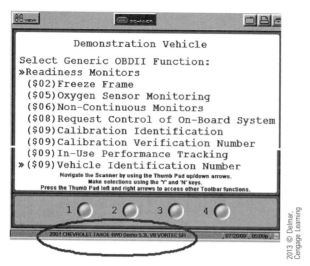

When the vehicle is identified under the actual Global OBDII option and not a VIN specific database, no specific vehicle information is listed at the bottom of the screen, as OBDII is a generic identification.

Global OBD Summary

When you are connecting to the vehicle's computer using a scan tool, you have the choice of two different areas to gather information. All of this information is accessed through the same data link connector (DLC) and comes from the same physical box (vehicle computer). It is easiest to picture the inside of the vehicle's computer, also called a Powertrain Control

Module (PCM), with two separate rooms in it; one being the VIN Specific or OEM room and the other being the Global OBDII room. The OEM room is accessed through the VIN specific databases on the scan tool while the Global OBDII is accessed by choosing this option from the Main Screen. Global OBDII is a valuable and often overlooked feature on scan tools. The information gathered here can make diagnosing a problem much easier and quicker, saving time and money. This is especially true when dealing with a substituted value situation but easily spotting it using Service $01. Repairs can be verified using Service $07 and Service $06 before releasing the vehicle back to the customer and risking a costly comeback repair, as well as a negative customer relations situation. Systematically performing a Global OBDII scan as the first step when using a scanner will save time and money over the long run by providing valuable diagnostic information to the technician not found when using a VIN specific database. Once the technician is in a VIN specific database, he or she still has access to some Global OBDII information through the Generic Functions menu option. This feature can also be a valuable time-saver when diagnosing vehicles. To truly be a power user of any diagnostic tool and get the most out of your diagnostic investment, it is imperative to practice and use the various Global OBDII functions and features.

Review Questions

1. Technician A states that there are separate vehicle data link connectors for accessing Global OBDII and VIN specific data. Technician B states that Global OBDII data is governed by the federal government. Who is correct?

    a. Tech A only
    b. Tech B only
    c. Both Tech A and Tech B
    d. Neither Tech A nor Tech B

2. Technician A states that a vehicle's computer can be described as having two rooms; one with the Global OBDII data and the other with the OEM or VIN specific data. Technician B states that the OEM room controls the MIL light. Who is correct?

    a. Tech A only
    b. Tech B only
    c. Both Tech A and Tech B
    d. Neither Tech A nor Tech B

3. Technician A states that manufacturers can substitute known good values for a faulty component into the vehicle's data stream by deriving that value based on the values of other good components. Technician B states that Global OBDII data is raw, unmanipulated data that contains no substituted values. Who is correct?

    a. Tech A only
    b. Tech B only
    c. Both Tech A and Tech B
    d. Neither Tech A nor Tech B

4. Technician A states that the dollar sign symbol ($) represents a hexadecimal number format used by the vehicle's computer. Technician B states that there are currently 10 Service Modes in Global OBDII. Who is correct?
   a. Tech A only
   b. Tech B only
   c. Both Tech A and Tech B
   d. Neither Tech A nor Tech B

5. Technician A states that Freeze Frame data is taken randomly during every drive cycle and stored into the computer memory. Technician B states that Freeze Frame data remains in the computer's memory even after the trouble codes are cleared. Who is correct?
   a. Tech A only
   b. Tech B only
   c. Both Tech A and Tech B
   d. Neither Tech A nor Tech B

6. Technician A states that any trouble codes found under Service $03 will cause the MIL to illuminate. Technician B states that trouble codes found under Service $07 will cause the MIL to illuminate. Who is correct?
   a. Tech A only
   b. Tech B only
   c. Both Tech A and Tech B
   d. Neither Tech A nor Tech B

7. Technician A states that to fully use Service $06, the technician will have to look up specific manufacturer information about Test and Component identifications. Technician B states that when using Service $06 each component test result is compared to a minimum and maximum value, so possible future problems can be more quickly identified as components approach their threshold values. Who is correct?
   a. Tech A only
   b. Tech B only
   c. Both Tech A and Tech B
   d. Neither Tech A nor Tech B

8. Technician A states that before returning a vehicle to a customer after a repair, it is best to test-drive the vehicle and then check Service $06 and Service $07 to verify the repair. Technician B states that Service $07 is useful in diagnosing intermittent problems. Who is correct?
   a. Tech A only
   b. Tech B only
   c. Both Tech A and Tech B
   d. Neither Tech A nor Tech B

9. Technician A states that Service $0A Permanent Codes can only be cleared by using a factory scan tool. Technician B states that Permanent Codes will *not* be erased if the battery is disconnected. Who is correct?

    a. Tech A only
    b. Tech B only
    c. Both Tech A and Tech B
    d. Neither Tech A nor Tech B

10. Technician A states that OBD Health Check displays the current status of the Readiness Monitors. Technician B states that printing a Full PID List under OBD Health Check will print out all of the available information gathered under this option. Who is correct?

    a. Tech A only
    b. Tech B only
    c. Both Tech A and Tech B
    d. Neither Tech A nor Tech B

11. Technician A states that the Generic Functions option is found under the VIN specific databases. Technician B states that all of the Global OBDII service modes can be accessed through the Generic Functions option. Who is correct?

    a. Tech A only
    b. Tech B only
    c. Both Tech A and Tech B
    d. Neither Tech A nor Tech B

chapter 8

Codes Menu

Upon completion of the Codes Menu module, you will be able to:

- Access Diagnostic Trouble Codes (DTCs) from the vehicle's computer using a scan tool
- Explain the difference between Current, Pending, and History trouble codes
- Clear Trouble Codes and reset Readiness Monitors
- Access Freeze Frame and Failure Records stored on the vehicle's computer
- Use the DTC Status feature to quickly access information stored on the vehicle's computer regarding a particular trouble code to aid in verifying vehicle repairs

REMINDER: This section is written so the reader can follow along using the Solus Pro or MODIS scan tools, if available. The Scanner software is identical on both of these tools, so all of the procedures will be the same. Most of the screenshots come from a MODIS, but the Solus Pro will be the same except for the screen size, so while the MODIS may show three trouble codes at a time the Solus Pro will only show two. Everything else will function exactly the same. You are strongly encouraged to follow along using the diagnostic tools, as this provides the hands-on knowledge necessary to pass the Certification Exams. The only required equipment is the Scanner unit itself and the AC power adapter cord. The demonstrations built into the Scanner software will provide the rest of the required material.

Codes Menu Overview

To fully understand this section, it is important to read and understand the Scanner Introduction section first. The Scanner Introduction section will explain the required steps to get to this point in the navigation of the software. Some of the codes menu navigation and features covered in this section were outlined and briefly discussed in the Global OBD section, so after completing this section you should be able to transfer some of this more detailed knowledge to Global OBDII when viewing trouble codes.

Diagnostic Trouble Codes (DTCs)

Diagnostic Trouble Codes, or DTCs, are stored by the vehicle's computer when a fault is discovered using the vehicle's own on-board diagnostic testing. On 1996 and newer vehicles equipped with OBDII, these codes are displayed as a five-digit alphanumeric numbering system set by the Society of Automotive Engineers (SAE). The first digit is a letter and designates the trouble code system. There are four possible letter designations, including P, B, C, and U. Powertrain is designated by the letter P, Body by the letter B, Chassis by the Letter C, and Network Communications by the letter U. The second digit breaks all DTCs into two broad categories by using either zero or one. Zero designates a SAE Generic code, while one designates a Manufacturer (OEM) Specific code. The third digit identifies the subsystem where the problem is occurring. Numbers 0 to 2 indicate a fault in the Fuel-Air Metering system, three indicates the ignition system or a misfire, four indicates the Auxiliary Emission Controls, five indicates the Vehicle Speed and Idle Control systems, six indicates the Computer Input and Output Circuits and finally seven and eight indicate the transmission. Retrieving DTCs from the vehicle is a key step in the diagnostic process but never confirms the failure of a component. After reading the DTCs, you should have some idea of where to start performing additional testing to confirm suspected problems before any new components are replaced on the vehicle. See Figure 8-1.

Codes Menu Overview **229**

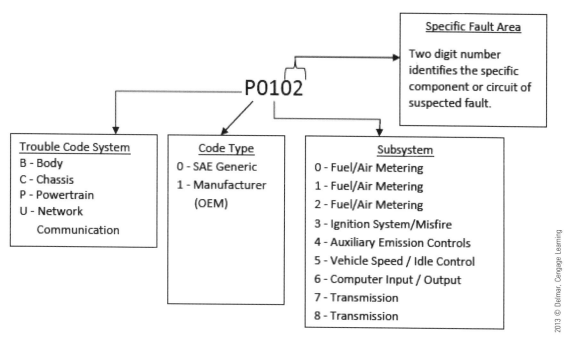

Figure 8-1 SAE has designated a specific format and meaning behind each digit in OBD-II trouble codes. These "generic" codes are common to all OEM's.

Please review the Scanner Introduction section to begin the Scanner demo, so that you can follow along with the rest of this section using the dignostic tool.

1. Here are the systems that this vehicle has available. Scroll to Engine and press Yes.

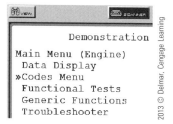

2. From the Main Menu (Engine) screen scroll and select Codes Menu and Press Yes.

Trouble Codes

1. This is the Manufacturer Specific Codes Menu. The demonstration will use GM as the example.
2. Press Yes to access Trouble Codes.

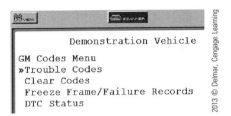

The "System" Codes Menu lists the various DTC designations. The system in this demonstration is the Engine. The names of these groupings can change between manufacturers. In this demonstration, all of the codes found under each category are the same, but this will not necessarily be the case when you are working on an actual vehicle diagnosing real problems. In a real time diagnostic situation, each DTC category has its own unique criteria for a DTC to be present. This can help isolate and find intermittent problems.

All of the DTC categories function the same way. The first category, Engine/Trans Trouble Code Information, will be detailed thoroughly: All scan tool functions within the category will be explained, but the additional categories will be explained as an overview. All of the scan tool functions within the categories are the same. Use this first category as the template to use the others.

3. Select "Engine/Trans Trouble Code Information" and Press Yes. This is also referred to as Current Trouble Codes.

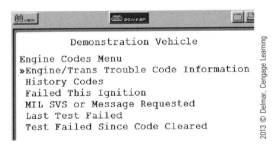

4. This screen lists all of the Current trouble codes in the vehicle's memory. Be sure to scroll down to view *all* the trouble codes, as they may not all fit on the screen.

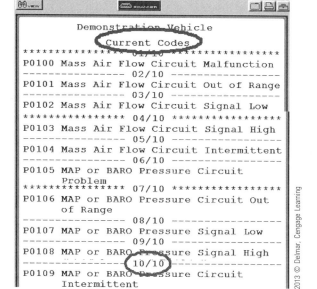

5. Scroll over to the right and highlight the File Folder icon and press Yes. Notice that when the Scanner button is no longer highlighted, the codes display box is not active and the codes cannot be scrolled through.

   Save: The Save Movie and Save Frame options are not functional when viewing codes. Save Image will take a picture of the current screen. To capture all of the codes using Save Image, you will have to take multiple screenshots and then scroll new codes into view each time. Refer to the Data Management section for more detailed information about these options and how to use them.

6. Scroll over to the right and highlight the Printer icon and press Yes.

   Print: "Full Screen" will perform its normal function of printing the screen just as you see it. This will include the visible codes on the screen at the time of the printing. "Full PID List" does not function in the codes menu. "Full Codes List" will print out all of the trouble codes, both on and off screen, in a simple text format. An example of Full Codes List Printout is shown.

```
Date/Time: 07/30/09 08:35a
              Current Codes
P0100 Mass Air Flow Circuit Malfunction
P0101 Mass Air Flow Circuit Out of Range
P0102 Mass Air Flow Circuit Signal Low
P0103 Mass Air Flow Circuit Signal High
P0104 Mass Air Flow Circuit Intermittent
P0105 MAP or BARO Pressure Circuit
      Problem
P0106 MAP or BARO Pressure Circuit Out
      of Range
P0107 MAP or BARO Pressure Signal Low
P0108 MAP or BARO Pressure Signal High
P0109 MAP or BARO Pressure Circuit
      Intermittent
```

Full Codes List Printout Example

7. Scroll over to the right and highlight the Tool Box icon and press Yes.
8. Press Yes to enter Custom Setup

Custom Setup: This is a shortcut back to the Scanner Units option under the main Utilities menu of the scan tool. Navigate the dialogue box to change between standard and metric units. Refer back to the Tool Setup and Utilities options explained under the specific section for the diagnostic tool you are using for more information.

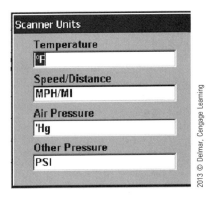

9. Press No to return to the Tool Box icon.
10. Press Yes again and scroll down to Save Data and press Yes.

Save Data: This is a shortcut back to the Scanner Units option under the main Utilities menu of the scan tool. Navigate the dialogue box to change the percent after trigger option, where data is stored, and what file type all of the screen images will be when saved. Refer back to the Tool Setup and Utilities options explained under the specific section for the diagnostic tool you are using or the Data Management section for more information.

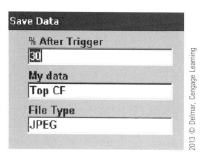

11. Press No to accept any changes and exit the Save Data menu box.
12. Scroll back over to the left and highlight the Scanner button.

13. Press No once.
14. Scroll down to "History Codes" and press Yes. History codes are problems that have happened in the past but may not be causing a problem at the current time. Use this menu option if the Malfunction Indicator Lamp (MIL) is on but there seem to be no problem symptoms.

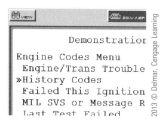

15. Press No to exit.
16. Scroll Down to "Failed This Ignition" and press Yes. This is sometimes referred to as "Pending Codes." These codes represent failures that occurred this ignition cycle but may not have triggered the MIL. So if there is no MIL on, but you suspect a problem, scan for codes using this option because some codes have to fail in multiple ignition cycles before triggering the MIL.
17. Press No to exit.
18. Scroll Down to "MIL, SVS, or Message Requested" and press Yes. This option lists the codes that are triggering the MIL, SVS (Service Vehicle Soon), or other indicator lamp to be turned on.

234 Chapter 8 Codes Menu

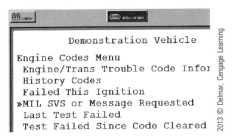

19. Press No to exit.
20. Scroll down to "Last Test Failed" and press Yes. This option allows you to see what codes have failed the last time their respective test has run. Remember, some problems will be intermittent, and the code could pass a test now and then.

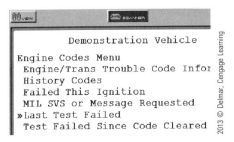

21. Press No to exit.
22. Scroll down to "Test Failed Since Code Cleared" and press Yes. If you encounter many codes and intermittent problems, it may be best to clear the codes and test drive the vehicle. After you drive the vehicle, this option will allow you to see what codes failed since you cleared them. This may give you an idea of what the higher priority or most active problems are.

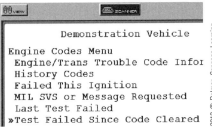

23. Press No to exit.
24. Press No again to return to the Codes Menu.

Clear Codes

1. Scroll Down to "Clear Codes" and press Yes.

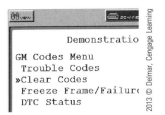

2. Press Yes again to clear the codes.

   You cannot actually clear the codes from the built-in demonstration, so it is OK to press Yes. Remember that when you do clear trouble codes that you clear all of the vehicle's computer memory, including Freeze Frame/Failure Records, and will reset all the OBDII Readiness Monitors.

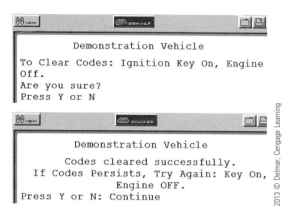

3. Press Yes to continue, and this will take you back to the Codes Menu.

Freeze Frame/Failure Records

1. Scroll down to Freeze Frame/Failure Records and press Yes. The next screen will list the codes that have Freeze Frame data associated with them. Select the code you wish to view and press Yes.

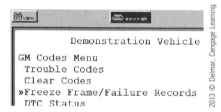

Freeze Frame Data is similar to a saved Movie. Many technicians think of this data as the black box recorder on airplanes that is always sought after a crash. When the MIL turns on, a snapshot of the data at the time of the problem is taken, this is known as Freeze Frame data. It can be very useful to see this saved data at the time the problem occurred. Navigate and manipulate the Freeze Frame screen just as you would a saved Movie or normal data gathering screen as we discussed in the Viewing and Interpreting Data section. Remember, you can change the view from Text to PID List or multiple Graphs to make diagnosing the problem much easier. If you want to save the Freeze Frame data for future reference, use the File Folder icon and save this as a separate Movie file. Otherwise, when the codes are cleared, the Freeze Frame data will be lost.

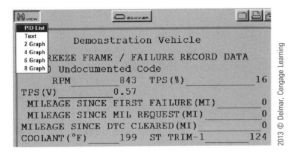

2. Press No two times to return to the Codes Menu.

DTC Status

1. Scroll Down to "DTC Status" and press Yes.

   DTC Status is used to gather data from the vehicle's computer about a specific trouble code. This option is very useful in verifying repair. Assume that we had a Mass Air Flow (MAF) sensor code P0102. Using that, we performed other tests and decided that the MAF was bad and replaced it. Before returning the vehicle to the customer, we would want to be sure that indeed the MAF sensor was bad and that the MIL would not come back on soon after the customer left. You should go and test-drive the repaired vehicle, and upon your return, use DTC status to see what the vehicle's computer has to say about code P0102.

2. The first digit in P0102 is 0, so press Yes to move to the second digit, leaving the first as a zero.

   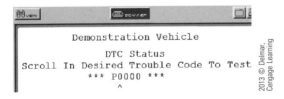

3. The second digit in P0102 is 1, so press the Up button to change the second digit to 1 and then press Yes to move to the third digit.

   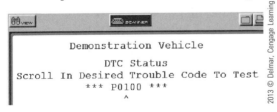

4. The third digit in P0102 is 0, so press Yes to move to the fourth digit leaving the third as a zero.

   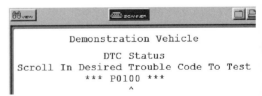

5. The fourth digit in P0102 is 2, so press the Up button to change the second digit to 2 and then press Yes to enter DTC Status.

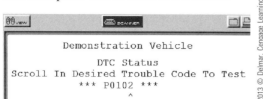

6. This quick readout gives you some valuable information regarding this specific code. You now know the tests that check for the proper function of this component have run and passed. There have also been no intermittent failures as there is no history code. This is a very good way to verify a repair before returning the vehicle to the customer. It can also be used to diagnose the initial complaints. After reviewing the codes, DTC status quickly gets all of the information about the specific codes on one screen instead of having to navigate through multiple screens and menus to gather all of this data.

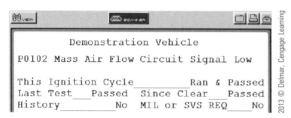

Codes Menu Summary

Retrieving Diagnostic Trouble Codes (DTCs) from the vehicle is a key step in the diagnostic process but never confirms the failure of a component. After reading the DTCs, you should have some idea of where to start performing additional testing to confirm suspected problems before any new components are replaced on the vehicle. Depending on the manufacturer, DTCs will be grouped into different categories to give insight into the possible problems happening in the vehicle. Current codes are problems that have failed and are continuing to fail to illuminate the MIL. History codes failed in the past and may have illuminated the MIL but the light may now be off. This may be indicating an intermittent problem. Pending Codes or Failed Last Ignition code indicate that a test has failed, but most tests will have to fail multiple drive cycles before the MIL is illuminated. Even if the MIL is not on, it is good practice to check for History and Pending codes because these will help guide you to problem areas. Before you return the vehicle to the customer, you must Clear the trouble codes. This will turn off the MIL and reset all of the OBDII Readiness Monitors. After Clearing the codes, perform a test-drive and then use DTC Status to check if your drive cycle has set any new codes. Be sure to check the Pending codes—if there are any present, then the vehicle should not be returned to the customer. These Pending codes could become current codes after the next drive cycle and can then illuminate the MIL. When the MIL turns on, a snapshot of the data at the time of the problem is taken, this is known as Freeze Frame data. Found under the Codes Menu, it can be very useful because it allows you to see the vehicle's data at the time the problem occurred. Navigate and manipulate the Freeze Frame screen just as you would a saved Movie or normal data-gathering screen as we discussed in the Viewing and Interpreting Data section. The information found under the Codes Menu should be some of the first information you access when starting a diagnosis on a vehicle.

Review Questions

1. Technician A states that DTC P0102 is a Generic SAE Powertrain code. Technician B states that a DTC usually confirms what component will need to be replaced. Who is correct?

    a. Tech A only

    b. Tech B only

    c. Both Tech A and Tech B

    d. Neither Tech A nor Tech B

2. Two technicians are discussing the screenshot below. Technician A states that pressing Yes now will display Pending DTCs. Technician B states that History Codes are helpful when looking for a problem with no current symptoms. Who is correct?

    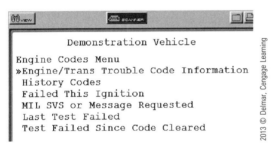

    a. Tech A only

    b. Tech B only

    c. Both Tech A and Tech B

    d. Neither Tech A nor Tech B

3. Two technicians are discussing the screenshot below. Technician A states that pressing Yes now will erase all the stored diagnostic trouble codes. Technician B states that pressing Yes now will also clear any saved Freeze Frame data and reset all OBDII Readiness Monitors. Who is correct?

    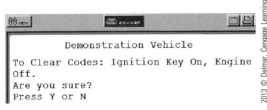

    a. Tech A only   c. Both Tech A and Tech B

    b. Tech B only   d. Neither Tech A nor Tech B

4. Two technicians are discussing the screenshot below. Technician A states that pressing Yes will print all of the available codes, including the ones not visible on the screen. Technician B states that there are 10 stored diagnostic trouble codes in this list. Who is correct?

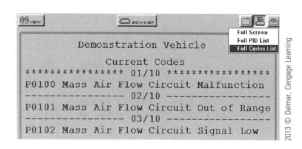

   a. Tech A only
   b. Tech B only
   c. Both Tech A and Tech B
   d. Neither Tech A nor Tech B

5. Technician A states that Freeze Frame Data is stored anytime an on-board diagnostic test is run. Technician B states that Freeze Frame Data can only be viewed in Text View. Who is correct?

   a. Tech A only
   b. Tech B only
   c. Both Tech A and Tech B
   d. Neither Tech A nor Tech B

6. Two technicians are discussing the screenshot below. Technician A states that this screen displays the DTCs that failed the current ignition cycle. Technician B states that pressing Yes now will take you to an intermediate codes menu with a printing option. Who is correct?

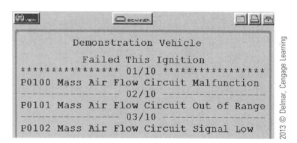

   a. Tech A only
   b. Tech B only
   c. Both Tech A and Tech B
   d. Neither Tech A nor Tech B

7. Two technicians are discussing the screenshot below. Technician A states that pressing Yes now will return you to the Engine Codes Menu. Technician B states that pressing No now will take you back to the codes list. Who is correct?

   a. Tech A only
   b. Tech B only
   c. Both Tech A and Tech B
   d. Neither Tech A nor Tech B

240 Chapter 8 Codes Menu

8. Two technicians are discussing the screenshot below. Technician A states that pressing the right arrow button will move you to the next digit. Technician B states that pressing up will scroll the zero to a one. Who is correct?

   a. Tech A only
   b. Tech B only
   c. Both Tech A and Tech B
   d. Neither Tech A nor Tech B

9. Two technicians are discussing the screenshot below. Technician A states that pressing Yes now will take you to the DTC Status information screen. Technician B states that to go back and change the fourth digit you need to press the left arrow button. Who is correct?

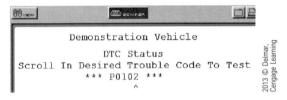

   a. Tech A only
   b. Tech B only
   c. Both Tech A and Tech B
   d. Neither Tech A nor Tech B

10. Two technicians are discussing the screenshot below. Technician A states that this is the History Codes display screen. Technician B states that this is the DTC Status information screen. Who is correct?

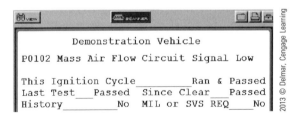

   a. Tech A only
   b. Tech B only
   c. Both Tech A and Tech B
   d. Neither Tech A nor Tech B

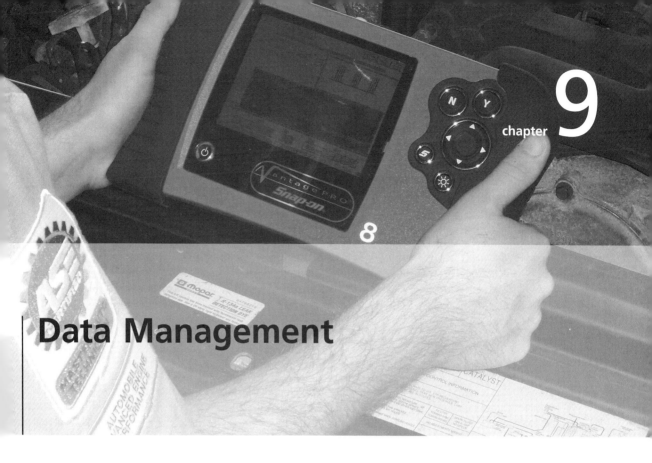

chapter 9

Data Management

Upon completion of the Data Management module, you will be able to:

- Explain how to use data management functions
- Identify different file types
- Load saved data files
- Demonstrate how to move and copy data from one data location to another
- Delete one file at a time or multiple files at once
- Edit file names for easier organization and identification
- Connect the diagnostic platform to a PC using the proper cables, software, and procedures to allow file swapping, updates, or other data transfers
- Use the free ShopStream Connect program to view diagnostic Movies and Snapshots and access the Ask-A-Tech forum, where experienced technicians share their knowledge and look for tips on tough diagnostic situations

Chapter 9 Data Management

REMINDER: This section is written so the reader can follow along using the Solus Pro, Vantage Pro, or MODIS tools if available. Please have a variety of saved files, such as Screenshots and Movies, on either a CF card or the internal memory so that you can manipulate these files in this section for practice. Please refer to the Viewing and Interpreting Scan Data section to learn how to save these files if you are unfamiliar with how to do it. Any file will work, so do not be concerned about what you are saving, just ensure that it is saved. Your instructor may have a CF card of practice files already made for you, or a coworker may be able to provide you with some saved files. The data management software is identical in function on all of these tools, so all of the procedures will be the same even if the icons and screen graphics are slightly different. Most of the screenshots come from a MODIS, but the Solus Pro and Vantage Pro will be identical except for the screen size and a few graphic icons. Everything else will function in exactly the same way. You are strongly encouraged to follow along using the diagnostic tools, as this provides the hands-on knowledge necessary to passing the Certification Exams. The only required equipment is the Scanner unit itself and the AC power adapter cord.

Data Management Overview

As vehicles become more complex and basic vehicle configurations diverge from traditional to hybrid and beyond, information will play an ever more important role in accurate diagnostics. Some of this information will be developed and saved by the technicians themselves. All of the diagnostic tools have the capability to save a tremendous amount of data with very little additional cost. Compact Flash (CF) cards and USB jump (flash) drives are both getting larger in capacity while coming down in price.

The Movies, Snapshots, and images that are easily created from the Scanner, Lab Scope, Ignition Scope, Power Graphing Meter, and other tool features are small in size and with a little planning can be archived on a personal PC with minimal effort. As you find glitches, bad waveforms, good waveforms, and other issues, you can quickly save this data and then build a file system on your PC to save it all. You can do this by Year, Make, Model, or customer. Take screenshots of the faulty component before and then after the repair, and then print these out for customers. True, they will not understand all of the parts of the waveform, but they will see a difference. This is one more step in justifying a repair and the costs associated with it. With the growing number of technician help websites and forums, like Snap-On's Ask-a-Tech, this saved data allows a much more detailed diagnosis of the vehicle from anywhere in the world. ShopStream Connect, a free download from Snap-On, allows others to view and manipulate Movies and Snapshots on their PC. Once you save the movie, which includes all of the information from all of the PIDs, it can be e-mailed or posted very easily, so anyone in the world can then open it up and view all of the data, change the views, use the cursors, and manipulate all of the other functions just as you would when using the actual diagnostic tool. The only limitation is not being able to collect any more live data because there is no further access to the vehicle. Even if you are completely stumped on the problem and have no idea of what data or screenshot to send, with ShopStream Connect it doesn't matter, as the assisting technician gets all the infromation and can work with it as they need.

Please have a variety of saved files, such as Screenshots and Movies, on either a CF card or the internal memory so you can manipulate these files in this section for practice in the next steps.

Data Management Features

1. From the Main Screen, scroll down to highlight the File Folder icon and then scroll to the right to highlight "Data Management" and press Yes.
2. From the Data Management screen, let's look at each button and its function.

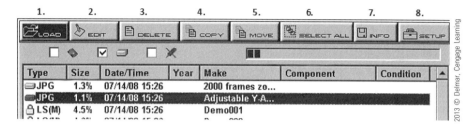

1. Load: When Yes is pressed with the Load button highlighted, the currently selected item in the file list will become active. If a picture file is selected, this allows the user to view the picture. If a movie file is selected, then the user can review and manipulate the data screens in movie mode. Load activates any saved file.

2. Edit: Edit is best used with a USB keyboard. This allows the user to rename files and also add additional information such as vehicle identification, component identification, and conditions found. Edit can also be used with the normal Yes and No button type navigation to more conveniently label files. The user can identify the year and make of the vehicle, along with the component being tested and its condition. The following illustrations show an example of this. Access a similar screen by highlighting (moving up or down in the list) a Movie (M) or Snapshot (P) file and then (moving left or right). Highlight the "Edit" button and press Yes. The screen below will come up and allow you to customize the file selected.

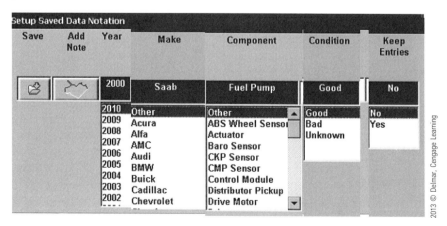

A dialogue box will appear and allow you to input information to identify the file you are about to save. Here is a breakdown of the features found in this box.

- **SAVE:** When the "Save" button on the far left is highlighted, pressing Yes will save the file. If the additional information boxes are not used, the file will be saved using a default file name that includes a date and time stamp. This is the quickest way to save a file that you do not need to identify right away. When you have finished inputting information and are ready to save the file, highlight this box and press Yes.

- **ADD NOTE:** If a USB keyboard is plugged into the top large USB port, then this button becomes active and can be highlighted. When you press Yes, you will be able to add 7 lines of additional information describing the file about to be saved. This is only available after a UBS keyboard has been plugged into the top large USB port.
- **YEAR:** When this box is highlighted, press Yes to open a drop-down menu in which the year of the vehicle being tested can be selected. Do this by scrolling down to the appropriate year and pressing Yes.
- **MAKE:** When this box is highlighted, press Yes to open a drop-down menu in which the make of the vehicle being tested can be selected. Do this by scrolling down to the appropriate make and pressing Yes.
- **COMPONENT:** When this box is highlighted, press Yes to open a drop-down menu in which the component being tested can be selected. Do this by scrolling down to the appropriate component and pressing Yes.
- **CONDITION:** When this box is highlighted, press Yes to open a drop-down menu in which the condition of the component being tested can be selected. Do this by scrolling down to the appropriate condition and pressing Yes.
- **KEEP ENTRIES:** When this box is highlighted, press Yes to open a drop-down menu to select either Yes or No to keep Entries. This feature allows you the option of keeping the last selected vehicle information in the tool's memory to make saving additional information from the same vehicle quicker and easier. If you are planning to save multiple movies, frames, or snapshots from the same vehicle, select Yes. The next time you save a file, the vehicle's information will already be selected. Selecting No will default all the information boxes to the generic settings each time a file is saved.

Using the Edit Feature

Here is an example of the Edit feature being used to customize a Lab Scope Preset. The procedure is similar for any type of file.

A. To Edit any File Name, make sure Edit is highlighted in the top tool bar. Then highlight the file you wish to edit in the lower display screen and then press Yes. This will open the Edit dialogue box.

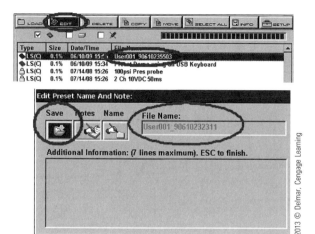

B. With the USB keyboard plugged in, scroll over to the right and highlight the Notes icon and press Yes. A typing cursor will appear in the Additional Information box, and you are free to type in any information you would like to help identify the Preset you just created. We suggest that you identify each channel, along with the year, make, model, and engine size of the vehicle. The Sweep and Trigger are the same for all the channels, so these are just listed under Channel one. This is a fictitious example illustrating the range of options that can be chosen. The extra minute it will take to save this the first time will save many minutes each time it is used, and the Lab Scope does not have to be reconfigured.

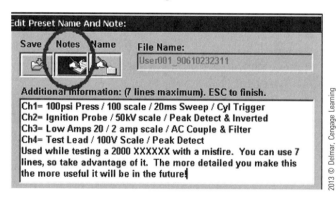

C. When you have finished typing additional information, press ESC on the keyboard to exit the typing box.
D. Scroll over to the right and highlight the Name box and press Yes.

E. Use the USB keyboard to delete the old file name and type in a new one that will easily and quickly identify this Preset to you in the future. Possible examples might be to label it by the component(s) being tested or the vehicle it is being connected to—whatever makes it easiest for you.

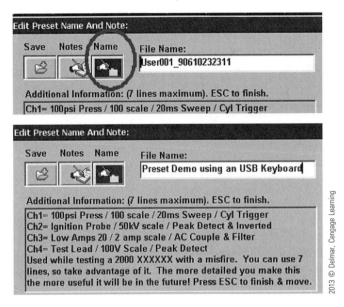

F. When you have finished typing the new file name, press ESC on the keyboard to exit the file name box.

G. Scroll over to the left until Save is highlighted and Press Yes.

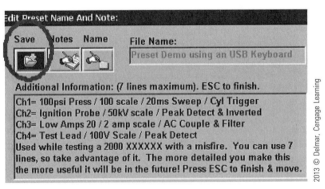

H. You will now see that the file name has changed and the additional information you added is visible when selecting presets.

Using the Edit Feature   **247**

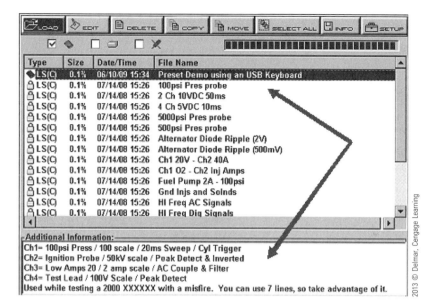

3. Delete: The highlighted file in the file list will be deleted if Yes is pressed when the Delete button is active. This will permanently remove any file from the storage disk.

4. Copy: The Copy button allows the user to move files to another data device while keeping the file on the current data device. Since a copy will be placed in the new destination location, the original source location will also retain the original file. If there is only one other possible data disk for the file to copy to, it will be done automatically. An example would be copying a file from the internal storage disk to the Top CF card. In this situation, if the user had connected both a Top CF card and USB storage disk to the tool, then a dialogue box would appear and ask the user to choose which destination disk the file should be copied to, either the Top CF card or the USB disk.

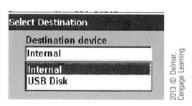

5. Move: This command is like combining the Delete and Copy commands into one function. When you use this command, the file will be copied to the designated destination storage device and then will automatically be deleted from the source storage device. An example would be a user who wishes to get a saved file from the internal storage device and archive it on his or her personal PC. An efficient way of doing this would be to use a USB storage device and Move the file from the internal storage device to the USB device. This will put a copy of the file on the USB drive and delete it from the internal device, so the internal device will now have room to save future data, and the USB device can easily be used to transfer the file to a PC.

6. Select All: This command is always used in conjunction with the other commands we've already discussed. To Delete, Copy, or Move multiple files at one time, use this command first to Select All of the files and then select Delete, Copy, or Move. Incorporating this feature into the example above helps illustrate the power and convenience of the data features incorporated into the various tools. A technician can easily save data to the internal storage device, then quickly Select All, and then Move that data to a USB device that will allow it to be easily transferred to a PC and this will automatically free up space on the internal device by deleting the source of the data. This way, the technician has kept and archived useful data to a PC and the tool is ready to save more data the next work day, week, or however often it needs to be emptied.

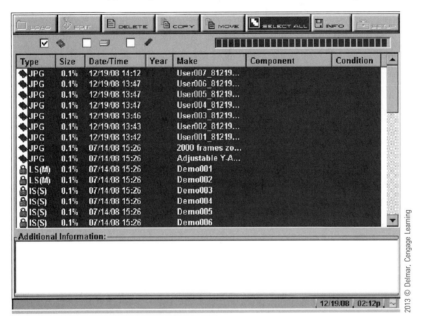

7. **Info:** This button gives the user more detailed information about the selected storage device's capacity. This will tell the user how many files have been saved, how many more can be saved, and then the total number of files able to be saved. In addition to this, it will also tell the user the amount of used and free disk space, as well as the total capacity. Information is presented both for users who prefer to think in terms of megabytes of space or in number of files.

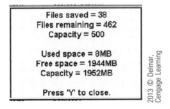

8. **Setup:** This feature allows the user to quickly change where the data is saved when using the tool, and this also then changes which files are being shown on the main file screen. The current file source is indicated by a check mark next to one of the three storage device icons.

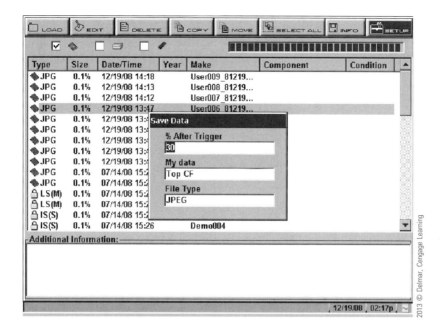

Data Storage Devices

Snap-On's diagnostic tools give you multiple places in which to store data. All tools have an internal storage device that is very convenient to use, but most users will also have other storage options that will expand the amount of files and data that can be saved. The active storage device is designated by a check mark in the box to the left of the device icon.

Top CF Card: This icon is to the far left of the screen and looks like a little red computer disk. This is an expandable storage device that is very easy and cheap to use. CF cards are readily available from large discount retailers and computer and office supply stores. At these same stores one may purchase a CF card reader that will connect to your PC using a USB connection and become an external disk drive. Removing the CF card from the diagnostic tool and inserting it into the card reader will allow a very quick and efficient way of transferring files from your diagnostic tool to your PC.

Internal Drive: The internal storage device icon resembles a grey computer disk drive and is the second icon from the left. This is the device that will store the tool's main software and any factory files or demonstrations. The factory files and software do not fill the entire drive, so the remainder is available for the user to store and save files. This device cannot be increased in size or removed by the user; it is set at the factory and is part of the tool. Although MODIS users can physically access the internal CF card, it is recommended that they only do this if directed to by a Snap-On representative, otherwise damage to the tool may result.

USB Storage Device: USB storage devices, also known as jump drives or memory sticks, are very common and have virtually replaced all other forms of compact mobile storage devices. This is the third icon from the left and is shaped like an USB memory stick. These are very affordable and anything over one gigabyte will easily store the amount of information needed by an average technician. This is the most efficient way of transferring files from the tool to a PC. The USB device will automatically be recognized by the tool, and then files can be Moved or Copied to it. The only disadvantage is that the device will protrude from the top of the tool and could be damaged. The CF card is flush with the tool and will eliminate this possibility. If a user does not want to purchase a CF card reader but still would like to build a library of information and archive this information to a PC, then it is recommended that, at the end of the day, all of the files from the CF card are moved to the USB drive and then that is taken and connected to the PC and the files are transferred.

Connect to PC: This option does not have an icon on the Data Management screen and will have to be accessed through the Utilities menu of the specific tool you are using, but it is another way of transferring files from the tool to a PC. The user will have to physically connect the tool to the PC by using the mini USB port on top of the tool and a standard USB port on the PC. To accomplish this, you will need a typical Mini-B USB cable. This probably came with your diagnostic tool but is also available from any office or electronics supply store. This is the same type of cable used to connect digital cameras and some external hard drives to PCs. To perform this task, from the Utilities menu select Connect to PC. A dialogue box will appear and indicate that the system will have to restart. After pressing Yes, the tool will reboot in Connect to PC mode, and the screen will have a different background. At this point, connect the mini USB end to the tool and the other end to your PC's USB port, and then follow the on screen instructions. Any data storage device you have connected to the tool such as the internal storage or the top CF card will now be recognized as eternal drives on your PC. These can be viewed under the My Computer option when using Windows. From here, you can cut, copy, paste, or drag and drop files between your PC and diagnostic tool. To make the transfer of files even easier, Snap-On has developed a program that can be loaded on your PC. This program is called ShopStream Connect, and it is a *free* download available on the Snap-On Diagnostics website at http://www1.snapon.com/diagnostics/us/InformationProducts/SSC. The Snap-On diagnostic website (www1.snapon.com/diagnostics/us) has detailed information on how to download and use this software.

Memory Indicator: The amount of available storage space left on the selected storage device is indicated by an increasing and decreasing blue bar. The blue bar represents available space. The amount of disk space will be updated as the user switches between storage devices using the Setup command. In the following illustrations, the top screenshot shows an EMPTY top CF card, while the bottom one shows a FULL internal drive. This looks counterintuitive, but just remember that the blue bar equals available space. The more blue bar, the more empty space.

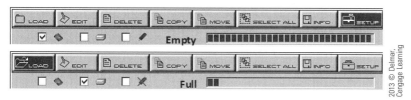

File Types

There are many different file types that one may see when looking through saved data files. It is important to understand what each of these file types are and how they are used.

BMP: An older image file type with very little file compression. It can be viewed on a PC very easily and edited with common PC photo editing software. This file is just a picture of what was on the screen at the time it was taken. None of the information on the screen can be moved, changed, or manipulated.

JPG: Joint Photographic Experts Group (jpeg or jpg) format. Jpeg is a newer file type with higher compression and is more easily integrated into newer PC programs. This file is just a screenshot, not a Movie or Snapshot. What you see on screen is captured, and nothing more. Other than basic photo editing, no information can be changed or manipulated. Both JPG and BMP file types are very PC friendly and can be viewed and edited using basic PC software and differ mainly in their ability to compress the image size to save memory. Either should work fine for the average technician wanting to post something online or e-mail the screenshot to a fellow technician.

SC(M): This a Scanner Movie file. While you are viewing data using the Scanner, a movie is *always* being recorded and this information is being saved to the memory buffer. The amount of frames or data saved in the buffer will change with software upgrades, but for example purposes let's say the buffer can hold 2000 frames of data. Once the buffer is full, the first frame of data that entered the buffer is erased and replaced with the 2001st frame of data. This continues so that most recent 2000 frames are always in the memory. When attempting to capture a glitch or event in the scanner data and archive it for later use, the user has two choices—either saving a Movie or creating a Snapshot. If a Movie is saved, then all of the information in the buffer is saved into a file with a SC(M) file type. When viewing this file, unlike a Screenshot, data can be moved and manipulated. PIDs can be rearranged and all frames of data can be viewed. This is a much more open file type that allows the user to see everything that was recorded. Realize that no information after the Movie was recorded is included in this file; only the information in the buffer is available for review.

SC(P): This is a Scanner Snapshot file type. A Snapshot is slightly different from a Movie; it can be made up of both past data stored in the buffer and new data that is continually being fed into the buffer. The moment the user activates or takes the Snapshot is called the trigger. The % after trigger option in the dialogue box allows the user to change how much old information is taken from the buffer and how much new information is collected. The default setting is typically 30% after trigger. Assuming a buffer of 2000 frames of data, this means that once the user triggers a Snapshot, the tool takes 1400 frames (70%) of the data from the buffer that was automatically recorded before the trigger and then continues to capture 600 frames (30%) of data after the trigger. These 2000 frames are bundled together and stored as a Snapshot file with the triggering event located between the beginning and the end of the entire Snapshot file and at the user specified percent. If the user wants to see more of the effect of the triggering event, then the percent after trigger should be increased; if the user wants to see more of the cause of the triggering event, then the percent after trigger should be decreased. When viewing this file, unlike a Screenshot, data can be moved and manipulated just like in a Movie, but now the user can also view the data that was recorded after the triggering event. PIDs can be rearranged and all frames of data can be viewed.

LS(M): This is a Lab Scope Movie file type. It has all the same functionality as a Scanner Movie file type but was recorded using the Lab Scope instead. Remember that the data can be reviewed, zoomed, and analyzed with cursors just as if the tool were hooked up live. Again, all of the recorded frames of data came from the buffer at the time the Movie was recorded.

LS(P): This is a Lab Scope Snapshot file type. It has all the same functionality as a Scanner Snapshot file type but was recorded using the Lab Scope instead. Remember that the data can be reviewed, zoomed, and analyzed with cursors, just as if the tool was hooked up live. Snapshots include a user defined percentage of frames that are recorded after the triggering event.

IS(M): This is an Ignition Scope Movie. This file is like all other Movie files but was recorded using the Ignition Scope.

IS(P): This is an Ignition Scope Snapshot. This file is like all other Snapshot files but was recorded using the Ignition Scope.

LS(C): This is a Lab Scope Preset saved file. Presets are saved configurations in either the Lab Scope or Ignition Scope. The user can go in and customize all the different configurations such as channels, test probes, scales, triggers, and other viewing criteria and then save this setup as a preset for future use. If a similar problem has to be tested again in the future, then the technician can simply open up the preset and get right to the testing because the scope will be configured and ready to go.

IS(C): This is an Ignition Scope Preset saved file. This is identical to the Lab Scope Preset except it was customized and saved using the Ignition Scope.

Here is a more complete list of the file types you may see in the data management section.

- **.bmp** – Bitmap of Saved Screen
- **.ism** – Ignition Scope Movie

- .isp – Ignition Scope Snapshot
- .iss – Ignition Scope Screen
- .lsc – Lab Scope Preset (File Storage Only)
- .lsm – Lab Scope Movie
- .lss – Lab Scope Screen
- .lsp – Lab Scope Snapshot
- .mmm – Multimeter Movie
- .mms – Multimeter Screen
- .pids – ShopStream Scanner Movie
- .scm – Scanner Movie
- .scp – Scanner Snapshot
- .scs – Scanner Screenshot
- .spm – ETHOS Scanner Movie
- .sps – Scanner Screen Bitmap

ShopStream Connect

ShopStream Connect is a free program that can be downloaded from the Snap-On Diagnostic website. The program emulates the tool that allows saved files to be viewed, just as you were using at the tool itself. Of course, because it is a program used to view saved data files, no new data can be collected from the vehicle. Now technicians have the ability to gather data from a vehicle and save it in a Movie or Snapshot file and view it without having to use the tool.

Advantages of using ShopStream Connect:

- View Information on a larger PC monitor instead of the smaller diagnostic tool screen.
- Shops with a limited number of diagnostic tool can more easily share them between technicians as after the data is captured that technician can review the data using a PC and allow another technician to gather data using the tool.
- ShopStream connect allows up to 16 PIDs to be viewed at one time so the programs can multiply the power of the original diagnostic tool.
- Plays back all Movies and Snapshots from the Scanner, Power Graphing Meter, and Lab Scope.
- Email a Movie file to a friend for review and he/she can now manipulate the data on the screen into a view that allows them to help you solve the diagnostic problem. You would not need to pick the information or view. Just send the entire Movie file and let them decide what to do with it from there.
- Post questions and files to the Ask-A-Tech forum that allows easy communication with other technicians in the field.

Downloading, Installing, & Updating ShopStream Connect

1. The *free* download is available on the Snap-On Diagnostics website at this address: www1.snapon.com/diagnostics/us/InformationProducts/SSC

2. You can also go to the diagnostics home page (http://www1.snapon.com/diagnostics/us) and then click the tab that says "Diagnostic Software."

3. From the Diagnostic Software page, the link to ShopStream Connect is in the lower middle of the page.

4. Click the link to Download ShopStream Connect.
5. Save the file to your computer in a place that is easy to remember, such as the desktop.
6. Once the download is complete, double click the Program icon to begin the installation.
7. Follow the on-screen instructions to complete the installation. From now on, you can use program icon on the desktop to start ShopStream Connect.

8. Once ShopStream Connect is started, it will check for updates and ask if you wish to download and install any that are available. Follow the on-screen instructions to accomplish any updates.

Navigating ShopStream Connect

1. The Main Screen is broken into various windows. There is a Windows-based file tree navigator to the far left with some premade links to the various diagnostic tools under that window. In this example, the MODIS link was clicked, which highlighted the

MODIS folder in the file tree. All the files located in that folder are displayed on the main window to the right. If an image file is selected, then a preview of that picture will be shown in the lower right corner window.

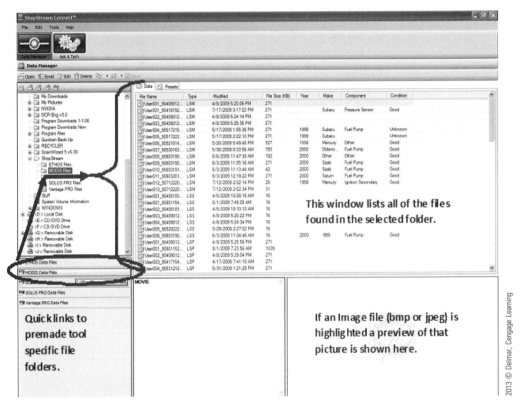

2. To add files, you can either use the PC Connect feature detailed in your specific tool's section of this book or save the files to a CF card or USE jump drive and transfer then to the PC that way. Having ShopStream Connect installed before you connect your tool via the mini USB cable to the PC and then using The PC Connect feature will make this whole process much easier. Once your tool is recognized, ShopStream Connect makes it easy to transfer the files to the proper preconfigured file folder. If using a CF card or USB jump drive, just follow normal PC procedures to cut or copy the files from your transfer storage device and then paste them into the proper folder in ShopStream Connect on your PC.

3. You can copy the Demo Scanner or Lab Scope (depending on your tool) files that come with your diagnostic tool to ShopStream Connect if you do not have any other files available. This will allow you to manipulate the control and navigate the tool. If the Demos are not available, you can quickly go out to the shop and capture a scanner Movie and upload that to ShopStream Connect. Please upload at least one Movie to ShopStream Connect before proceeding.

4. Here are two Movies captured using a Solus Pro scan tool. We know these are scanner movies because of the "SCM" type designation. They are found in the Solus Pro file folder. Either double click the highlighted Movie file or click the "Open" command near the middle left of the screen.

Navigating ShopStream Connect   **257**

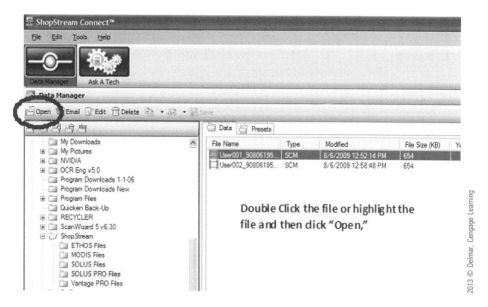

Double Click the file or highlight the file and then click "Open,"

5. The Movie will now open up in the scanner viewer. You may need to maximize the screen to see all of the information.

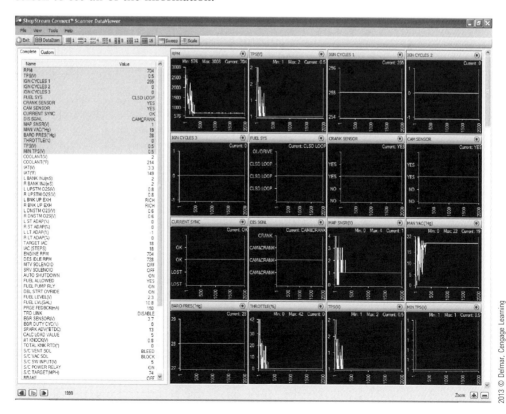

Reviewing a Scanner Movie File

1. First, let's look at our display options. All of these icons across the top toolbar are similar to on/off buttons. Many of theses options can be used in different combinations together.

   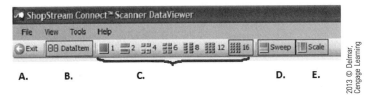

   A – **Exit:** Clicking this will exit you out of the Scanner DataViewer.

   B – **Data Item:** This controls the PID List display. Clicking this option will turn the visibility of the PID List on and off.

   C – **Different Graph Views:** These are on and off buttons, but only one can be on at a time. You have the choice of viewing one graph all the way up to viewing 16 graphs. If you just want to view graphs and no PID List data items, click the Data Item button to turn it off and then select the number of graphs you wish to view. Likewise, if you wish to view no graphs and just the PID List, turn off the graphs by clicking the graph button that is currently highlighted (on) to turn it off.

   D – **Sweep:** This turns on and off the visibility of the Sweep display along the horizontal axis for each graph. When you are viewing a scanner movie, this identifies the number of frames taken and which frame is currently being viewed. Turn this off to increase the size of the graph because it frees up more screen space.

   E – **Scale:** This turns on and off the visibility of the Scale factor on the vertical axis of each graph. The Scale display shows the minimum and maximum value recorded in the data stream and also displays the current frame value in the middle of the scale.

2. Take a moment to turn on and off each of these options and see how they can help you build a screen that allows you to see the information you need to make a diagnostic decision.

3. There are some icons across the bottom toolbar that also help us customize the display settings and allow us to see the information we need to make a diagnostic decision.

   A – **Rewind:** This moves the cursor on the graphs one frame to the left or rewinds the movie one frame at a time.

   B – **Play:** This will play the Movie one frame at a time. You can watch the cursor travel along the graphs as it indicates what frame is currently being viewed.

C – **Fast Forward:** This moves the cursor on the screen one frame to the right or moves the movie forward one frame at a time.

D – **Frame Number:** This indicates the current frame being identified by the cursor on the screen.

E – **Zoom In (+):** This will change the number of data frames able to be viewed on the screen at one time for each graph. Because less information is being viewed but the screen size stays the same, the data is more spread out, allowing you to more easily see changes in the data values.

F – **Zoom Out (-):** This allows you to view more data frames at one time. If you continue to click on this and zoom out far enough, you will be able to view the entire movie on one screen. The data becomes very bunched together and sometimes hard to read, but you have a bird's-eye view of all the data.

4. Let's put these features together and review what they can do.

   Your Movie screen will display different data, and that is OK because we are just looking for the navigation of the program.

5. Select 2 Graphs with the Data List, Sweep, and Scale also displayed. Right now we'll just view the first two default graphs. Your screen should resemble the example.

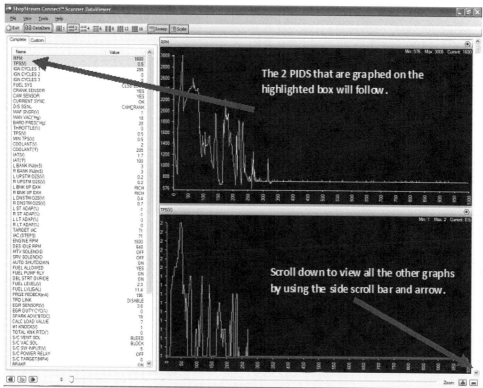

260 Chapter 9 Data Management

6. This example is displaying 2000 frames of data and most of the "action" is to the left side of the screen. To zoom in on these frames, we have a couple of options.

7. We first need to move the Cursor to the part of the graph that we want to Zoom In on. This can be done by using the Fast Forward button because our current position is Frame 0.

8. Click on the Fast Forward button to move the Cursor. Notice the new position of the Cursor (highlighted with a blue circle). Your Cursor color may be different from the one in the example. We'll explain this later.

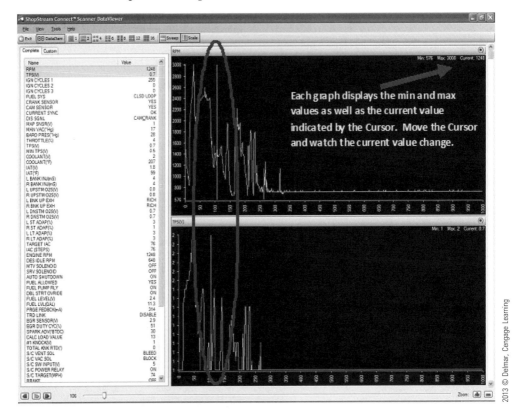

9. Notice you can use the mouse pointer to grab the Cursor line right on the graph and then, while holding the mouse button, drag the cursor to the desired position.

10. Press the Zoom In (plus sign +) button to decrease the number of frames we are looking at and allow for a more detailed view of the graphed data.

11. Continue to Zoom in until you get a good look at the data. The example screenshot has about 100 frames of data on the screen. The Cursor is more easily seen when zoomed in.

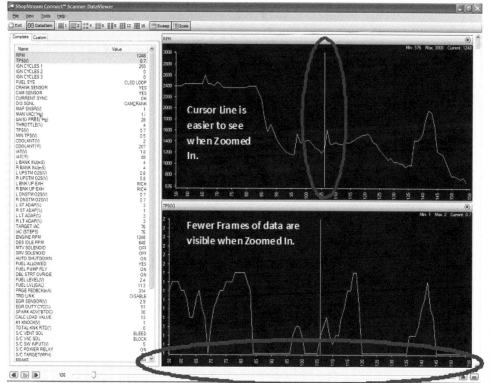

Graph Properties

1. The graphs can be modified in a number of ways. To view both the Individual and Shared Graph Properties, click the little hex head in the upper right hand corner of any graph.

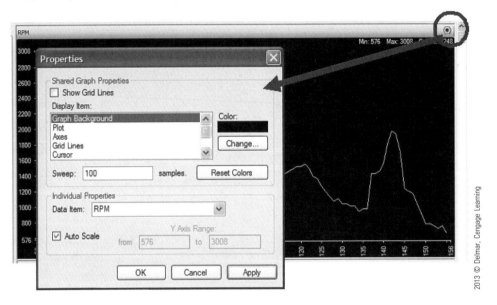

2. The Properties dialogue box is divided into two categories: "Shared Graph Properties" and "Individual Properties." As implied by the names, any options selected or changed under the Shared Properties heading will change *all* the graphs. The options under the Individual Properties will be local to the selected graph. Different graphs can be selected by using the drop-down menu.

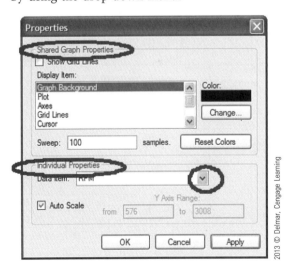

3. Let's explore what some of these features are. Here is a screenshot of the RPM graph before changing the properties.

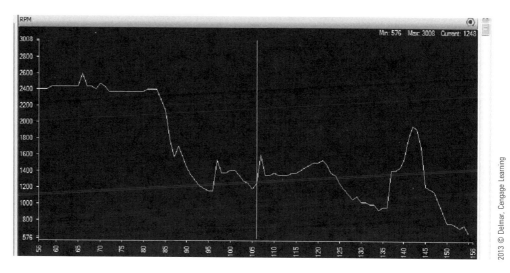

4. Change the following items for practice using the Properties dialogue box.

   Shared Graph Properties
   - Check "Show Grid Lines"
   - Graph Background = Dark Red
   - Plot = Yellow
   - Axes = White
   - Grid Lines = Grey
   - Cursor = Blue
   - Sweep = 150 Samples

   Individual Properties
   - Un-Check Auto Scale and set the range from 1000 to 2500.

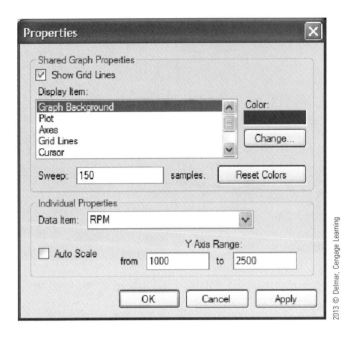

5. When you have finished, click Apply and see the results.
6. The graphs are fully customizable, and you have complete control over what you see and how you see it.

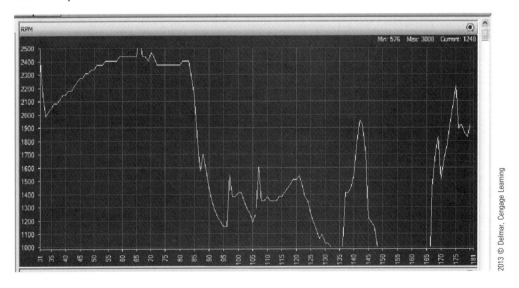

7. We can easily undo this by going back to properties, un-checking the options, and clicking "Reset Colors."

Custom Data Views

1. There are usually many PIDs collected by the scan tool, and the odds of the PIDs we want to look at being next to each other is low, so we have to have a way of changing the location of the different PIDs.

2. First, let's look at the PID list. To change arrangement of these PIDs, simply click on the PID you wish to move and, while holding the mouse button, drag the PID to the desired location. You will notice a small line appear between the other PIDs as you drag the mouse pointer over the list. This indicates the position of where the PID will be placed when the mouse button is released. Be sure to keep the mouse cursor over the text of the other PIDs, or the program will not allow you to drop the PID into a new position.

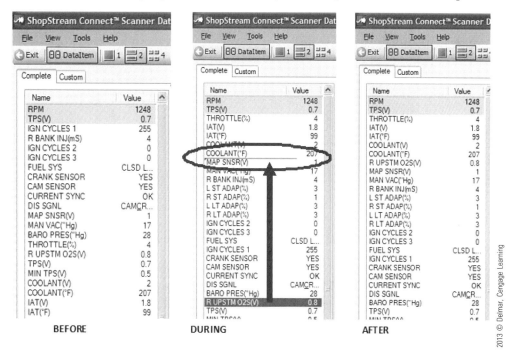

BEFORE   DURING   AFTER

3. As you move the PIDs in the Data List, the PIDs in the graph view also move accordingly.

4. Rearranging the graphs is very similar. Here is a before screenshot of an 8 PID view.

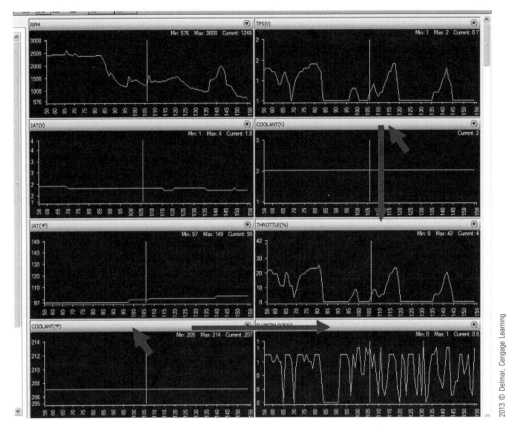

5. In this example, let's move the TPS PIDS next to each other as well as group the Coolant PIDs. To accomplish this, click and hold left mouse button in the PID ID banner bar and then drag the PID to a new location. This new location has to be the PID ID banner bar of another PID. In this example, click on the banner bar of the Coolant (V) and drag it to the banner bar of the Throttle (%) and they will switch places. The TPS PIDs will now be next to each other.

6. Next, click and drag the banner bar of Coolant (°F) over to the right to R UPSTM O2S (V) banner bar and release the mouse button to have these PIDs switch places.

7. The result would look like this, with the TPS and Coolant PIDs next to each other.

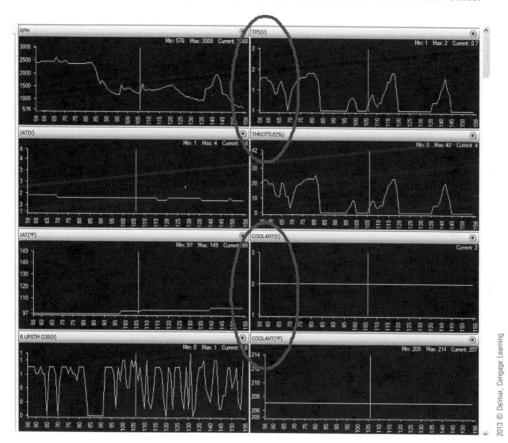

Custom Data List

When you are connected to a live vehicle, a Custom Data List will increase the sampling rate of data coming from the vehicle's computer because some PIDs are turned off and *not* collected. We can only review saved files using ShopStream Connect. However, the Custom Data List can still be used to focus attention on a certain hand-selected group of PIDs. This will not change the data at all, only the number of PIDs visible in the PID List.

1. Click the "Custom" tab next to the "Complete" tab at the top of the PID List window.

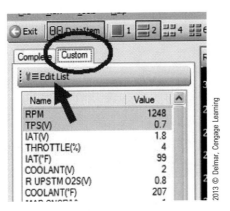

2. Click on "Edit List" just underneath the Custom tab.
3. Click "Select None" to remove all the check marks from the PIDs.

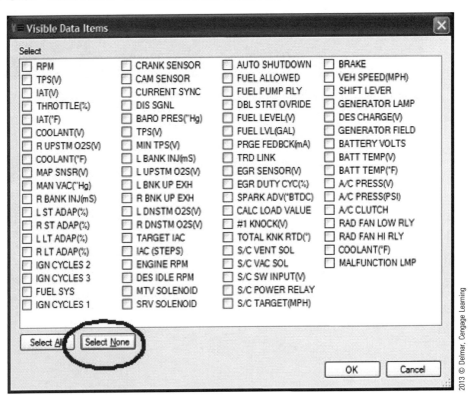

4. Assume that all we want or need to look at are the fuel control PIDs, such as the oxygen sensors, pulse widths, and fuel trims.

5. Go through the list and select only the PIDs you need to make a diagnostic decision.

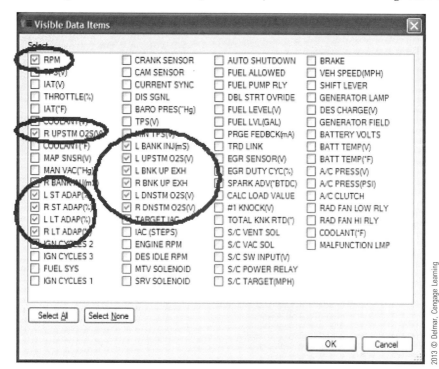

6. Click OK to view the information.
7. With 13 PIDs selected, switch to a 16 Graph view to see all of the relationships. If necessary, use the procedures for Custom Data View to further rearrange this list. This will allow you to see the information and relationships you need to make an accurate diagnostic decision.

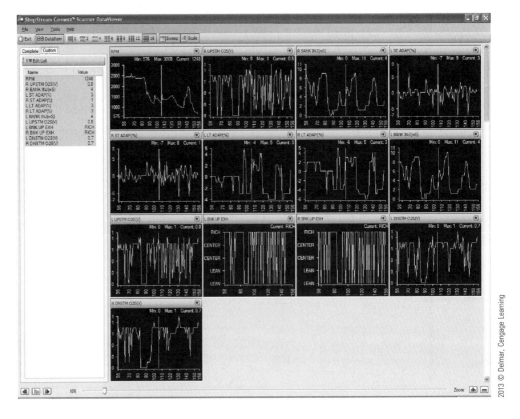

Reviewing a Lab Scope Movie File

8. From the Main Screen, look for an LSM file Type, as this is a Lab Scope Movie. Double click this file to open it. This example will use Demo 1 that comes with the MODIS, but Demo 1 from the Vantage Pro will also work to navigate the screen.

9. First, let's look at the upper toolbar options on the left side of the screen.

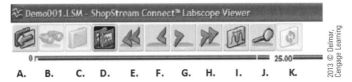

A – **Exit:** Click here to exit the Lab Scope viewer.

B – **View:** This is not used in the Lab Scope viewer.

C – **Play:** This is not used in the Lab Scope viewer.

D – **Auto Scroll:** Clicking here will allow you to have the tool automatically scroll (play) through the saved movie either in a fast or slow speed.

E – **Fast Rewind:** Each click of this button will move the screen one complete frame. In this example, the frame would move from 25.00 to 24.00.

F – **Slow Rewind:** Clicking this button will move the screen through each data point. In this example, the screen would move from 25.00 to 24.99.

Custom Data List **271**

G – **Slow Forward:** Clicking this button will move the screen through each data point. In this example, the screen would move from 25.00 to 25.01.

H – **Fast Forward:** Each click of this button will move the screen one complete frame. In this example, the frame would move from 25.00 to 26.00.

I – **Cursors:** Clicking this icon will turn the Cursors on and off. The Cursors allow you to gather numerical data from the waveforms.

J – **Zoom:** Clicking this icon will open a drop-down menu, allowing you to choose the zoom level. The zoom out feature increases the number of frames on the screen so you can see more of the data at one time.

K – **Reset:** This is not used in the Lab Scope viewer.

10. Next, we'll look at the options in the upper tool bar near the right side of the screen.

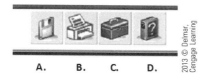

A – **Save:** This feature is not used in the viewer.

B – **Print:** This will print out the current screen. What you see is what you get. The background will *not* be black because it defaults to white when printing.

C – **Utilities:** This is where you can access some display options of the lab scope.

- **Units:** Not used in the viewer
- **Grid:** Can be used to turn on and off the background grid.
- **Trigger Display:** Not used in the viewer.
- **Scales Display:** Turns on and off the numeric identification on both the vertical scale and the horizontal sweep.
- **Inverse Colors:** Will switch to a white/grey background. This was formerly used for printing but is more of a user display preference now because all printing automatically defaults to a white background.

11. The bottom toolbar only has two options available in the viewer. The two options are Channel Selection and Zero Offset. You can reposition the waveforms on the screen in the vertical axis using Zero Offset. Once the correct channel has been clicked and activated, then click the up and down arrows next to the Zero Offset button to move the waveform on the screen.

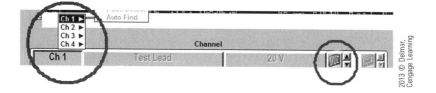

Sample Lab Scope Navigation

1. First, Zoom out to 16x to get a better picture of the events and how they are related to each other.

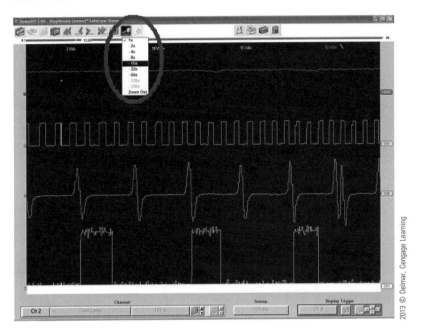

2. Once zoomed out, use the Review icons to scroll through the waveform movie. Choose a point of interest and move the vertical white Zoom line as close to that point as possible.

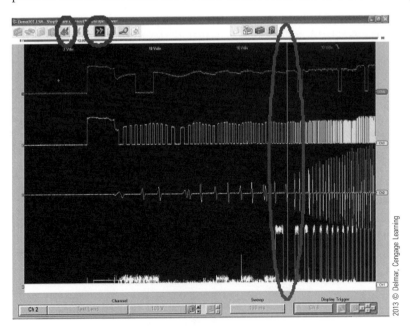

3. Zoom back in to 1x on that point of interest and turn on the Cursors. Cursor 1 is highlighted first, and clicking the left and right arrow icons will move the Cursor across the screen. Clicking the Cursor 1& 2 icon will highlight (activate) the opposite Cursor so it can then be moved by using the arrow icons. Here is an example of the cursors on a four channel trace. Notice that the numbers are color coded to the channels on the Lab Scope module. The exact value of where each cursor hits the waveform is displayed in the information box. The Delta (triangle) column still calculates the difference between Cursor 1 and Cursor 2 for each channel. The amount of time between the two cursors is displayed at the bottom of the box and is labeled Delta Time. Use the cursors to get any value needed from the waveforms, as well as to measure time and frequency. The cursor function is a very powerful tool that should not be overlooked. Give the picture on the screen real meaning by attaching accurate and precise numerical data to it and quickly calculate differences using the Delta column.

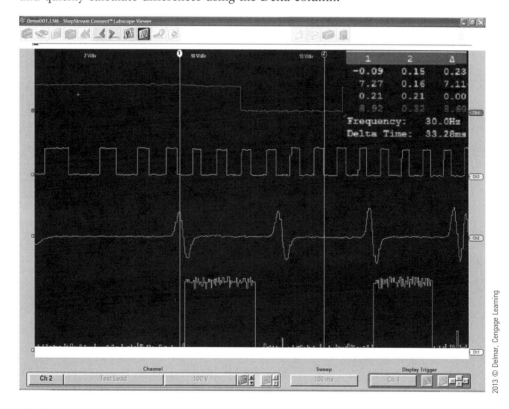

4. Click the Cursor button again to turn the Cursors off. You will be asked if you would like to keep the Cursors on for reference. Click OK to leave them on or *cancel* to turn them off.

5. Using the Zero Offset option, move Channel 1 up and superimpose it on Channel 2. Then move Channel 3 and superimpose it over Channel 4.

274 Chapter 9 Data Management

6. To change the active channel, click on the Channel icon that will open up a small dialogue box that lists all of the available channels. Click on the channel you wish to activate, then go back and click on the channel button to confirm your selection.
7. Here is what the superimposed patterns look like.

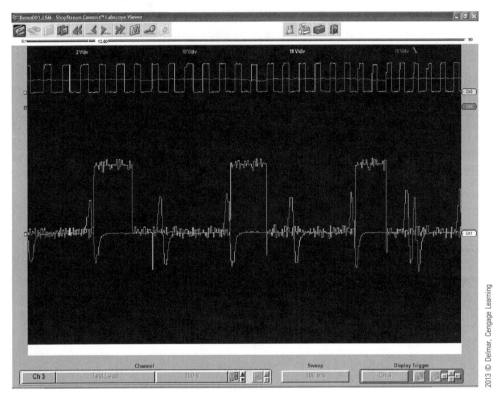

8. When you have finished viewing the Lab Scope Movie, click the Exit icon.
9. From the ShopStream Connect Main Screen, click on the Ask-A-Tech icon near the upper left of the screen.

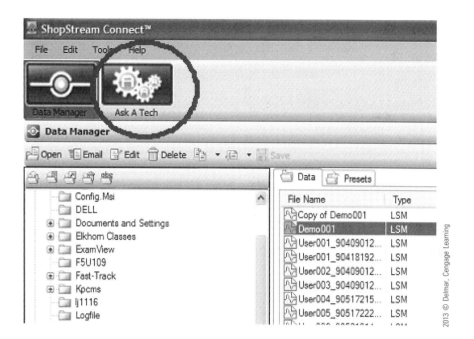

10. Ask-A-Tech is a paid subscription service, but you get the subscription as part of your diagnostic tool's software upgrade. With this being a paid service, and not open to anyone online, the information is being generated by dedicated individuals who see the value in keeping their equipment up-to-date. You can also access all of the Fast-Track Troubleshooter information through the Ask-A-Tech website, so now you would not have to install the Troubleshooter DVD that comes with your software upgrade. Because this program is bundled with ShopStream Connect, it is very convenient to upload Scanner and Lab Scope Movie for other members to view in order to help you with your problem. Now technicians from around the world can basically look over each others' shoulders and see the information from the diagnostic tool and not have to solely rely on the descriptive information provided by the person asking the question. This makes posting a question quicker, and because the technicians who reply have all the information from your diagnostic tool, the answers tend to be more accurate.

11. From the Ask-A-Tech home page, you need to log in.

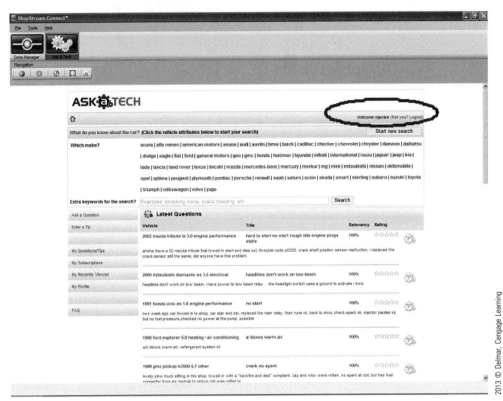

12. When you are facing a tough problem and looking to Ask-A-Tech for some guidance, the first thing to do is look and see if your problem is perhaps a fairly common one or has been asked about before. It is helpful to narrow your search results by entering the vehicle information and looking at what has been discussed or asked about in regard to that particular vehicle in the past. Begin this process by first selecting the vehicle make.

13. Continue entering information about the vehicle until it is identified. Then select the system you are working on or browse the results for the vehicle itself, including all the systems.

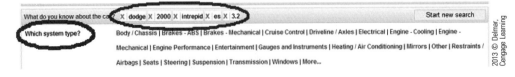

14. You can also use the keyword search to find the information you are looking for.

15. Here is an example of the list of tips about the example car we entered.

16. For more information about the Ask-A-Tech website, use the Frequently Asked Questions (FAQs) found in the side menu options.

17. Ask-A-Tech is a community of professional technicians that are looking to collaborate to solve tough problems more quickly and increase productivity. The questions, responses, and tips that are created using the program are updated instantly so that the latest information is always available. This subscription comes as part of your software upgrade, so use all of the resources that come with that purchase to maximize its cost benefit.

Summary

Data Management is a key part of using today's sophisticated diagnostic tools but this does not have to be a complex problem. Saving and storing information can help increase productivity because there is no reason to resolve a problem that has already been diagnosed, but there is too much information out there to accurately remember it all. All of the diagnostic tools have the capability to save a tremendous amount of data with very little additional cost. The Movies, Snapshots, and images that are easily created from the Scanner, Lab Scope, Ignition Scope, Power Graphing Meter, and other tool features are small in size and, with a little planning, can be archived on a personal PC with minimal effort.

To reassure customers and give them confidence in your business, take screenshots of the data stream or waveform of the faulty component before and then after the repair. Print these screenshots out and explain them to the customer. They will appreciate your time and feel more respected that you offered to explain this "complex information" instead of the normal, "It's fixed, Have a nice day," which can translate into, "You are too stupid for me to waste my time even attempting to explain anything to you." True, the customer will not understand all of the parts of the data or what you are saying, but they will see a difference and the vehicle will no longer have the problem. This can be just one more step is building trust with the customer—Can your competition provide the same service quality or customer care? Saving this same type of information with the customer's repair order history can also make future repairs easier for the technician because one can see what was done and then double check to be sure everything is still working accordingly or if the same or similar problem has returned.

With the growing number of technician help websites and forums like Snap-On's Ask-a-Tech, this saved data allows a much more detailed diagnosis of the vehicle from anywhere in the world. ShopStream Connect, a free download from Snap-On, allows others to view and manipulate Movies and Snapshots on a PC. If you build a database or library of not just personal customer information but diagnostic information about customer vehicles, it will help you find small problems before they become big ones. Taking time to archive waveforms and Movies that show both good and bad examples can help train new technicians, and these can be shared on forums such as Ask-A-Tech. The more industry professionals who share their experience by distributing their data, the greater the collective knowledge of the group. The ability to gather, store, send, and receive data will play a bigger role in the future as automotive technology advances.

Review Questions

1. Technician A states that today's diagnostic tool can easily capture and save different data files. Technician B states that the data files saved on the diagnostic tool can *not* be easily transferred to a PC. Who is correct?
    a. Tech A only
    b. Tech B only
    c. Both Tech A and Tech B
    d. Neither Tech A nor Tech B

2. Two technicians are discussing the screenshot below. Technician A states that pressing Yes now will open the highlighted file. Technician B states that the highlighted file is an image or picture file. Who is correct?

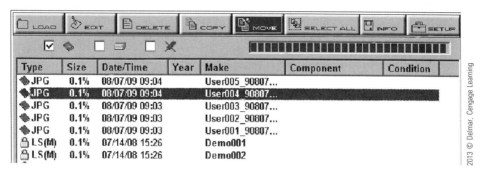

a. Tech A only
b. Tech B only
c. Both Tech A and Tech B
d. Neither Tech A nor Tech B

3. Two technicians are discussing the screenshot below. Technician A states that the Internal Storage device is currently selected. Technician B states that the currently selected storage device's file capacity is near empty so there is plenty of free space to save more files. Who is correct?

a. Tech A only
b. Tech B only
c. Both Tech A and Tech B
d. Neither Tech A nor Tech B

4. Two technicians are discussing the screenshot below. Technician A states that pressing Yes now will delete the highlighted file from the Top (extra) CF card and save it to the Internal (main) storage device. Technician B states that the files currently being viewed are stored on the Top (extra) CF card storage device. Who is correct?

280 Chapter 9 Data Management

   a. Tech A only
   b. Tech B only
   c. Both Tech A and Tech B
   d. Neither Tech A nor Tech B

5. Two technicians are discussing the screenshot below. Technician A states that pressing Yes now will allow you to change which data device is currently active. Technician B states that there is an USB jump drive connected to this diagnostic tool. Who is correct?

   a. Tech A only
   b. Tech B only
   c. Both Tech A and Tech B
   d. Neither Tech A nor Tech B

6. Technician A states that the Copy command deletes the file from the source drive and saves it to the destination drive. Technician B states that the Move command keeps the file on the source drive but also saves it to the destination drive. Who is correct?

   a. Tech A only
   b. Tech B only
   c. Both Tech A and Tech B
   d. Neither Tech A nor Tech B

7. Two technicians are discussing the screenshot below. Technician A states that the Select All command could have been used to highlight all of the files. Technician B states that pressing Yes now will delete all of the files from the currently active drive. Who is correct?

   a. Tech A only
   b. Tech B only
   c. Both Tech A and Tech B
   d. Neither Tech A nor Tech B

8. Two technicians are discussing the screenshot below. Technician A states pressing Yes now will open the highlighted file. Technician B states that the Add Note button is only active when an USB keyboard is plugged into the diagnostic tool. Who is correct?

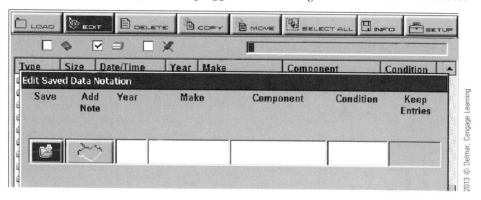

a. Tech A only
b. Tech B only
c. Both Tech A and Tech B
d. Neither Tech A nor Tech B

9. Technician A states that the SC(P) file type designates a Scanner Preset. Technician B states that the IS(M) file type designates an Ignition Scope Movie. Who is correct?
a. Tech A only
b. Tech B only
c. Both Tech A and Tech B
d. Neither Tech A nor Tech B

10. Technician A states that the JPG file type designates a picture file. Technician B states that the BMP file type designates a picture file. Who is correct?
a. Tech A only
b. Tech B only
c. Both Tech A and Tech B
d. Neither Tech A nor Tech B

11. Technician A states that ShopStream Connect is a free download from the Snap-On Diagnostic website. Technician B states that ShopStream allows saved Movies and Snapshots from Snap-On's diagnostic tools to be viewed on your PC. Who is correct?
a. Tech A only
b. Tech B only
c. Both Tech A and Tech B
d. Neither Tech A nor Tech B

282 Chapter 9 Data Management

12. Technician A states that ShopStream Connect makes it easier to transfer files directly from your diagnostic tool to your PC. Technician B states that when reviewing Movie files using ShopStream Connect the PIDs *cannot* be rearranged and display options are locked to the options selected when the Movie was recorded. Who is correct?

    a. Tech A only
    b. Tech B only
    c. Both Tech A and Tech B
    d. Neither Tech A nor Tech B

13. Technician A states that if a Movie was recorded on a Solus Pro that can display a maximum of 4 graphs at one time, then when reviewing the Movie on ShopStream Connect, only 4 graphs are allowed to be viewed at one time. Technician B states that the ShopStream Connect Lab Scope viewer does *not* allow you to move the waveforms using the Zero Offset command. Who is correct?

    a. Tech A only
    b. Tech B only
    c. Both Tech A and Tech B
    d. Neither Tech A nor Tech B

14. Two technicians are discussing Ask-A-Tech. Technician A states that it is an online forum where technicians can post questions and look up time saving tips to help with complicated automotive problems. Technician B states that you get a subscription to it when purchasing a software upgrade for your diagnostic tool. Who is correct?

    a. Tech A only
    b. Tech B only
    c. Both Tech A and Tech B
    d. Neither Tech A nor Tech B

15. Technician A states that access to Ask-A-Tech is done through a separate download from the Snap-On website. Technician B states that when using Ask-A-Tech you have online access to all the information stored on the Fast-Track Troubleshooter DVD. Who is correct?

    a. Tech A only
    b. Tech B only
    c. Both Tech A and Tech B
    d. Neither Tech A nor Tech B

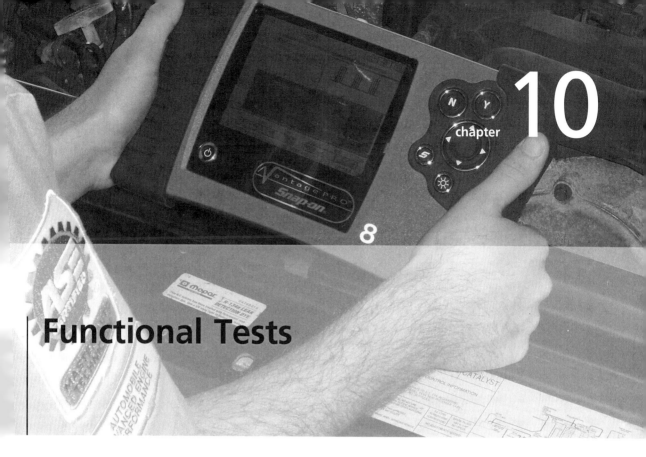

Functional Tests

Upon completion of the Functional Test module, you will be able to:

- Define the four types of functional tests
- Identify various tests available under each of the four functional test categories
- Use Functional Tests to aid in the diagnostic process
- Demonstrate the use of at least one test under each category
- Use Fast-Track Troubleshooter to learn about and successfully perform unfamiliar Functional Tests

REMINDER: This section is written so the reader can follow along using the Solus Pro or MODIS scan tools, if available. The Scanner software is identical on both of these tools, so all of the procedures will be the same. Most of the screenshots come from a MODIS, but the Solus Pro will be identical except for the screen size, so the MODIS may show more information per screen while the Solus Pro user may have to scroll down to view all of the information. Everything else will function in exactly the same way. You are strongly encouraged to follow along using the diagnostic tools, as this provides the hands-on knowledge necessary to pass the Certification Exams. The only required equipment is the Scanner unit itself and the AC power adapter cord. The demonstrations built into the Scanner software will provide the rest of the required material.

Functional Tests Overview

Functional Tests are sometimes called bi-directional control because the scan tool is now sending a request to the vehicle's computer and taking control of various components rather than just reading information. Functional Tests are a great diagnostic tool. Using them can quickly narrow problem areas. With so many vehicle components being controlled by an on-board computer, it is necessary to check and see that these computers are functioning properly and are able to perform the desired task. A simple example could be a radiator cooling fan controlled by a temperature sensor. The temperature sensor monitors the engine temperature and continually sends this information to the Powertrain Control Module (PCM). When a specified temperature is reached, the PCM will command the cooling fan on until the temperature cools below a specified point. Assume the cooling fan is not working. There are many possible reasons for this. Whenever you are diagnosing a problem, it is helpful to narrow down the problem area. This system has an input, the temperature sensor, and an output (the cooling fan) with the PCM in the middle. Using a Functional Test, we can quickly cut the potential problem area in half. We use a Functional Test to command the cooling fan on. The scan tool now takes control of the PCM and activates the cooling fan circuit, and, sure enough, the fan begins to operate without any problems. What does this tell you? The PCM itself and the cooling fan operational circuit, including all of the wires, connectors, relays, and motor must be functioning. This leaves the input side or temperature sensor and wiring left to diagnose. Using one Functional Test on a scan tool, we eliminated the many tests needed to verify the electrical circuit and components of the cooling fan system.

One of the biggest challenges of using any Functional Test is knowing that it existed in the first place. Factory scan tools will have unlimited access to their model vehicles, but aftermarket scan tools usually have fewer functional tests available. Technicians may not take the time to go and check if a test is available because it may be a waste of time, and time is very valuable. Taking the time to find out what types of Functional Tests are available will save you time and equate to more profit in the future. (After all, of course, the time to learn about functional tests is *not* when the customer is standing over you and the time crunch is on.) One of the main objectives of this section is to familiarize you with the various functional tests available. This is just a foundation or starting point because as new updates come out, additional functional tests will be added, so you always need to refresh your mental list of available tests.

Functional Test Descriptions

To help you understand and remember the various Functional Tests available, we will group them into four categories: Information, Toggle, Reset, and Variable Control tests.

Information Tests

These are read-only tests that return back information about the vehicle. Examples include the VIN and Calibration P/N test. These tests are simply used to help identify required information about the vehicle.

Toggle Tests

This type of test has two distinct settings, such as on or off and open or closed. Usually this type of test controls a switch, solenoid, relay or similar device. Examples of this test include A/C Relay on or off, Fuel Injector Enable or Disable, Vent Solenoid open or closed, or a Warning Lamp on or off. This type of test checks to see if a component is functioning and able to be controlled by the PCM through the control circuit.

Variable Control Tests

This type of test allows a specific component to be set at a certain value or percentage. These components are not just on or off but have an operating range or duty cycle. As with all Functional Tests Variable Control tests check for proper operation of the specified component. However, many times you look at other PID values (components) while you are manipulating the one being tested and then look for desired reactions from other vehicle circuits and systems. Examples of this type of test include Idle Air Control (IAC) Motor, Exhaust Gas Recirculation (EGR) valve, and Fuel Gauge Sweep.

Reset Tests

This type of test resets adaptive or learned strategies and other saved values in the vehicle's computer. This type of test has to be performed after completing certain repairs or used to reset a computer to a known good factory setting for diagnostic purposes. The Crankshaft Position Variation Learn is a good example of this Functional Test.

Performing Functional Tests is a key step in the diagnostic procedure, especially when you are working on computer-controlled vehicle systems. To effectively use Functional Tests, you need to understand the different types of tests available and the diagnostic advantages of each. As stated earlier, the most important part of using a Functional Test is knowing the test is available in the first place, which involves developing a diagnostic mindset that has you think of helpful Functional Tests while diagnosing a vehicle, and then taking the time to go and search for that test. The number of times you find the test, as well as the time you save diagnosing the vehicle, will greatly exceed the amount of time you "waste" looking for a nonexistent test. Follow a systematic diagnostic procedure that uses all of the tool's capabilities at your disposal. Let's begin to look at specific Functional Tests in more detail.

Please review the Scanner Introduction section to begin the Scanner demo, so you can follow along with the rest of this section using the dignostic tool.

1. Here are the systems that this vehicle has available. Scroll to Engine and press Yes.
2. From the Main Menu (Engine) screen scroll and select Functional Tests and Press Yes.

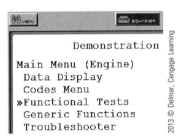

3. This is the Functional Test Menu for the Engine System. Each type of test will be shown in more detail in the coming section.

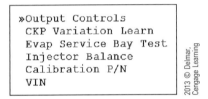

While navigating and learning about the various Functional Tests always be thinking about what type it is, Information, Toggle, Variable Control, or Reset and also think of a diagnostic situation in which this test would be useful. These two attachments will make it much easier for you to remember the different tests in the future when they will actually be needed to help diagnosis a vehicle.

Task: Your instructor may have you take Screenshots (Save Images) of the following screens as proof that you are navigating the tool, finding the actual test, and getting familiar with the test procedures. The easiest way to do this is to program the S-button (Brightness/Contrast button for MODIS) to "Save Image." This procedure is detailed in your scan tool's specific section. The other option is to scroll up to the file folder in the upper right corner of the screen and select Save Image, but this will be more cumbersome and time consuming. Obtain a CF card or jump drive from your Instructor and set the scan tool to save to that device using the Save Data option under the Toolbox icon in the upper right corner of the screen.

Some technicians find it helpful to print out the various screenshots and build a quick reference index of the available Functional Tests for the common automobile makes that come into their shops. This can be done by simply transferring all of the screenshot pictures from your jump drive or CF card to a PC and using a program such as Microsoft Word to paste/insert all of the pictures files into. This is essentially the same procedure used to develop all of the pictures in this book. Spending a little time now to create a quick reference index can equate to saving a lot of time later.

Because we are using a Demonstration Program and are not connected to a Live vehicle, there will be limitations to the depth at which we can explore some of the tests. The data will not always match the actual test results.

CKP Variation Learn Test

1. Scroll down to select CKP Variation Learn and press Yes.

```
Output Controls
»CKP Variation Learn
 Evap Service Bay Test
 Injector Balance
 Calibration P/N
 VIN
```

What type of Functional Test is this?

Answer at the end of this example.

2. Read the on screen directions and take screenshots of each.

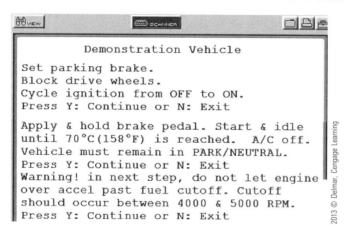

3. Press No to exit at the end.

Answer: This is a Reset test.

Injector Balance Test

1. Scroll down to select Injector Balance Test and press Yes.

```
Output Controls
 CKP Variation Learn
 Evap Service Bay Test
»Injector Balance
 Calibration P/N
 VIN
```

What type of Functional Test is this?

Answer at the end of this example.

288 Chapter 10 Functional Tests

2. Read the on-screen directions and take screenshots of each.

```
Select Test Mode.  Press Y to Continue.
»A/C RELAY (ON/OFF)
 AIR PUMP RELAY -IF EQUIPPED (ON/OFF)
 CANISTER PURGE(%)
 EVAP PURGE/SEAL
 FUEL CLOSED LOOP (CLSD/OPEN)
 FUEL GAUGE ENABLE (%)
 FUEL PUMP RELAY (ON/OFF)

 FUEL TRIM ENABLE (ON/OFF)
 FUEL TRIM (RESET)
 IAC CONTROL
 INJECTOR #1 DISABLE (YES/NO)
 INJECTOR #2 DISABLE (YES/NO)
»INJECTOR #3 DISABLE (YES/NO)            [v]

 INJECTOR #4 DISABLE (YES/NO)
 INJECTOR #5 DISABLE (YES/NO)
 INJECTOR #6 DISABLE (YES/NO)
 INJECTOR #7 DISABLE (YES/NO)
 INJECTOR #8 DISABLE (YES/NO)
»LINEAR EGR(%)
 MALFUNCTION INDICATOR LAMP (ON/OFF)
 O2 HEATER 1 *IF EQUIPPED* (ON/OFF)
»VENT SOLENOID (ON/OFF)
```

3. Press No to exit at the end.

   Answer: This is a Toggle test.

4. Scroll down to Select Calibration P/N and press Yes.

```
 Output Controls
 CKP Variation Learn
 Evap Service Bay Test
 Injector Balance
»Calibration P/N
 VIN
```

What type of Functional Test is this?

Answer at the end of this example.

5. Press No to exit at the end.

   Answer: This is an Information test.

Output Controls

1. Scroll back up and select Output Controls and press Yes.

```
»Output Controls
 CKP Variation Learn
 Evap Service Bay Test
 Injector Balance
 Calibration P/N
 VIN
```

2. This is another long list of potential time saving Functional Tests.

```
Select Test Mode.   Press Y to Continue.
»A/C RELAY (ON/OFF)
 AIR PUMP RELAY -IF EQUIPPED (ON/OFF)
 CANISTER PURGE(%)
 EVAP PURGE/SEAL
 FUEL CLOSED LOOP (CLSD/OPEN)
 FUEL GAUGE ENABLE (%)
 FUEL PUMP RELAY (ON/OFF)

 FUEL TRIM ENABLE (ON/OFF)
 FUEL TRIM (RESET)
 IAC CONTROL
 INJECTOR #1 DISABLE (YES/NO)
 INJECTOR #2 DISABLE (YES/NO)
»INJECTOR #3 DISABLE (YES/NO)           [v]

 INJECTOR #4 DISABLE (YES/NO)
 INJECTOR #5 DISABLE (YES/NO)
 INJECTOR #6 DISABLE (YES/NO)
 INJECTOR #7 DISABLE (YES/NO)
 INJECTOR #8 DISABLE (YES/NO)
»LINEAR EGR(%)
 MALFUNCTION INDICATOR LAMP (ON/OFF)
 O2 HEATER 1 *IF EQUIPPED* (ON/OFF)
»VENT SOLENOID (ON/OFF)
```

3. Capture a screenshot of all of the tests by taking an initial screenshot and then scrolling down until there is a new list of Functional Tests visible. Then take another screenshot. Continue this until all of the tests have been documented.

Idle Air Control Test

1. Scroll down to select IAC Control and press Yes.

```
Select Test Mode.   Press Y to
 FUEL CLOSED LOOP (CLSD/OPEN)
 FUEL GAUGE ENABLE (%)
 FUEL PUMP RELAY (ON/OFF)
 FUEL TRIM ENABLE (ON/OFF)
 FUEL TRIM (RESET)
»IAC CONTROL
 INJECTOR #1 DISABLE (YES/NO)
```

What type of Functional Test is this?

Answer at the end of this example.

2. This is one of the most complicated screens to navigate on the scan tool. The good news is that navigation is not that hard once a few details are explained. The first objective is to recognize some physical limitations of the scan tool. Snap-On has purposely kept its scan tool buttons and controls virtually the same since the MT2500 Red Brick Scanner. This includes a Yes button, a No button, and a directional control device. This has made it easier and more intuitive to move up and learn the new scan tools but still limits all of our navigation to three basic main buttons: Yes, No, and directional control.

This makes it necessary for these buttons to perform various tasks, even some nonintuitive tasks, depending on the screen layout.

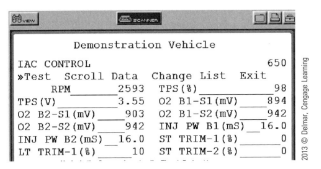

3. The Idle Air Control Motor can be set to a variable range of positions indicated by a value on the far right of the screen. In this case, the initial value is 650.

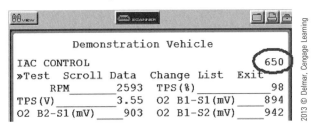

4. To take control of the IAC, press Yes with "Test" selected by the double arrow and watch as it changes to the current value of 650 with an asterisk next to it.

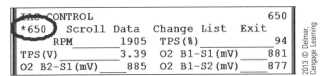

5. At this point, we can dial up or dial down the IAC and watch the values in the data list change accordingly. We would obviously expect to see some correlation in the RPM reading. NOTE: Since this is a demo, the readings may not give you expected results but rather only an illustration of what would happen if connected to a live vehicle.

6. With "Test" active, press the Up arrow button to scroll up the value of the IAC to 1250. Note the current reading on the far right of the screen change with the "Test" value. You know you have active control of the component when the asterisk is next to the test value.

7. Note the PID readings and how they change. Assume you cannot see the PID you wish because it is off of the screen. In the next step, we'll scroll the data.

8. Press No, to deactivate the "Test" control. The double arrow cursor will appear again. NOTE: The current tested value that you set remains active. You know this by the

1250 indicated on the far right of the screen. Even though you cannot change the test value right now, you did *not* lose control of the component.

9. WARNING: Even though your instincts tell you to press Yes, to scroll right and to select "Scroll Data," *don't* do it. That's because, if you do this, what actually happens is the highlighted Scanner button in the upper tool bar moves over to the right and highlights the File Folder icon. So, again, do *not* do this.

10. This is where having a limited number of control buttons requires a little training because navigation becomes nonintuitive. With control of the "Test" deactivated (double arrow cursor visible), press the Up arrow button to scroll right and select "Scroll Data." Again, press UP to move RIGHT.

```
IAC CONTROL                           1250
  1250  »Scroll Data  Change List  Exit
     RPM           1984   TPS(%)          93
TPS(V)             0.04   O2 B1-S1(mV)   903
```

11. Press Yes to select "Scroll Data." Notice that, when active, the double arrow cursor disappears, and now pressing the Up and Down arrow buttons allows you to scroll the data list. All this is the case while you still have control of the component because it is set to the 1250 we chose earlier.

```
IAC CONTROL                           1250
   Test  Scroll Data  Change List  Exit
LT TRIM-1(%)      12   ST TRIM-2(%)       0
LT TRIM-2(%)      11   ST TRM AVG1 %      0
LT TRM AVG1 %      9   ST TRM AVG2 %      0
LT TRM AVG2 %      9   FT LEARN          NO
```

12. Assume that the PID you wish to view is still absent from the list. Perhaps it is in another PID group. To check other groups, we'll use the "Change List" option.

13. Press No to deactivate "Scroll List." The double arrow cursor will appear once again.

14. Just as before, press the Up arrow button to scroll right and select "Change List."

```
IAC CONTROL                              1250
  Test  Scroll Data >Change List< Exit
     RPM         1374   TPS(%)             75
  TPS(V)          2.71  O2 B1-S1(mV)      712
  O2 B2-S1(mV)     794  O2 B1-S2(mV)      781
```

15. Press Yes to change the current PID list.
16. Scroll down to "Sensor Data" and press Yes to select this PID list group.

```
Data Menu
  ENGINE DATA 1
  MISFIRE DATA
  EGR, EVAP, ACC
 »SENSOR DATA
```

17. Notice how the PID list has changed. As a review, and so we can see the new PID values, press the Down arrow button to scroll Left and then select Scroll Data by pressing Yes.

```
IAC CONTROL                              1250
  Test »Scroll Data  Change List  Exit
     RPM         1729   TPS(%)             52
  MAP("Hg)        27.5
  MILEAGE SINCE DTC CLEARED(MI)             0
  O2 B1-S1(mV)     660  O2 B2-S1(mV)      304
  O2 B1-S2(mV)     777  O2 B2-S2(mV)      773
  MAF(Hz)         6141  MAF(gm/Sec)     59.96
```

18. Again, the double arrow cursor will disappear, and pressing the Up and Down arrow buttons will scroll through the data list. All this is the case while we have maintained control of the IAC component as indicated by the 1250 value we chose earlier.

```
IAC CONTROL                              1250
  Test  Scroll Data  Change List  Exit
  TPS(V)          3.69  MAP(V)           4.75
  BARO("Hg)      29.9   BARO(V)          4.80
  COOLANT(°F)     199   INTAKE AIR(°F)   117
  START CLNT(°F)  196   DES EGR(%)         0
  EGR POS(V)      0.86  CAM HI TO LO       0
  CAM LO TO HI      0
```

19. Because the purpose of this and many other Functional Tests is to see the effects of the different PID values, it is important for us to see the values we desire and, more importantly, see how they react. To accomplish this, it may be necessary to "Change List" first, "Scroll Data" second, and finally, when the data view is correct with all of the desired PIDs visible, then activate the Functional Test. Even at this point, it may be hard to pinpoint the effects and the relationships between the various PIDs.

20. One of the best ways to see the relationships between PIDs is to use the Graphing capabilities of the scan tool. To use this feature, first enable the Functional Test and set your component to the value you desire. In our example, this was 1250 for the IAC.

21. Scroll over to the left to highlight the View button and then scroll Down to select 4 Graph and press Yes.

22. The scan tool began to collect and record data frames to the memory buffer as soon as we entered this screen. When we activated the Functional Test, the PID values hopefully showed some response. This response will be shown on the graph. You will have to do a little estimating about the time at which the Functional Test was activated. This is when the rough average of 10 frames per second can be used. Depending on the vehicle and your Custom Data List, the actual number of frames can be different. Assuming you navigated the screen and activated the test within the first 15 to 30 seconds of entering the Functional Test screen, the response seen in the graphs should fall between frames 150 and 300. Since we can store 2000 frames, this is easily captured and able to be reviewed. All of the functions discussed in the Viewing and Interpreting Data section are available here. Choose the number of Graphs you wish to view and then Lock to rearrange the PIDs, Scale, Sort, and Zoom PIDs to build the best possible view and finally use the Cursor function to gather numeric data from the graph. *NOTE:* Because this is a demo, the readings may not give you expected results, but rather only an illustration of what would happen if you were connected to a live vehicle.

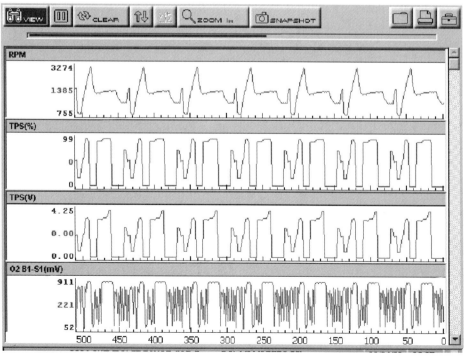

23. The PID Trigger option can be very useful in this circumstance. If you know what values you would like to see, then set the PID Trigger levels at those thresholds *first*, then go back to the Functional Test and activate it. Assuming the test has its desired reaction, a Snapshot will automatically be taken, confirming the action of the Functional Test. The colored line indicating the trigger point on the graph now also represents the point at which the functional test was activated. This reference point can also be used to look for other relationships between various PIDs. PID Triggers are explained in more detail in the Viewing and Interpreting Data section. Please refer to that section for more information.

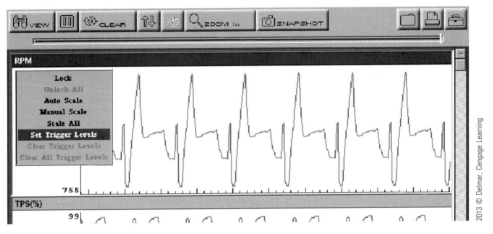

24. To return to the IAC Functional Test screen press No, if necessary, to return to the top toolbar. Next, scroll over to the View button, press Yes to open up the viewing options, select Text, and press the Yes button.

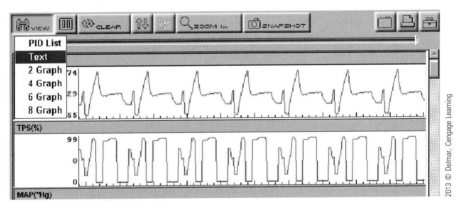

25. You will return back to the Functional Test screen. Notice that you still have control of the component being tested, as indicated by the 1250 value we chose earlier.

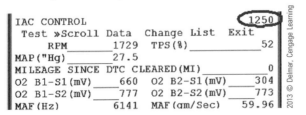

26. When you have finished performing the Functional Test, you can exit by again pressing the Up arrow button to scroll right until Exit is selected by the double arrow and then press Yes.

    Another way to exit this screen is to press No when the double arrow cursor is visible on the screen and the Scanner button is highlighted at the same time. This will bring you back to the Output Control Functional Test list.

    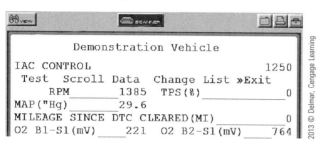

27. Upon exiting, you will lose control of the component and it will be returned to its normal value as determined by the vehicle's computer.

    Answer: This is a Variable Control Test

1. Scroll down to select Injector #1 Disable (Yes/No) and press Yes.

    ```
    Select Test Mode.   Press Y to C
      FUEL CLOSED LOOP  (CLSD/OPEN)
      FUEL GAUGE ENABLE (%)
      FUEL PUMP RELAY   (ON/OFF)
      FUEL TRIM ENABLE  (ON/OFF)
      FUEL TRIM         (RESET)
      IAC CONTROL
    »INJECTOR #1 DISABLE (YES/NO)
    ```

    What type of Functional Test is this?
    Answer at the end of this example.

2. Read the on-screen directions and take screenshots of each.

    ```
    Set parking brake, block drive wheels.
    Start and idle engine. A/C should remain
    OFF for test.
    Press Y: Continue or N: Exit
    ```

3. This screen is very similar to the IAC Functional Test screen, and it is navigated in the same way. It is important for you to keep clear in your mind what the test is actually performing and when it is being performed. The name of the test is "Injector #1 Disable," and currently the double arrow cursor is next to the Yes option on the screen, so is the injector disabled or is it active? The answer is <u>active</u>, as we never disabled it. To keep the current state of any Functional Test clear in your mind, always look towards the upper right corner of the screen—this will indicate the current state. When you first enter the Functional Test, the default setting is whatever the vehicle computer is currently calling for or a normal state. There is no value in the upper

right corner because we know the vehicle is running, and so the injector is supposed to be functioning.

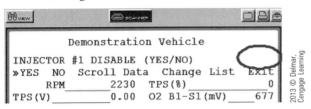

4. Press Yes to select "Yes" using the double arrow cursor. Notice that Yes now appears in the upper right corner of the screen indicating the current status. Because the Functional Test is named Injector #1 Disable, add this name to the answer found in the upper right part of the screen to conclude the current status of the component. In this example, the answer is: Yes, Injector #1 is disabled at that current time. Notice that the double arrow cursor moves to No.

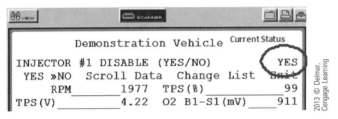

5. The easiest way to keep the current status of the component clear is to always look to the upper right corner of the screen. This is the current status. Looking at the options under the name of the Functional Test is similar to viewing the buttons in the scanner data section. The double arrow cursor is not indicating what is happening right now, but instead is indicating what button is selected and what is going to happen if the button is pushed.

6. In this screen, if we wished to enable Injector #1, we would have to select No by pressing the Yes button. The No command is selected by the double arrow cursor. The question to ask yourself is: Is Injector #1 Disabled? The answer is No.

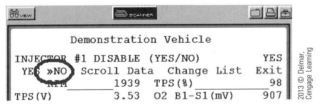

7. Notice that the current status in the upper right corner area has now been updated to No, so Injector #1 is not disabled and the double arrow cursor moves to Yes, in case we want to disable the injector in the future.

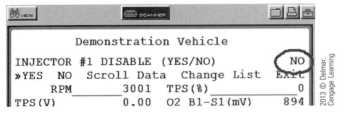

8. When you are performing this or any Functional Test, use the procedures and ideas presented under the IAC Functional Test (described earlier) to easily view the data PIDs you need to make an accurate diagnostic decision.

9. Press No to exit or press the Up arrow button to scroll over to the right until Exit is selected by the double arrow cursor. Then press Yes.

Answer: This is a Toggle test.

Transmission Functional Tests

1. Keep pressing No until you get back to the Select System Menu.
2. Scroll down to select Transmission and then press Yes.

```
Select System:
  Engine
 »Transmission
  Antilock Brakes
  Airbag
  Transfer Case
  BCM
  IPC
```

3. Press Yes to continue after the connector location and key identification screen.

```
Connect: OBD-II Adapter, no key
required.
Location: 16 pin connector under left
side of dash. Press Y: Continue
```

4. Scroll down to select Functional Tests and press Yes.

```
Main Menu (Trans)
  Data Display
  Codes Menu
 »Functional Tests
  Troubleshooter
```

Task: Capture a screenshot of all of the Functional Tests available for this system by taking an initial screenshot and then scrolling down until there is a new list of Functional Tests visible, then take another screenshot. Continue this until all of the tests have been documented.

5. Your screenshots should include all of the tests in this example.

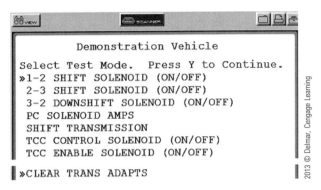

Task: Can you identify each test by type? Print each screenshot and then, next to each Functional Test description, write either Information, Toggle, Variable Control, or Reset.

This can also be done electronically. Put the screenshots onto a PC and then, using Microsoft Word or a similar program, type either Information, Toggle, Variable Control, or Reset next to each Functional Test description.

Your instructor may assign this as an actual assignment, or you may do this on your own to build a quick reference chart of the types of Functional Tests available on a GM SUV. Remember, it is very difficult to use a Functional Test to diagnose a vehicle when you do not know what specific Functional Tests exist.

6. Keep pressing No until you get back to the Select System Menu.

ABS Functional Tests

1. Scroll down to select Antilock Brakes and then press Yes.

```
Select System:
  Engine
  Transmission
 »Antilock Brakes
  Airbag
  Transfer Case
  BCM
  IPC
```

2. Press Yes to continue after the connector location and key identification screen.

```
Connect: OBD-II Adapter, no key
required.
Location: 16 pin connector under left
side of dash. Press Y: Continue
```

3. Do you see any Functional Tests available? Although there is no menu option labeled "Functional Tests," we do see the "Automated Bleed" option. This is a Function Test.

4. Scroll down to select Automated Bleed and press Yes.

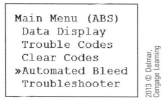

Task: Capture a screenshot of all of the Automated Bleed procedure screens.

5. Your screenshots should include all of the procedures in this example.

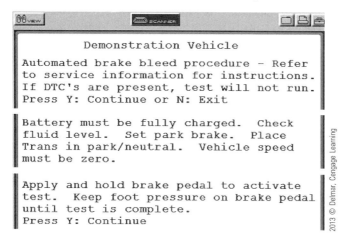

6. Keep pressing No until you get back to the Select System Menu.

Airbag Functional Tests

1. Scroll down to select Airbag and then press Yes.

```
Select System:
  Engine
  Transmission
  Antilock Brakes
»Airbag
  Transfer Case
  BCM
  IPC
```

2. Press Yes to continue after the connector location and key identification screen.

```
Connect: OBD-II Adapter, no key
required.
Location: 16 pin connector under left
side of dash. Press Y: Continue
```

3. Do you see any Functional Tests available? Obviously, there are no Functional Tests found when working on the Airbag System.

```
Main Menu (Airbag)
»Data Display
  Trouble Codes
  Clear Codes
```

Task: Capture a screenshot to confirm there are no Functional Tests available in the Airbag System.

4. Your screenshots should include all of the tests in this example.
5. Keep pressing No until you get back to the Select System Menu.

Transfer Case Functional Tests

1. Scroll down to select Transfer Case and then press Yes.

```
Select System:
  Engine
  Transmission
  Antilock Brakes
  Airbag
»Transfer Case
  BCM
  IPC
```

2. Press Yes to continue after the connector location and key identification screen.

```
Connect: OBD-II Adapter, no key
required.
Location: 16 pin connector under left
side of dash. Press Y: Continue
```

3. Scroll down to select Functional Tests and press Yes.

```
Connect: OBD-II Adapter, no key
required.
Location: 16 pin connector under left
side of dash. Press Y: Continue
```

Task: Capture a screenshot of all of the Functional Tests available for this system by taking an initial screenshot and then scrolling down until there is a new list of Functional Tests visible. Then take another screenshot. Continue this until all of the tests have been documented.

4. Your screenshots should include all of the tests in this example.

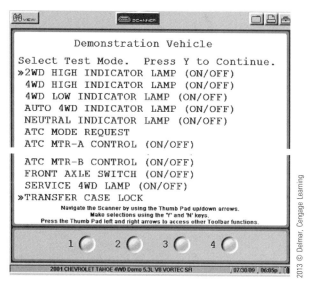

Task: Can you identify each test by type? Print each screenshot and then, next to each Functional Test description, write either Information, Toggle, Variable Control, or Reset.

This can also be done electronically. Put the screenshots onto a PC and then using Microsoft Word or a similar program, type either Information, Toggle, Variable Control, or Reset next to each Functional Test description.

5. Keep pressing No until you get back to the Select System Menu.

Body Control Module (BCM) Functional Tests

6. Scroll down to select BCM (Body Control Module) and then press Yes.

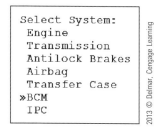

7. Press Yes to continue after the connector location and key identification screen.

```
Connect: OBD-II Adapter, no key
required.
Location: 16 pin connector under left
side of dash. Press Y: Continue
```

302 Chapter 10 Functional Tests

8. Scroll down to select Functional Tests and press Yes.

```
Main Menu (BCM)
  Data Display
  Trouble Codes
  Clear Codes
 »Functional Tests
```

Task: Capture a screenshot of all of the Functional Tests available for this system by taking an initial screenshot and then scrolling down until there is a new list of Functional Tests visible. Then take another screenshot. Continue this until all of the tests have been documented.

9. Your screenshots should include all of the tests in this example.

Task: Can you identify each test by type? Print each screenshot and then, next to each Functional Test description, write either Information, Toggle, Variable Control, or Reset.

This can also be done electronically. Put the screenshots onto a PC and then, using Microsoft Word or similar program, type either Information, Toggle, Variable Control, or Reset next to each Functional Test description.

10. Keep pressing No until you get back to the Select System Menu.

Instrument Panel Cluster (IPC) Functional Tests

1. Scroll down to select IPC (Instrument Panel Cluster) and then press Yes.

```
Select System:
  Engine
  Transmission
  Antilock Brakes
  Airbag
  Transfer Case
  BCM
 »IPC
```

Instrument Panel Cluster (IPC) Functional Tests **303**

2. Press Yes to continue after the connector location and key identification screen.

   ```
   Connect: OBD-II Adapter, no key
   required.
   Location: 16 pin connector under left
   side of dash. Press Y: Continue
   ```

3. Scroll down to select Functional Tests and press Yes.

   ```
   Main Menu (IPC)
     Data Display
     Trouble Codes
     Clear Codes
   »Functional Tests
   ```

   Task: Capture a screenshot of all of the Functional Tests available for this system by taking an initial screenshot and then scrolling down until there is a new list of Functional Tests visible. Then take another screenshot. Continue this until all of the tests have been documented.

4. Your screenshots should include all of the tests in this example.

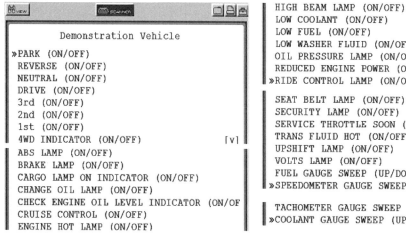

5. Keep pressing No until you get back to the Select System Menu.
6. Go back to the Engine System. Scroll to Engine and press Yes.

   ```
   Select System:
   »Engine
    Transmission
    Antilock Brakes
    Airbag
    Transfer Case
    BCM
    IPC
   ```

7. From the Main Menu (Engine) screen, scroll and select Functional Tests and Press Yes.

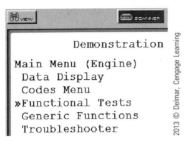

8. We'll continue from this point in Functional Test Training.

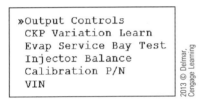

Functional Test Training

Some Functional Tests may be unfamiliar to you, especially if you are a new technician. There is built-in training on the procedures of how and why you would perform many of the available Functional Tests built into the Fast-Track Troubleshooter. This feature is discussed in the Fast-Track Troubleshooter section but will be reviewed here also. The scan tool is here to help you diagnose the vehicle correctly the first time. Use all of the available help and training features it offers to accomplish this. We are going to look at two examples. The two Functional Tests we need help with are the CKP Variation Learn and the Injector Balance.

1. Press No to get to the Main Menu Engine.
2. Scroll down and select Troubleshooter and press Yes.

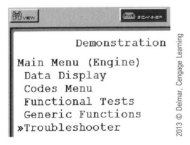

3. With Engine selected, press Yes.

Functional Test Training **305**

4. Then scroll down to select Test and Procedures.

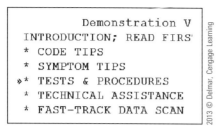

5. Scroll down to select Functional Test Descriptions and press Yes.

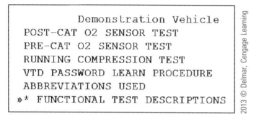

6. Scroll down to select CKP Variation Learn and press Yes.

```
»CKP VARIATION LEARN
 EGR(%)
 FUEL CLOSED LOOP (CLSD/OPEN)
 FUEL PUMP RELAY (ON/OFF)
 FUEL TRIM (RESET)
 FUEL TRIM ENABLE (ON/OFF)
```

Task: Capture a screenshot of all of the CKP Variation Learn procedure screens. Your screenshots should include all of the procedures in this example.

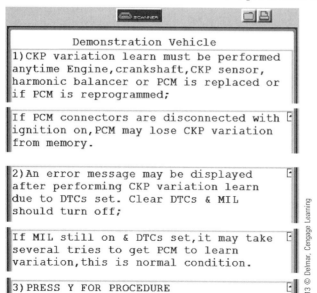

306 Chapter 10 Functional Tests

7. The Troubleshooter description explains why the Functional Test needs to be performed, as well as some procedures for accomplishing the job. At the end of the description, there is a hyperlink straight to the Functional Test itself, so there is no need to manually navigate back to the test.

8. Press No to return to the Functional Test description list.

9. Scroll down to select Injector Balance and press Yes.

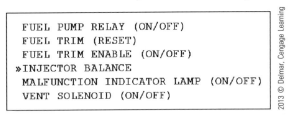

Task: Capture a screenshot of all of the Injector Balance procedure screens.

Your screenshots should include all of the procedures in this example.

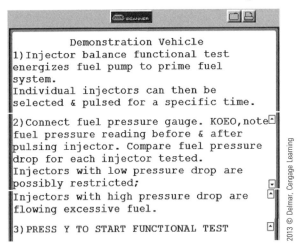

10. In this example, Trouble shooter explains the reason for the test, steps to perform the test, tools required to perform the test, and test results along with probable causes. Again, there is a hyperlink straight to the Functional Test itself.

11. Keep pressing No until you get to the Main Menu Engine.

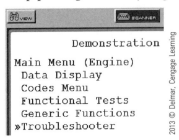

12. Scroll to highlight the View button and press Yes to return to the Main Screen.

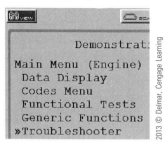

Researching other Functional Tests

It is important for you to build a mental list of available Functional Tests so that, when you are faced with a diagnostic problem, you'll be able to use the Functional Test feature of the tool as part of your diagnostic routine. Functional Tests in an aftermarket scan tool can vary widely by year, make, and model of vehicle but, generally speaking, most manufacturers will have similar tests available for their models. By exploring the Functional Test menus, you begin to know what tests are commonly available for the different manufacturers. This is experience-based knowledge because only by using the tool and navigating the menus do you gain this insight. The purpose of this section is to provide some directed time to explore and navigate different Functional Tests.

Many times, you can access the Functional Test list for a vehicle without having to be connected to the vehicle. You will *not* be able to perform any of the Functional Tests or even see the test activation screen, as was allowed when viewing the demonstration vehicle. These tasks are aimed at building your mental list of available Functional Tests.

To help facilitate this experience, we are going to look at a variety of Functional Test menus. Your instructor will explain how he/she will want you to record and turn in your assignment information. You may be given a blank CF card or asked to use a jump drive to save screenshots and turn that in for grading. After collecting the screenshots, you may be asked to put those pictures into a PowerPoint or Word file and print that out to turn in for credit. Everyone in the class could be assigned a couple of different manufacturers, to research the available Functional Tests for different years and models from that manufacturer, and then to present their findings as a PowerPoint presentation to the class, with a handout, so at the end of the class everyone leaves with a stack of quick reference handouts detailing the available Functional Tests from a variety of manufacturers. You can also review the example screenshots as a quick tour of the Functional Test available for different manufacturers, but it is highly recommended that you don't just review the screenshots but instead actually navigate to them and then try to find Functional Tests on a few vehicles of your own choosing. This hands-on navigational exercise is very important for becoming a Power User of the tool and passing the Snap-On Diagnostic Certification Exam.

Directions: Complete the following tasks by taking identical screenshots of the ones given. This exercise will check your ability to identify vehicles, navigate the tool, find Functional Tests and capture screenshots. Very little navigation detail is included, as that is part of the assignment.

1. **Task:** Collect identical screenshots from this Chrysler vehicle that has Federal Emissions, air conditioning, and an automatic transmission. You do *not* need to connect to the vehicle to gather these screenshots.

```
             Chrysler   1983-2008
Current Vehicle Identification Is:
Vehicle: 2000 DODGE INTREPID
Engine: 3.2L V6 MPI
Press Y: For Same Vehicle N: For New ID.
```

```
Main Menu (ENG)
  Codes & Data Menu
 »Functional Tests
  System Tests
  Generic Functions
  Troubleshooter
```

```
Functional Tests:
»ATM Tests
 AIS Motor Test
 Fuel Pressure Test
 MIN Airflow RPM
 Module Info
```

```
             Chrysler   1983-2008
Select A Test (Key On, Engine Off):
»Stop Current Actuator Test
 A/C Clutch Relay
 Auto Shutdown Relay
 EGR Solenoid
 Fuel Pump Relay Circuit
 Generator Field
 High Speed Fan Relay

 IAC (AIS) Motor
 IAC Motor Open 1step
 IAC Motor Shut 1step
 Injector #1
 Injector #2
»Injector #3

 Injector #4
 Injector #5
 Injector #6
 LDP Solenoid
 Low Speed Fan Relay
»MTV Solenoid

 O2 Heater Test
 O2 Sensor Bias
 Purge Solenoid
 S/C Power Relay
 S/C Vacuum Solenoid
»S/C Vent Solenoid
 Speed Control Servo Solenoids
 SRV Solenoid
»All Solenoids and Relays
```

```
              Chrysler
Main Menu (ENG)
  Codes & Data Menu
  Functional Tests
 »System Tests
  Generic Functions
  Troubleshooter
```

```
              Chrysler   1
Systems Tests:
 »EGR Systems Test
  EVAP Monitor Test
  Generator Field Test
  INJ. Kill Test
  Misfire Counters
  Purge Vapors Test
  Read VIN
```

```
              Chrysler
Main Menu (Body)
  Codes & Data Menu
 »ATM Tests
```

```
              Chrysler   1983-2008
Select A Test!
 »All External Lights
  Chime
  Courtesy Lamp
  Decklid (Trunk) Release
  Door Lock Relay
  Door Unlock Relay
  Driver Door Unlock Relay

  Front Fog Lamps
  Headlamp Relay
  Horn Relay
  Park/Tail Lamp Relay
  Wiper Deicer Output
 »Wiper Motor
```

2. **Task:** Collect identical screenshots from this Ford vehicle that has air conditioning and an automatic transmission. You do *not* need to connect to the vehicle to gather these screenshots.

```
              Ford   1981-2008
VIN: -F--X18L-3-------
Vehicle: 2003 FORD F-150 4WD
Engine: 5.4L V8 DOHC EEC-V SEFI
Press Y: Continue.  N: for New ID.
```

Chapter 10 Functional Tests

```
                 Ford
Main Menu (PCM)
  Data Display
  Codes Menu
 »Functional Tests
  Generic Functions
  Troubleshooter
```

```
            Ford  1981-2008
Select Test:
»Idle Speed Command
 EGR Vacuum Regulator ON/OFF Command
 EVAP Canister Vent Sol ON/OFF Command
 EVAP System Cold Soak Time Bypass
 EVAP System Test
 EVAP Vapor Mgmt Valve ON/OFF Command
 Module ID

 Output State Test
 Reset Keep Alive Memory (KAM)
»Transmission Functional Tests
```

```
            Ford  1981-2008
Select Test:
»Auto Trans Bench Mode
 Auto Trans Drive Mode
 Clear Trans Tables
 Enable Trans Tables
 Stop Trans Adaptive Learning
 Stop Use Of Trans Adaptive
```

3. **Task:** Collect identical screenshots from this Hyundai vehicle. You do *not* need to connect to the vehicle to gather these screenshots.

```
            Asian 1983-2008
VIN: ---W---H-4-------
Vehicle: 2004 HYUNDAI SONATA
Engine: 2.7L V6 DOHC
Press Y: Continue.  N: for New ID.
```

```
                 Asian 1!
Main Menu (Engine)
  Codes & Data Menu
 »Functional Tests
  System Tests
  Generic Functions
  Troubleshooter
```

```
              Asian 1983-2008
Select Test:
»A/C Compressor Relay
 Canister Purge Valve
 Cooling Fan Relay-Hi
 Cooling Fan Relay-Lo
 Diagnostic Lamp
 Fuel Pump
 Idle Speed Actuator
 Ignition Coil - Cylinder 1, 4
 Ignition Coil - Cylinder 2, 5
 Ignition Coil - Cylinder 3, 6
 Injector #1 Disable
 Injector #2 Disable
»Injector #3 Disable
 Injector #4 Disable
 Injector #5 Disable
 Injector #6 Disable
»Main Relay
```

```
              Asian 1!
Main Menu (Engine)
 Codes & Data Menu
 Functional Tests
»System Tests
 Generic Functions
 Troubleshooter
```

```
              Asian 1
Systems Tests:
»EVAP Monitor Test
 End of list
```

4. **Task:** Collect identical screenshots from this Toyota vehicle. You do *not* need to connect to the vehicle to gather these screenshots. Notice that Functional Tests is replaced with Actuator Tests—a different name, but the same basic features and navigation.

```
              Asian 1983-2008
VIN: ----F--K-5-------
Vehicle: 2005 TOYOTA CAMRY
Engine: 3.0L V6 MFI (1MZ-FE)
Press Y: Continue.  N: for New ID.
```

```
Main Menu (Engine)
 Data Display
 Codes Menu
»Actuator Tests
 Generic Functions
 Troubleshooter
```

```
              Asian 1983-2008
Select Test:
»A/C MAG Clutch
 A/F Lean/Rich (%)
 Canister Press CTRL VSV (On/Off)
 EVAP VSV Alone (On/Off)
 FC IDL Prohibit (On/Off)
 Fuel Pump (On/Off)
 Injector Volume (%)

 Intake Control VSV (On/Off)
 Intake Control VSV2 (On/Off)
 Tank Press Bypass VSV (On/Off)
 TC/TE1 (On/Off)
 VVT Control Bank 1 (On/Off)
»VVT Control Bank 2 (On/Off)
```

```
              Asian
Main Menu (Trans)
  Data Display
  Codes Menu
 »Actuator Tests
  Troubleshooter
```

```
              Asian 1983-2008
Select Test:
»Line Press Up Solenoid (On/Off)
 Lock Up Solenoid (On/Off)
 Shift
```

5. **Task:** Collect identical screenshots from this Nissan vehicle. You do *not* need to connect to the vehicle to gather these screenshots. Notice that Functional Tests is replaced with Actuator Tests—a different name, but the same basic features and navigation.

```
              Asian 1983-2008
VIN: ---BA----4-------
Vehicle: 2004 NISSAN MAXIMA
Engine: 3.5L V6 MFI (VQ35DE)
Press Y: Continue.  N: for New ID.
```

```
              Asian 1!
Main Menu (Engine)
  Data (No Codes)
  Codes Menu
 »Actuator Tests
  Generic Functions
  Troubleshooter
```

```
              Asian 1983-2008
 Select Test:
»Coolant Temperature (°C)
 EGR Vol Cont/V Range (Steps)
 Engine Mounting (Trvl/Idle)
 Fuel Injection (%)
 Fuel Pump (On/Off)
 Ignition Timing (°)
 Injector #1 Disable (Yes/No)

 Injector #2 Disable (Yes/No)
 Injector #3 Disable (Yes/No)
 Injector #4 Disable (Yes/No)
 Injector #5 Disable (Yes/No)
 Injector #6 Disable (Yes/No)
»Intank Fuel Temperature Sensor (°C)

 Purge Vol Cont/V (%)
 Radiator Fan Output HI (On/Off)
 Radiator Fan Output LO (On/Off)
 Vent Control/V (On/Off)
 VIAS Solenoid (On/Off)
»V/T Assign Angle (°)
```

6. **Task:** Select a vehicle of your choice and connect to it. Take a screenshot of the System Menu to indicate all of the systems, Engine, Transmissions, ABS, and so on, available on your vehicle. Go through and select each System and look for Functional Tests, remembering that they will not always be labeled as functional tests. Take multiple screenshots while scrolling through the Functional Test list to capture all available tests, just as we did in the previous tasks. If the System has no Functional Tests available, take a screenshot of the Main Menu for that System to prove to your instructor that you did look. Your instructor will explain how he/she will want you to record and turn in your assignment information. You may be given a blank CF card or asked to use a jump drive to save screenshots and turn that in for grading. After collecting the screenshots, you may be asked to put those pictures into a PowerPoint or Word file and print that out to turn in for credit or present your findings to the rest of the class or co-workers.

7. **Task:** Select a vehicle of your choice, but it has to be a different manufacturer from the vehicle you used in the previous task. Connect to the vehicle. Take a screenshot of the System Menu to indicate all of the systems, Engine, Transmissions, ABS, and so on, available on your vehicle. Go through and select each System and look for Functional Tests, remembering that they will not always be labeled as functional tests. Take multiple screenshots while scrolling through the Functional Test list to capture all available tests, just as we did in the previous tasks. If the System has no Functional Tests available, take a screenshot of the Main Menu for that System to prove to your instructor that you did look. Your instructor will explain how he/she will want you to record and turn in your assignment information. You may be given a blank CF card or asked to use a jump drive to save screenshots and turn that in for grading. After collecting the screenshots, you may be asked to put those pictures into a PowerPoint or Word file and print that out to turn in for credit or present your findings to the rest of the class or co-workers. Your Instructor will provide further details.

8. **Task:** Select a vehicle of your choice and then look up Functional Test descriptions in the Troubleshooter. From the Functional Test description menu, select a test of your choice. Read and become familiar with the test and the possible results. Perform the Functional Test and, using the graphing feature, collect screenshots to show or prove that the Functional Test was performed and the vehicle system acted accordingly. If there is a malfunction in how the vehicle system responded, then justify your conclusion using Troubleshooter Fast-Track Data Scan (Normal Values) or another reference source. Compile your findings (before and after Functional Test activation screenshots, reference sources, etc.) into a short PowerPoint that can either be printed and turned in for credit or presented to the class. Your Instructor will provide further details.

9. **Task:** Select a vehicle and Functional test of your choice. Perform the Functional Test and, using the graphing feature, collect screenshots to show or prove that the Functional Test was performed and the vehicle system acted accordingly. If there is a malfunction in how the vehicle system responded, then justify your conclusion using Troubleshooter Fast-Track Data Scan (Normal Values) or another reference source. Compile your findings (before and after Functional Test activation screenshots, reference sources, etc.) into a short PowerPoint that can either be printed and turned in for credit or presented to the class. Your Instructor will provide further details.

Functional Test Summary

Functional Tests are sometimes called bi-directional control because the scan tool is now sending a request to the vehicle's computer and taking control of various components, rather than just reading information. Functional Tests are a great diagnostic tool. Using them can quickly narrow problem areas. With so many vehicle components being controlled by an on-board computer, it is necessary to check and see that these computers are functioning properly and are able to perform the desired tasks. One of the biggest challenges of using the Functional Tests feature is knowing what tests are available because aftermarket scan tools have fewer tests available than factory scan tools. To help you understand and remember the various Functional Tests available, we group them into four categories: Information, Toggle, Reset, and Variable Control. Information tests display information about a vehicle. Toggle tests have two distinct settings such as on/off or open/closed. Reset tests remove or erase adaptive or learned strategies from the vehicle's computer and return it to a default setting. Variable Control tests allow a user to adjust a component's operation through a wide range of settings, such as choosing a specific duty cycle percent. When performing a Functional Test, it is important to be able to capture the responses or effects from the other vehicle components, and this is best done through the Graphing view and by employing a capturing technique, such as PID Trigger. Experience is the best teacher when using Functional Tests because the more tests you perform, the better you'll grasp what tests are available, what the normal response of the test should be, and what test(s) work best for diagnosing specific situations. Explore what the tool has to offer, and practice the different tests you find.

Review Questions

1. Technician A states that Functional Tests allow a user to override the vehicle computer and take control of a component. Technician B states that Functional Tests are sometimes referred to as bi-directional control. Who is correct?
   a. Tech A only
   b. Tech B only
   c. Both Tech A and Tech B
   d. Neither Tech A nor Tech B

2. Technician A states that Functional Tests are useful in the diagnostic procedure by helping isolate the problem area. Technician B states that after performing a Functional Test reactions can be viewed through the data stream. Who is correct?
   a. Tech A only
   b. Tech B only
   c. Both Tech A and Tech B
   d. Neither Tech A nor Tech B

3. Technician A states that a Functional Test that reads and reports the VIN number is an example of a procedural test. Technician B states that controlling a fuel injector is an example of a toggle test. Who is correct?
   a. Tech A only
   b. Tech B only
   c. Both Tech A and Tech B
   d. Neither Tech A nor Tech B

4. Technician A states that adjusting the idle air control motor is an example of a variable control test. Technician B states that CKP Variation Learn is an example of a factory test. Who is correct?
   a. Tech A only
   b. Tech B only
   c. Both Tech A and Tech B
   d. Neither Tech A nor Tech B

5. Technician A states that a toggle test can incrementally change the percentage at which a component is either opened or closed. Technician B states that you must perform a reset test after performing any Functional Tests on a vehicle. Who is correct?
   a. Tech A only
   b. Tech B only
   c. Both Tech A and Tech B
   d. Neither Tech A nor Tech B

6. Technician A states that reset tests can erase adaptive strategies and return settings to a factory default. Technician B states that information tests allow you to see reactions in the data stream. Who is correct?

   a. Tech A only
   b. Tech B only
   c. Both Tech A and Tech B
   d. Neither Tech A nor Tech B

7. Two technicians are discussing the screenshot below. Technician A states that there are only six available Functional Tests on this vehicle. Technician B states the selected test is an example of an information test. Who is correct?

   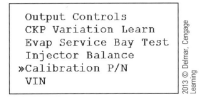

   a. Tech A only
   b. Tech B only
   c. Both Tech A and Tech B
   d. Neither Tech A nor Tech B

8. Two technicians are discussing the screenshot below. Technician A states that pressing the Right arrow button will move the cursor to "Scroll Data." Technician B states that pressing the Up arrow button will increase the value of the IAC. Who is correct?

   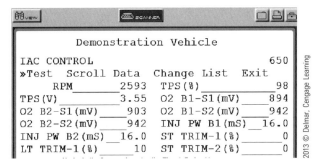

   a. Tech A only
   b. Tech B only
   c. Both Tech A and Tech B
   d. Neither Tech A nor Tech B

9. Two technicians are discussing the screenshot below. Technician A states that pressing Yes now will allow the user to change to a different grouping of PIDs. Technician B states that, even though "Test" is not selected, the user can still have control of the IAC. Who is correct?

```
IAC CONTROL                              1250
  Test  Scroll Data »Change List  Exit
      RPM            1374   TPS(%)            75
  TPS(V)             2.71   O2 B1-S1(mV)     712
  O2 B2-S1(mV)        794   O2 B1-S2(mV)     781
```

   a. Tech A only
   b. Tech B only
   c. Both Tech A and Tech B
   d. Neither Tech A nor Tech B

10. Two technicians are discussing the screenshot below. Technician A states that injector #1 is currently disabled. Technician B states that pressing No now will enable the injector #1. Who is correct?

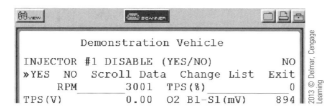

   a. Tech A only
   b. Tech B only
   c. Both Tech A and Tech B
   d. Neither Tech A nor Tech B

11. Two technicians are discussing the screenshot below. Technician A states that the fuel pump relay is currently off. Technician B states that pressing Yes now will allow the user to choose either a Movie, Frame, or Image file to save. Who is correct?

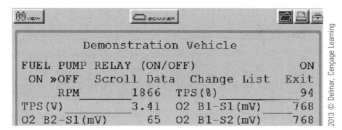

   a. Tech A only   c. Both Tech A and Tech B
   b. Tech B only   d. Neither Tech A nor Tech B

12. Technician A states that going to Troubleshooter and looking under Functional Test descriptions will give the user a brief overview of how to perform the test and what to look for in the effects of it. Technician B states that the Automated Bleed found under the ABS system of many vehicles could be considered a Functional Test. Who is correct?

    a. Tech A only
    b. Tech B only
    c. Both Tech A and Tech B
    d. Neither Tech A nor Tech B

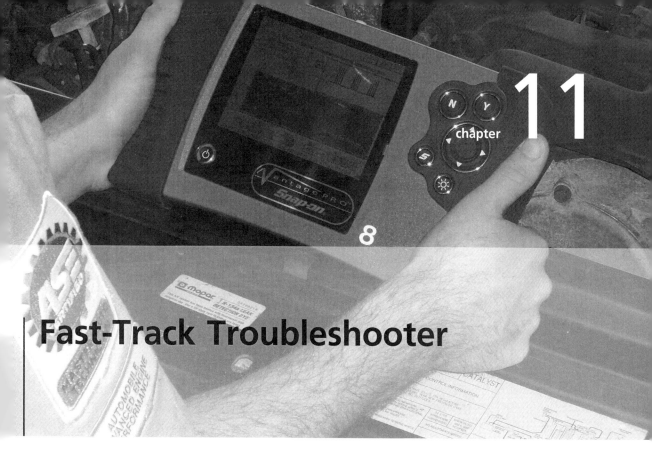

Fast-Track Troubleshooter

Upon completion of the Fast-Track Troubleshooter module, you will be able to:

- Navigate through the menus and access the Fast-Track Troubleshooter
- Explain the importance of Experience-Based information that is VIN specific
- Clarify the differences between Code Setting Conditions, Code Tips, and Symptom Tips
- Review diagnostic strategies under Code Tips
- Retrieve Normal Values for common components
- Reflect on the amount of time-saving, vehicle-specific diagnostic information at one's fingertips, while never having to walk away from the vehicle

Overview

Fast-Track Troubleshooter is one of the most powerful yet underused features of the diagnostic tools. This is a database collection of diagnostic tips and fixes that dates back to the late 1980s and is developed by actual technicians in the field. The diagnostic research facility is located in San Jose, California, and is staffed by field technicians by day and Troubleshooter diagnostic technicians in the evenings. These technicians have actual service center jobs during the day and help develop this database after hours, so they are seeing the same problems and having the same frustrations as you are. They are building a database with fixes and tips that they hope will lessen the frustration for the end user. All of the information found in the Fast-Track Troubleshooter is truly experience based, coming from real technicians who are working on real vehicles, solving real problems. Tips are verified at least five times before being added to the database, so you can rely on its accuracy. Frustrating and tough problems can be more quickly diagnosed, assuming the owner of the diagnostic tool understands and uses the features of Fast-Track Troubleshooter in an efficient manner. It is good business to not let this resource go unused as it can increase productivity. Troubleshooter will allow a novice technician to work more independently and learn diagnostic procedures while progressing to the next level. For experienced and veteran technicians, Troubleshooter is a quick reference tool that allows more problems to be diagnosed quicker, which ultimately improves the bottom line, increasing profitability and thus justifying part of the tool's cost.

Here are the key points to remember about Fast-Track Troubleshooter:

- Information and tests come from real technicians working real jobs on real vehicles. Truly Experience Based Information involving actual repairs.
- Information can be found on the Engine and Transmission systems of most vehicles and newer versions of the software also include information on other systems, including antilock brakes.
- All information and tests are VIN specific to the identified vehicle so component locations, wire colors, and other specifications are specific to the problem you're diagnosing.
- Fast-Track Troubleshooter is similar to having an automotive expert for Engine, Transmission, and ABS systems at your side, helping you with trouble codes, symptoms (with no codes present), diagnostic tests and procedures, current Technical Service Bulletins, and a library of component and sensor specifications. All of this is accessed through the tool without leaving the vehicle.
- Troubleshooter comes with a DVD full of specific vehicle reference information that is easy to use. Once it is installed on a personal computer or laptop, simply search by using the reference numbers provided in the Troubleshooter diagnostic procedures. All of this information is also available on the Ask-A-Tech website (www.askatech.com). OEM electrical diagrams, component locations, and other valuable reference information is readily available and easily searchable. You *do not* have to have an Ask-A-Tech account login to access this information. Use the search references provided for you by Troubleshooter when using the tool.

Please review the Scanner Introduction section to begin the Scanner demo, so that you can follow along with the rest of this section using the dignostic tool.

1. Here are the systems that this vehicle has available. Scroll to Engine and press Yes.

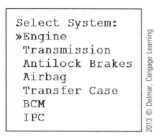

2. From the Main Menu (Engine) screen scroll and select Troubleshooter and Press Yes.

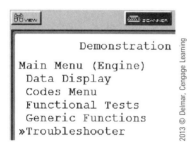

Fast-Track Troubleshooter Features

The following screenshots and features can be viewed by using the demonstration vehicle built into the software. Please follow the demonstration instructions and screenshots to access the Troubleshooter information and see the features and benefits firsthand. No vehicle is needed so this can be done at any time and anywhere. To truly become a power user of the diagnostic equipment, it is important for you to perform the hands-on navigation of the menus and see the information on your specific tool.

1. Press Yes to select the desired system (we'll use Engine) and this will bring you to the Troubleshooter main screen that has the introduction material and the five main features of the Troubleshooter.

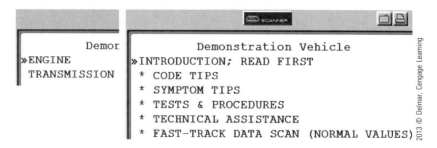

322 Chapter 11 Fast-Track Troubleshooter

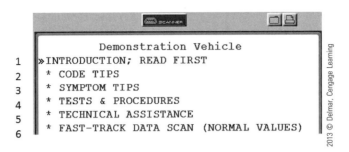

1. **Introduction:** The introduction gives an overview of the Troubleshooter features and how information in Troubleshooter is organized. It also outlines a simple, yet effective diagnostic strategy using Troubleshooter and has links to more detailed information, for example, OBD-II trouble codes, how to interpret them, how to use freeze-frame data, and the proper order of what codes should be diagnosed first. This is an excellent place to learn the basics of scanner diagnostics or review some fundamental skills.

2. **Code Tips:** This is where a technician goes to find out more vehicle-specific information on a particular trouble code. It will have code setting conditions and a step-by-step testing procedure to find the root cause of the failure to help ensure a quick and accurate diagnosis and repair. Be sure to scroll all the way to the bottom to see all available tests and procedural tips.

3. **Symptom Tips:** Even if there is no trouble code stored in the vehicle's computer, Troubleshooter can still be a big help. Problems that do not produce a code can be more difficult to diagnose because there is no clear starting point. Many times, a problem is one that has been diagnosed before and may be a fairly common occurrence with a particular vehicle. Symptom tips list common problems with the identified vehicle and outlines step-by-step how to correct these problems. This is where Troubleshooter can save hours of time and frustration. Quickly scroll through the list of symptom tips to see if the problem you're facing is similar to ones that have been fixed in the past and are part of a known issue. Obviously, a technician will not find the fix every time, but by getting in the habit of checking this on a routine basis, the times a solution is found will more than make up for the extra time you spend searching for solutions. Be sure to scroll all the way to the bottom to see all available symptoms.

4. **Tests & Procedures:** This section explains how to perform tests on some of the more common sensors and components. Scrolling all the way to the bottom will show a useful feature when performing functional tests while diagnosing a problem There is a detailed description of most of the functional tests the diagnostic tool can do. Any questions about how to set up or perform any functional tests should be directed here.

5. **Technical Assistance:** It is important for you to remember that all of this information is specific to the vehicle the user identified by the VIN number when connecting the scanner. This section contains any Technical Service Bulletins, OBD-II Drive Cycle information, as well as other important information necessary for completing repairs on this vehicle. Be sure to scroll all the way to the bottom to see all available options.

6. **Fast-Track Data Scan (Normal Values):** This is a library of known good specifications for nearly all of the data parameters displayed by the diagnostic tool. This quickly and efficiently displays the current data from the vehicle, as well as explaining what the values should be. All that is needed from the technician is a judgment based on the

current data from the vehicle compared to the known good normal values. All of this instantly on one screen makes it very easy for even new technicians to make accurate diagnostic decisions.

Troubleshooter in Detail

The next sections will look at each Troubleshooter in more detail, as well as outline the procedure for accessing the information. To fully understand and appreciate the features, it is best to follow along using your own diagnostic tool.

Task: Your instructor may have you take screenshots (Save Images) of the following screens as proof you are navigating the tool, finding the trouble code specific information, looking for symptoms, and finding specifications for PID values. The easiest way to do this is to program the S-button (Brightness/Contrast button for MODIS) to "Save Image." This procedure is detailed in your scan tool's specific section. The other option is to scroll up to the file folder in the upper right corner of the screen and select Save Image, but this will be more cumbersome and time consuming. Obtain a CF card or jump drive from your Instructor and set the scan tool to save to that device using the Save Data option under the Toolbox icon in the upper right corner of the screen.

Code Tips

The demonstration vehicle has multiple trouble codes stored in its memory. For practice, we are going to look at the P0102 Mass Air Flow code. Troubleshooter provides a detailed diagnostic flow chart on how to repair most trouble codes. It is important to remember that this diagnostic flow chart was developed by seasoned technicians in the field and is meant for the specific vehicle you identified and are communicating with. The steps are numbered and have a rationale to their sequence. The first screen explains the conditions under which the code will set. The next steps will help walk the technician through a series of tests and inspections to help find the root cause of the problem. The earlier steps will be simpler and easier to perform inspections that can commonly cause the problem. This is the most efficient place to start. Assuming those inspections pass the procedure, we will get into more detail and more extensive testing will be explained. Let's see what this looks like when using the demonstration vehicle and trouble code P0102.

How, and what, do we test in order to repair a P0102 trouble code?

1. From the Troubleshooter Main Screen, scroll down to Code Tips and press Yes.

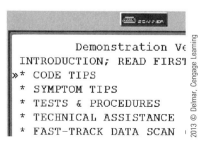

2. Identify the trouble code P0102 by selecting the appropriate range of code numbers, ultimately finding and selecting code P0102.

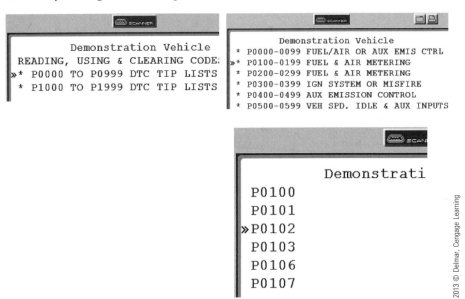

3. Read through the diagnostic procedure and always be sure to scroll all the way to the bottom. Even if the text doesn't go to the bottom of the screen, there may be more data on the next page. Keep scrolling!

4. The first page/screen will usually give a brief description of the code-setting conditions. This may be helpful in determining the cause of the problem.

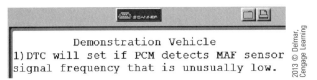

5. The next steps include some quick and easy visual checks. This won't require a lot of disassembly when working on a vehicle, but these are also commonly found problems that can cause this code to set. Don't overlook the quick and easy fix. Only dig deeper if necessary.

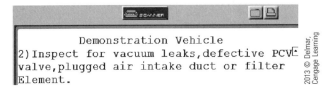

6. If the first inspections don't solve the problem, then continue on with more detailed tests to verify the problem and find the root cause. In this next test, the diagnostic tool explains the recommended specification for this vehicle's MAF at idle and also goes out and collects the real-time data from the PCM. In a real diagnostic situation, the tool would be connected to the data link connector and the vehicle idling. As you can see

on the screen, the MAF frequency should be above 10Hz, and the actual MAF Hz reading from the vehicle is displayed at the bottom of the screen. As the technician, all you need to do is make a judgment call of Pass or Fail.

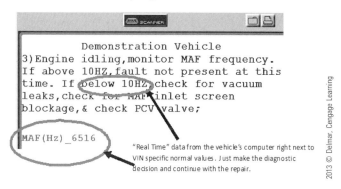

7. If the problem persists and no cause has been identified, continue following the diagnostic procedure. The next few screens have the technician perform some electrical tests. On the last screen of this particular test a reference number is given, REF G119. All of the references found in Troubleshooter are located on the Troubleshooter DVD disk that comes with your diagnostic tool of software update. This DVD needs to be installed on a computer and will allow the technician to quickly find vehicle specific information for the particular test being described in Troubleshooter.

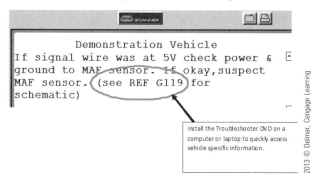

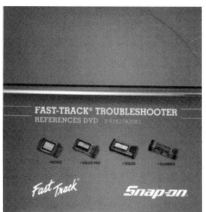

Once the Troubleshooter DVD is installed and the program opened, the user can enter the reference number in the search box. For this demonstration, the reference number is G119. After searching for this reference number, the Troubleshooter brings up the following diagram. Remember, this is vehicle-specific information and helpful in completing the diagnostic procedure outlined in Troubleshooter. This information is also available on the Ask-A-Tech website (www.askatech.com). OEM electrical diagrams, like the one in the example, is easily searchable. You *do not* have to have an Ask-A-Tech account login to access this information. Use the search references provided for you by Troubleshooter, in this case G119, when using the tool. Go to the website, type G119 into the search box, and follow the instructions to easily access the information. This is a much quicker and more efficient way of finding pertinent reference information of the specific vehicle you are working on. If you use other computer based or online based references, it will take you longer to search and navigate.

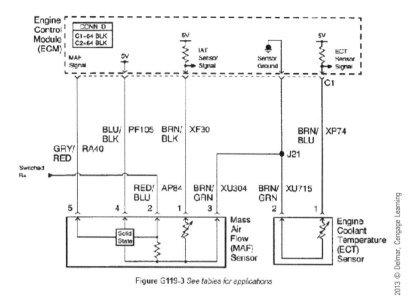

Figure G119-3 See tables for applications

Symptom Tips

Symptom Tips is where a technician can go for help when there are no trouble codes present. This is a valuable tool when you are working on intermittent problems found in Engine, Transmission, and ABS systems. Many of the symptoms under the Engine system deal with no start conditions due to various reasons. For practice, we'll look at how to diagnosis a No Start, No Spark symptom.

What do you do if the vehicle will not start, and you suspect it has no spark?

1. From the Troubleshooter Main Screen, scroll down to Symptom Tips and press Yes.
2. Then scroll down until you select NO Start, NO Spark and then press Yes.

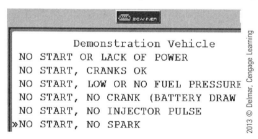

3. As with Code Tips, begin to read and follow the diagnostic procedure. See how the directions are straight forward, easy to follow, and include the part numbers of any special tools needed to complete the diagnosis.

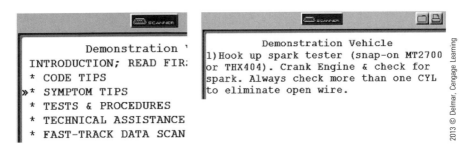

4. Scroll down to the next step and again see how Troubleshooter will give you the specification to look for. Go out and gather that specific data from the vehicle and display it at the bottom of the screen. This screen also includes a reference number that should be searched for using the Troubleshooter Program installed on a shop computer using the Troubleshooter DVD. Again, this information is also available on the Ask-A-Tech website (www.askatech.com). Reference information is readily available and easily searchable. You *do not* have to have an Ask-A-Tech account login to access this information. Use the search references provided for you by Troubleshooter when using the tool and type them into the search box on the website.

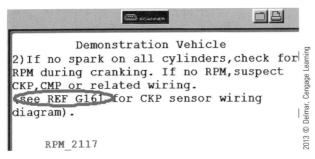

5. Scrolling down to step 3 provides you with directions for the final test and another reference item.

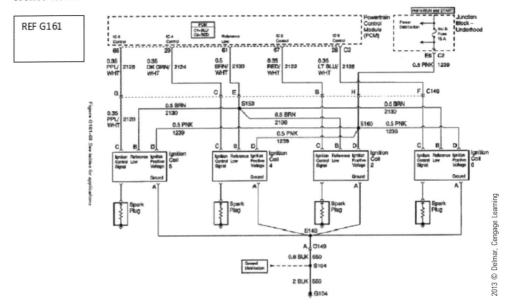

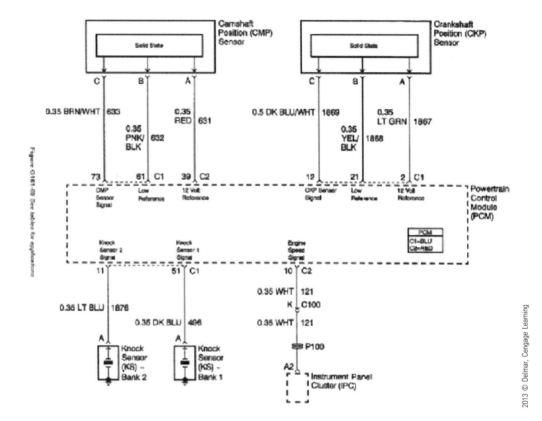

Tests & Procedures

This section explains, step by step, how to perform common diagnostic tests on the vehicle. It also includes a subsection that provides details and procedures about the Functional Tests (bidirectional control) that are capable of being performed using the diagnostic tool. For practice, we'll look at how to perform the Injector Balance test using the diagnostic tool to control the pulsing of the fuel injectors.

1. From the Troubleshooter Main Screen, scroll down to Tests and Procedures and press Yes.

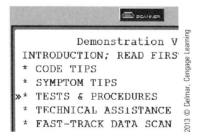

2. Then scroll down until you select Functional Test Descriptions and then press Yes.

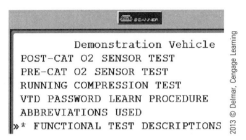

3. Continue to scroll down on the new screen until you select Injector Balance and then press Yes.

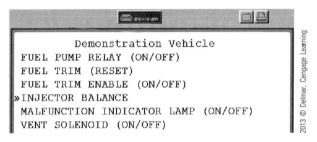

4. The following screens explain the purpose of the Injector Balance test, the tools needed to perform the test, possible results of the test and their diagnostic meanings, as well as a hyperlink to the Functional Test itself.

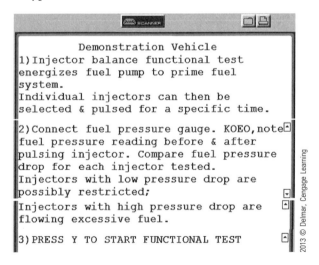

Technical Assistance

This feature includes pertinent Technical Service Bulletins, OBD-II Drive Cycle procedures, OBD-II Readiness Monitor Information, and other key information field technicians found important when diagnosing and repairing the identified vehicle. Here are the steps to access the Technical Assistance information and a list of the information available for the demonstration vehicle. Please take the time to navigate through the information and read a couple of topics to get a feel of what type of information is available right on your diagnostic tool.

1. From the Troubleshooter Main Screen, scroll down to Technical Assistance and press Yes.
2. The following is a couple of screenshots combined to show all of the informative topics available under this feature. Each topic has multiple screens of information. Please take a few moments to review a few topics to get a feel for the basic navigation and kind of information available.

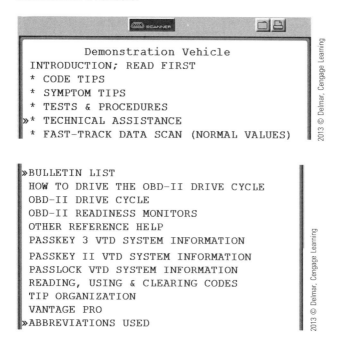

Fast-Track Data Scan (Normal Values)

This feature is a helpful time-saver because it allows the technician to quickly find the normal value of a given parameter without leaving the vehicle to look it up. The diagnostic tool will also go and gather the current value from the vehicle and display it on the screen. Once again, the technician is left with making the judgment of whether the value is within the given range or requires further testing. While many of the values are not absolute values, they do give a good baseline for making a quick diagnostic decision. The ease with which this information can be used makes it a very good first step towards focusing in on the root cause of the problem. For practice, we'll look at the normal value of the Mass Air Flow Sensor (MAF).

What should the value of the MAF sensor be?

1. From the Troubleshooter Main Screen, scroll down to Fast-Track Data Scan (Normal Values) and press Yes.

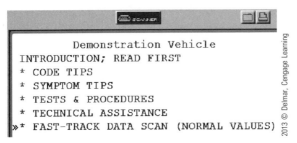

2. Then scroll down until you select MAF (gm/Sec) and then press Yes.

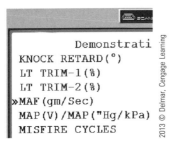

3. The next few screens explain what the normal value of the MAF should be and display the current value of this parameter. Based on this information, the technician makes a judgment whether the value is acceptable or not. The biggest advantage of this feature is that the technician does not have to leave the vehicle to find this information because it is all part of the diagnostic tool.

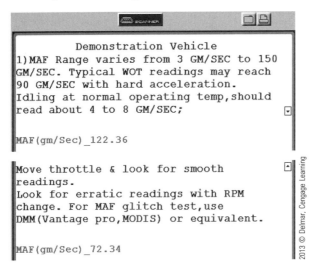

Fast-Track Troubleshooter Research

Directions: Complete the following tasks by taking identical screenshots of the ones given. This exercise is going to check your ability to navigate the tool, find Troubleshooter information, and capture screenshots. Very little navigation detail is included, as that is part of the assignment. You may be given a blank CF card or asked to use a jump drive to save screenshots and turn that in for grading. After collecting the screenshots, you may be asked to put those pictures into a PowerPoint or Word file and print that out to turn in for credit or present your findings to the rest of the class. Your Instructor will provide further details.

1. **Task:** Collect identical screenshots using the demonstration vehicle. After scanning the vehicle, assume that a C0300 transmission code was found. Check Troubleshooter for some tips on where to begin diagnosing this problem. Collect the following screenshots from both the scan tool and shop computer (using the Troubleshooter DVD). To capture screenshots on the computer, press the "Print Screen" button and then, in Microsoft Word or a similar program, either right click and "Paste" or use the keyboard shortcut of "Ctrl V" at the same time to paste the screenshot into the document.

```
                    Demonstration Vehicle
»C0300      1)DTC sets with loss of signal from rear
 C0305      propshaft speed sensor. If rear
 C0308      propshaft speed drops below 16 RPM for
 C0309      30 seconds with Engine running. Front
 C0310      AXLE engaged & vehicle speed above 10
 C0315      MPH,DTC will set.

            2)VSS also located on rear output shaft
            near rear propshaft speed sensor. Do
            not confuse this during testing (see
            REF G1038 for component locations);

            3)After repairs have been completed,it
            is essential to clear DTCs or symptoms
            will remain. (see REF G1039 for wiring
            diagram).
```

2013 © Delmar, Cengage Learning

334 Chapter 11 Fast-Track Troubleshooter

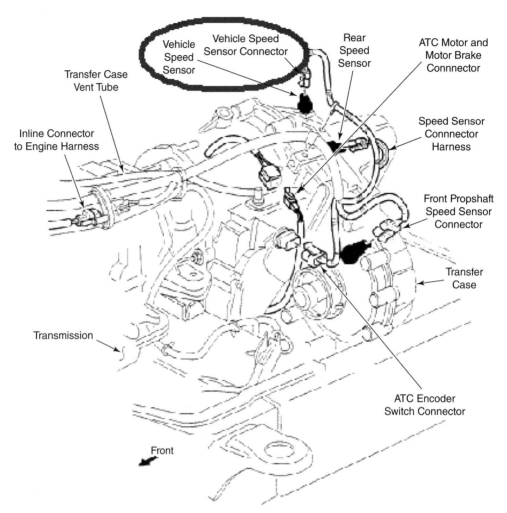

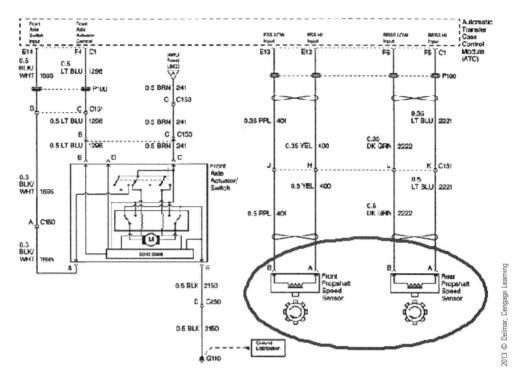

2. **Task:** Using the demonstration vehicle, select a common problem symptom found on this vehicle under the Symptom Tips menu option. There are 11 Engine tips and 2 Transmission tips. So, depending on class size, you may have to work in groups. Be sure as a class that all the Symptom Tips are covered. Your Instructor will give you further details. Take screenshots of each procedural screen, collect any reference materials given in the procedure, then summarize the common cause and repair procedure for this Symptom Tip. After collecting the screenshots, develop a short PowerPoint presentation explaining your findings and share it with the class. Your Instructor will provide further details.

3. **Task:** Using the demonstration vehicle, select 4 PIDs under the Fast-Track Data Scan (Normal Values) and take screenshots of all available information. Summarize your findings. After collecting the screenshots, you may be asked to put those pictures into a PowerPoint or Word file and print that out to turn in for credit or present your findings to the rest of the class. Your Instructor will provide further details.

4. **Task:** Using the vehicle of your choice, select one Code Tip, one Symptom Tip, one Technical Service Bulletin, and 2 PIDs under the Fast-Track Data Scan (Normal Values) and take screenshots of all available information. Summarize your findings. After collecting the screenshots, you may be asked to put those pictures into a PowerPoint or Word file and print that out to turn in for credit or present your findings to the rest of the class. Your Instructor will provide further details.

Summary

Fast-Track Troubleshooter is a great asset to any technician. Not having to leave the vehicle to find technical service bulletins, component specifications, and other service information will increase productivity for veteran and new technicians alike. The benefits of this feature can only be realized if put to use in a consistent fashion. Troubleshooter will not solve every problem every time, but if used consistently the amount of time it saves will far outweigh the amount of time spent using it. With the vast amounts of information available today, there is no reason to re-diagnose a problem that already has a known successful fix, performed and verified by a team of actual industry technicians, and the quickest place for a technician to look for this answer would be in the Fast-Track Troubleshooter feature of the diagnostic tool they're already using.

Key points to remember:

- All the information is from technicians currently working in the industry.
- All the information is VIN specific to the vehicle you are currently diagnosing.
- Use Troubleshooter not just for diagnosing trouble codes, but also for diagnosing by symptom, learning about tests and procedures, retrieving technical information, and verifying specifications.
- Install the Fast-Track Troubleshooter Reference DVD that comes with your diagnostic tool and subsequent software updates. This is an excellent quick reference. All of this information is also available on the Ask-A-Tech website (www.askatech.com). OEM electrical diagrams, component locations, and other valuable reference information is readily available and easily searchable. You *do not* have to have an Ask-A-Tech account login to access this information. Use the search references provided for you by Troubleshooter when using the tool.

Review Questions

1. Technician A states that Troubleshooter information comes from experienced technicians currently working in an automotive service center. Technician B states that Troubleshooter will provide repair information going back to the early 1980's. Who is correct?
    a. Tech A only
    b. Tech B only
    c. Both Tech A and Tech B
    d. Neither Tech A nor Tech B

2. Technician A states that Troubleshooter information is VIN-specific information focused on the vehicle you identified. Technician B states that Troubleshooter information is only found under Engine Systems. Who is correct?
    a. Tech A only
    b. Tech B only
    c. Both Tech A and Tech B
    d. Neither Tech A nor Tech B

3. Technician A states that the Troubleshooter DVD that comes with your diagnostic tool and software upgrades should be installed on a PC and used to quickly look up reference information given throughout the Troubleshooter procedures. Technician B states that experienced technicians will find little useful information in Troubleshooter. Who is correct?

   a. Tech A only
   b. Tech B only
   c. Both Tech A and Tech B
   d. Neither Tech A nor Tech B

4. Technician A states that Troubleshooter is only useful if there is a trouble code found in the vehicle's memory. Technician B states that Technical Service Bulletins can be accessed from Troubleshooter. Who is correct?

   a. Tech A only
   b. Tech B only
   c. Both Tech A and Tech B
   d. Neither Tech A nor Tech B

5. Technician A states that Code Tips will give generic procedures on how to generally diagnose a trouble code. Technician B states that the procedures found under Code Tips are in a specific diagnostic order, starting with easy checks of common problems and working towards more time-consuming diagnostic procedures. Who is correct?

   a. Tech A only
   b. Tech B only
   c. Both Tech A and Tech B
   d. Neither Tech A nor Tech B

6. Technician A states that Symptom Tips are diagnostic and repair procedures for known common problems found on the specific vehicle you have identified. Technician B states that even if there is *no* trouble code found on the initial scan, Troubleshooter should be used to look for Symptom Tips to help diagnosis the problem. Who is correct?

   a. Tech A only
   b. Tech B only
   c. Both Tech A and Tech B
   d. Neither Tech A nor Tech B

7. Technician A states that to quickly find what the normal value for any given PID is supposed to be, you should look under Tests and Procedures. Technician B states that drive-cycle procedures are also found under Tests and Procedures. Who is correct?

   a. Tech A only
   b. Tech B only
   c. Both Tech A and Tech B
   d. Neither Tech A nor Tech B

8. Two technicians are discussing the screenshot below. Technician A states that the value in red is the current "live" value being sent from the vehicle's computer. Technician B states that it is common to find "live" PID values throughout the Troubleshooter procedures. Who is correct?

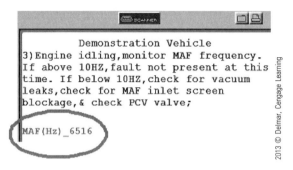

   a. Tech A only
   b. Tech B only
   c. Both Tech A and Tech B
   d. Neither Tech A nor Tech B

9. Two technicians are discussing the screenshot below. Technician A states that REF G119 refers to a wiring diagram found on the Troubleshooter Reference DVD. Technician B states the Troubleshooter Reference DVD can only be searched by reference numbers identified in the Troubleshooter procedure. Who is correct?

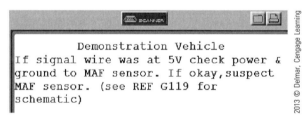

   a. Tech A only
   b. Tech B only
   c. Both Tech A and Tech B
   d. Neither Tech A nor Tech B

10. Two technicians are discussing the screenshot below. Technician A states that this screen is found under the Technical Assistance menu option. Technician B states that pressing Yes now will provide access to information on how to perform a variety of functional tests. Who is correct?

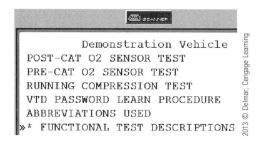

   a. Tech A only
   b. Tech B only
   c. Both Tech A and Tech B
   d. Neither Tech A nor Tech B

11. Technician A states that using Troubleshooter in a consistent systematic fashion will lower a technician's overall productivity. Technician B states that Code Tips only provide information on generic OBDII trouble codes. Who is correct?

   a. Tech A only
   b. Tech B only
   c. Both Tech A and Tech B
   d. Neither Tech A nor Tech B

12. Two technicians are discussing the screenshot below. Technician A states that Pressing Yes now will open list of Technical Service Bulletins (TSBs). Technician B states that this list of information is found under the Technical Assistance menu option. Who is correct?

   a. Tech A only
   b. Tech B only
   c. Both Tech A and Tech B
   d. Neither Tech A nor Tech B

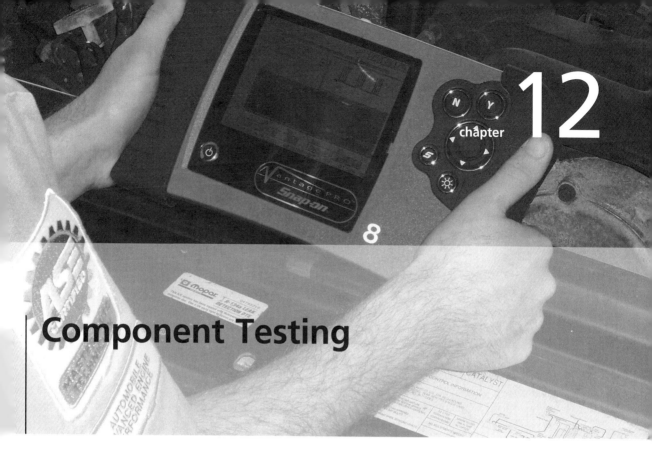

chapter 12

Component Testing

Upon completion of the Component Testing module, you will be able to:

- Identify a new vehicle to use in the Component Testing Meter
- Find and select a component test from the menu
- Diagnose a faulty component using the Component Test Meter
- Learn the basics of Lab Scope diagnostics, while being guided with step-by-step directions and a library of good and bad waveform examples
- Explain how to use the built-in automotive reference features that aid in the continued training of technicians

Component Test Meter Overview

The Component Test Meter (CTM) is similar in function to the Fast-Track Troubleshooter. The Troubleshooter helps technicians diagnose problems while interpreting back-door data through the data link connector using the Scanner. The Component Test Meter (CTM) helps technicians diagnosis problems interpreting front-door data through the use of the Lab Scope or Multimeter functions. The CTM database is a collection of specific component tests, information, and specifications that dates back to the early 1980s. The specific tests, procedures, and specifications come from the same industry technicians located in San Jose, California, that provide the information for the Troubleshooter. These technicians have actual service center jobs during the day and help develop this database after hours, so they are seeing the same problems and having the same frustrations as you. However, they are building a database with tests and tips in order to help lessen frustration for the diagnostic tool's end user. All of the information found in the CTM is truly experience based, coming from actual technicians, working on real vehicles, solving real problems. These frustrations and tough problems can be more quickly diagnosed assuming the owner of the diagnostic tool understands and utilizes the features of the CTM in an efficient manner. It is good business to not let this resource go unused as it can increase productivity. The CTM allows a novice to intermediate technician to work more independently and learn diagnostic test procedures and general use of a Lab Scope. For veteran technicians, the CTM is a quick reference tool that allows more problems to be diagnosed quicker, which ultimately improves the bottom line, increasing profitability and thus justifying part of the tool's cost.

The CTM menu offers some useful information links. It is best to view this section as a continually updated automotive textbook. This electronic textbook has detailed procedures on how to use some of the diagnostic tool's accessories such as pressure transducers, amp clamp, and ignition adapters, as well as an overview of the features and benefits of the diagnostic tool itself. In addition, you will find a quick reference A-Z automotive index and built-in automotive training lessons that take a technician through some basic automotive knowledge and diagnostic procedures. It is important to understand the amount of information available in these diagnostic tools because this understanding can lead to higher skilled technicians, quicker repairs, and ultimately greater profits for the company.

REMINDER: The Component Testing section will provide the information for both the MODIS and Vantage Pro diagnostic tools. The software running these two diagnostic tools is virtually identical. Some of the icons on the Vantage Pro may look a bit different on the screen due to a resolution difference, but the functions are the exactly same. You will more clearly understand the information presented in this section if you follow along using the diagnostic tool. Using the tool in conjunction with the book helps you build the hands-on knowledge required to pass the Snap-On Certification Exams.

Component Testing Menu

The Component Testing menu lists all of the features you need to identify vehicles, select component tests, and find other automotive related information. Each menu option will be detailed in the following section. Follow along with your diagnostic tool to gain greater familiarity with the navigation.

Component Testing Menu

1. Scroll to the Component Testing Icon. A menu box will appear to the right.

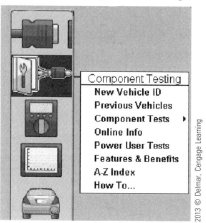

2. To begin testing a vehicle, a technician will have to identify that vehicle using the diagnostic tool. The tool will have a vehicle currently identified, and this is shown at the bottom of the screen. If the vehicle shown here is not the vehicle you are working, on a New Vehicle will have to be identified.

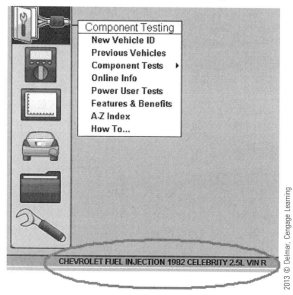

3. Scroll over to the right and select New Vehicle ID by pressing the Yes button.

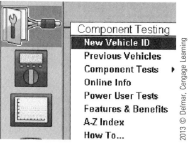

For demonstrations purposes, we'll use our standard 2001 Chevrolet Tahoe 5.3L VIN T.

4. Select US Domestic.

5. Scroll and select the Manufacturer: Chevrolet.

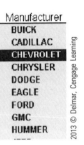

The next selection you need to make is System Type. This is dependent on what component you will be testing. Depending on the vehicle make, the system options will change. Common options included on most vehicle makes include ABS, Carbureted, Charging System, Fuel Injection, and Transmission. The largest list of components is found under the Fuel Injection selection and includes most of the sensors and components used to diagnose drivability concerns. For the demonstration, select Fuel Injection.

6. Scroll and select the System Type: Fuel Injection.

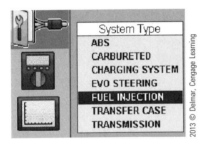

7. Scroll and select the Year: 2001.

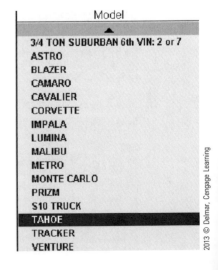

8. Scroll and select the Model: Tahoe.

Component Testing Menu 345

9. Scroll and select the Engine: 5.3L VIN T.

10. A dialogue box will appear and ask you to accept the identification. Review the information at the bottom of the screen first. If that information matches the vehicle you are currently working on, then press Yes to accept the identification.

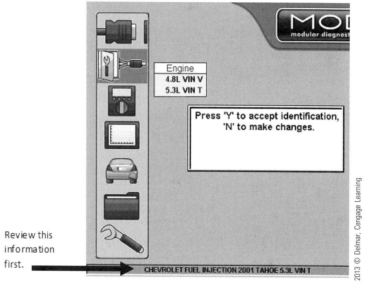

Review this information first.

The next screen lists the available Component Tests of the system you selected for the vehicle you identified. For the demonstration, the Component Tests for the Fuel Injection system for the 2001 Chevy Tahoe are listed. Notice the scroll arrow at the bottom of the list, indicating there are more tests that can be selected. It is important for you to understand that this is the same screen for the menu option Component Tests. To quicken the diagnostic process, the tool automatically jumps to Component Tests and skips over Previous Vehicles.

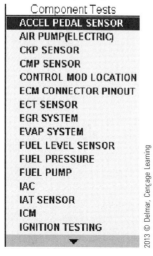

346 Chapter 12 Component Testing

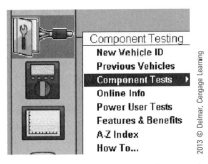

You do not need to follow along with the Previous Vehicles menu option explanation. Leave your tool set to the Component Tests screen, and we'll pick that up in a minute. It is important to explain the Previous Vehicle option because it will help you to save time in the future. The Previous Vehicles option allows you to choose one of the last 10 identified vehicles. If you have already identified a vehicle, and it is still stored in the list of 10, then simply scroll down to it and press Yes.

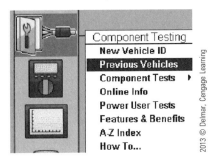

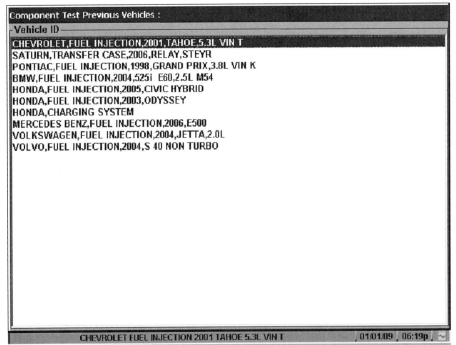

Let's now get back to our Component Testing. Component Testing is similar to basic training for the Lab Scope. The technician will be accessing front-door data using the lab scope and analyzing waveforms but will need to do little, if any, lab-scope manipulation. All of the settings have been preset, and all of the connection information has been included. This is excellent for newer technicians still learning to use lab scopes, but it is also helpful for veteran technicians who don't want to waste time setting up the parameters of the lab scope and looking up wiring diagrams for connection information. For this technician, all of the parameters are set, and all of the electrical connector information is included, so all that needs to be done is analysis of the waveform. The tool will provide good samples if needed. We are going to look at a couple of different examples, so you can see all of the help and guidance that the tool can offer.

Let's first look at the Knock Sensor. Assume we used our scanner to first look at the trouble codes, and maybe even used Troubleshooter to help in our diagnosis. Troubleshooter suggests that we need to perform a DC voltage test on the Knock Sensor to check for proper operation. From this point, you would identify the vehicle under the Component Testing menu, as we detailed above, and then select the Knock Sensor from the Component Tests box.

11. Scroll and select the component you wish to test: Knock Sensor.

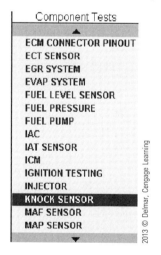

Let's first learn more about the component we're about to test.

12. Select Component Information.

Each of the four boxes explains key points that will help the technician test the selected component. Remember, all of this information is VIN specific to the vehicle you identified.

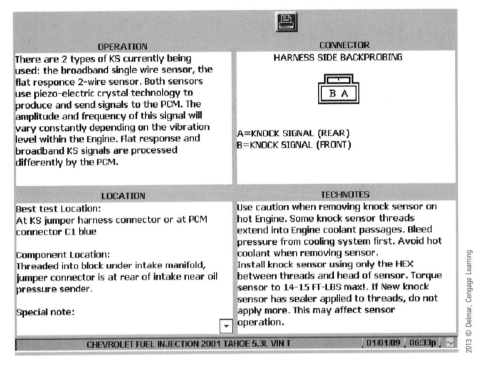

Operation: The operation box will give a brief description of how the component functions and how it fits into the automotive system. This is very basic, but very important information when trying to diagnosis problems.

Connector: This is a connector view that identifies the pins and sometimes includes wire colors. This information is needed to properly back probe the harness to get a waveform on the lab scope.

Location: This box explains the location of the component itself so that it is easier for the technician to find on the vehicle. It also includes test locations because the component itself might *not* be the best place to connect and get a reading. Remember this information comes from technicians in the field, so they want to do it as quickly and easily as you do.

Tech Notes: This box will not be found for every component, but if it is an option, it is very important to read it. This is a spot for important information that the veteran technicians want to pass on to other technicians. It usually includes special tips, test procedures, or things to watch out for or be careful of. Reading and following the information found in this box will save a lot of mistakes and unnecessary frustration.

To properly navigate the Component Information screen, it is very important to be aware of what is currently active on the screen. The active function is always highlighted in blue. The printer at the top of the screen is currently the active function. Pressing Yes now would print out all of the information on this screen. It is important to always look for arrows in the lower right corners of all the boxes to see if there is more information available. From our demonstration screen, we can see that the Location box has more information to view.

13. Press Down to move the active feature on the screen to the Operation box. You will notice that the blue highlight is now not on the printer but instead is highlighting the outline of the Operation box.

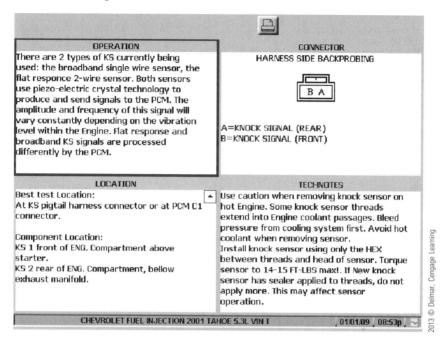

14. Press down one more time to highlight the Location box. In order for you to scroll the rest of the information, the Location box must not just be highlighted but also activated.

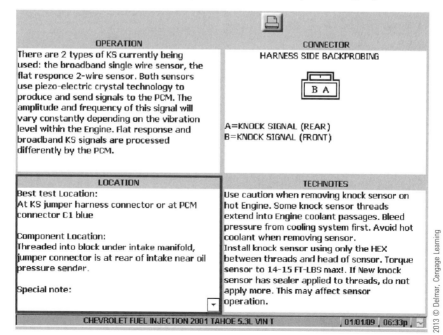

350 Chapter 12 Component Testing

15. Press Yes to activate the Location box. You will notice a thick blue banner appear at the top of the box, indicating that it is activated. The box will sometimes enlarge to show information previously off the screen. Notice the arrow in the lower right corner. It indicates that you will still need to scroll to read all of the information. Press Down to read and review all of the information until the arrow moves to the top of the box.

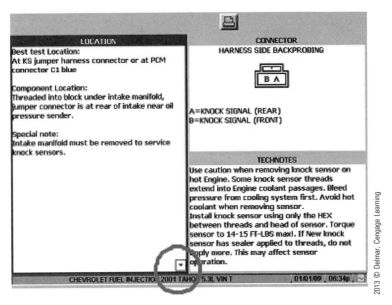

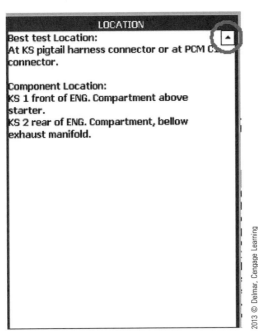

16. Press No to deactivate the box. This allows you to move around the screen.
17. Press No one more time to return to the Printer icon at the top of the screen.

18. Press No one more time to exit to the Knock Sensor component menu.
19. Scroll down and select Tests.

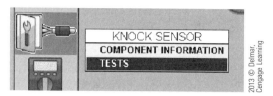

There may be more than one test available for each component. The tests are listed in the order in which they should be performed. Each test takes you further down the diagnostic procedure to accurately condemn the component if all the tests fail. The test specifications are included with each test, so just connect, read, evaluate, and decide. For this demonstration, there is only one test available, the DC Voltage Test.

20. Select DC Voltage Test.

Your screen should look like this.

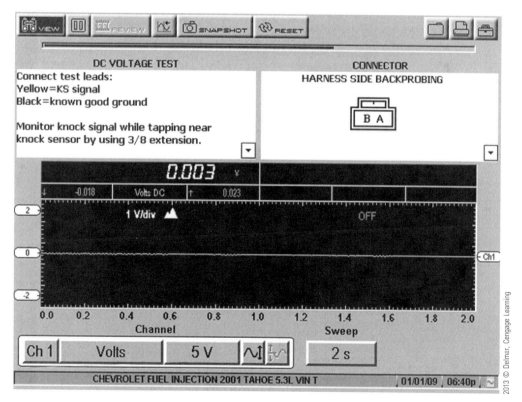

Notice that the Component Test Meter has decided that the Power Graphing Meter was the best tool for the job and not the Lab Scope. One way you can tell this is because there is no trigger option in the lower right corner. All of the functions, tweaks, of the Power Graphing Meter or the Lab Scope if that is the tool of choice, are accessible from this screen but do not need to be manipulated very much. Much of the time your waveform will fit on the screen with the preset adjustments. Sometimes minor adjustments will have to be made to manipulate the view so you can see what you are looking for. All of these tweaks and adjustments will be covered in detail in the Multimeter and Lab Scope sections. Right now we are going to focus on navigating the procedure boxes and learning to use the information provide by the Component Test Meter.

For the demonstration we are not going to hook the diagnostic tool up to an actual vehicle but we are going to look at how the tool explains exactly how to perform this test. This is just-in-time training at its best and explains exactly what you need to know, right when you need it. At this point to perform the Knock Sensor DV Voltage Test all you need to do is read and follow directions.

First connect (back probe) the yellow test lead to the Knock Sensor (KS) signal wire. To find out which wire is the signal wire look at the harness information in the Connector information box. Notice that there is an arrow indicating more information.

21. Scroll down and then over in order to highlight the Connector box.

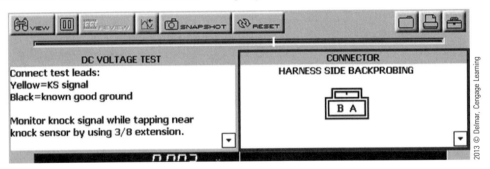

22. To see the rest of the harness information, activate the box by pressing Yes. Look for the thick blue banner at the top of the box, indicating it has become active. The box will then enlarge, and you will be able to read the rest of the connector information.

It is very easy to see that you want to use the yellow test lead to connect (back probe) pin A for the rear Knock Sensor and pin B for the front Knock Sensor.

The next step says to connect the back test lead to a known good ground. This can be any clean metal engine ground or the negative battery terminal.

Finally, it says to monitor the signal while tapping near the sensor by using a 3/8" extension. The rest of the information in the DC Voltage Test box is cut off, but, again, there is an arrow indicating there is more to be viewed. To view the last piece of this procedure, follow the next set of steps.

23. Press No to deactivate the Connector box.
24. Scroll over to highlight the DC Voltage Test box.

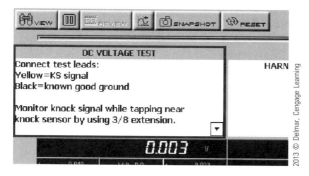

25. To see the rest of the procedural information, activate this box by pressing Yes. Look for the thick blue banner at the top of the box, indicating it has become active.

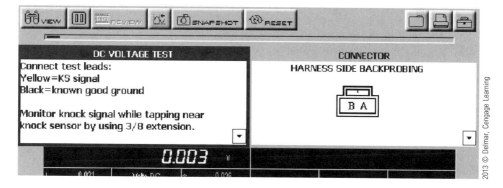

26. Now press down to scroll to the next informational screen.

Chapter 12 Component Testing

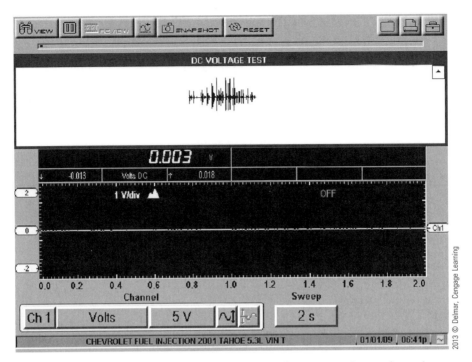

As you can see, the last piece of information is a known good waveform that you can use to match with what you are seeing on the screen. We are not actually connected to a vehicle, so no signal will be generated, but in a real diagnostic situation all that is left to do is to make a judgment about the quality of the signal coming from the Knock Sensor. If the pattern you see on the screen does not match the known good waveform, then you have found your problem and have now confirmed that the sensor is bad. This type of testing helps eliminate guessing and the replacing of good parts with newer good parts.

One other feature that is helpful when using the Component test meter is the Full Screen option. There may be times when the waveform you are viewing is covered by the informational boxes at the top of the screen. After you have read the procedure and have the signal on the screen, you can hide the information boxes and focus on the waveform itself. The next steps will accomplish this.

27. Scroll up to close the sample waveform for the Knock Sensor.
28. Press No to deactivate the DC Voltage Test box.
29. Scroll up to the View button and press Yes.

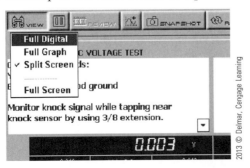

30. Scroll down to Full Screen and press Yes.

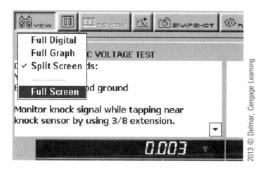

Your screen should now look like this:

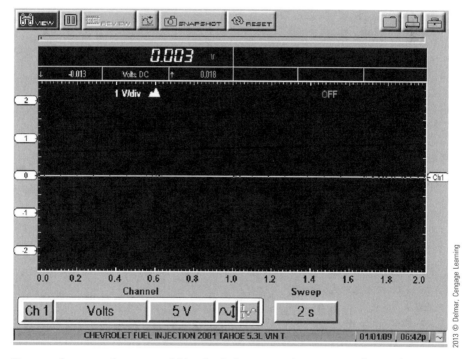

Repeat the procedure to unhide the information boxes. You'll see that the Full Screen option is an option that is either on or off. If it is on, there is a check mark next to Full Screen. To turn it off, select View, scroll down to Full Screen, and then press Yes. This will put the screen back to a normal view.

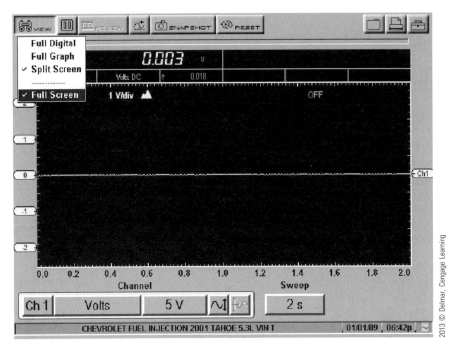

Before we go on to our next example, remember that all of the other function buttons across the top and bottom of the screen for both the Power Graphing Meter and the Lab Scope are dealt with in great detail in their respective chapters. The full functionality of these tools is accessible from the Component Test Meter in addition to the additional informational screens.

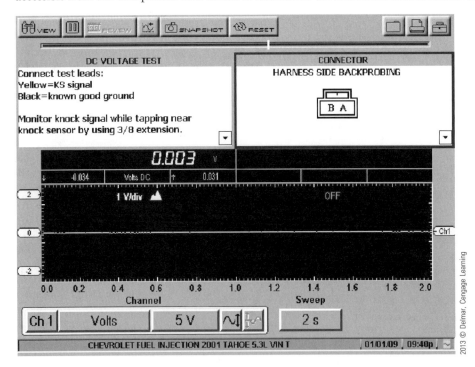

Component Testing Menu 357

The next example will incorporate and review all of the above navigational information that we used in the Knock Sensor example above but will also show some of the more detailed information available from the Component Test Meter. The vehicle identification will remain the same, 2001 Chevy Tahoe 5.3L, but this time our Component Test will be on the Fuel Injector.

1. Press No until you exit out back to the main Component Testing Menu and then scroll to Component Tests. Be sure the proper vehicle is identified at the bottom of the screen. If it is not, perform the Vehicle Identification sequence as described above or use the Previous Vehicles feature. Your screen should look like the following.

2. Next, scroll over to the right and then down to Injector and press Yes.

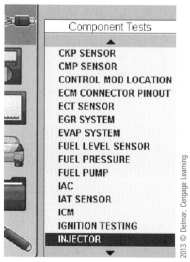

3. Press Yes with Component Information highlighted. This will review the component information that is available to the technician. Notice that there is no Tech Notes box for this component. This means there is no known special procedure, tip, specification, or circumstance that you need to be aware of at this time. The standard Operation, Connector, and Location informational boxes are available.

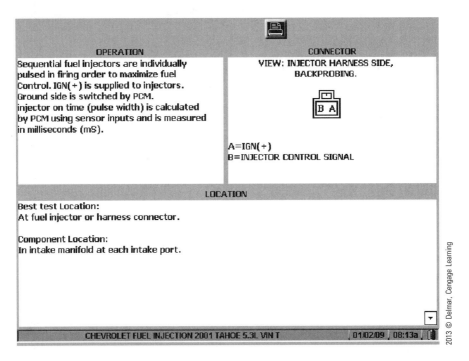

4. There is more information available in the Location box as designated by the scroll arrow in the lower right corner. Press down once to highlight the Operation box, then again to highlight the Location box.

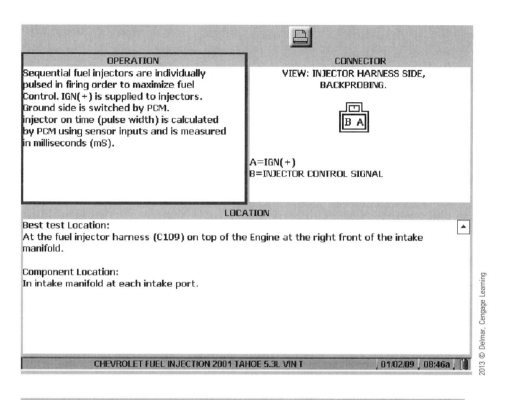

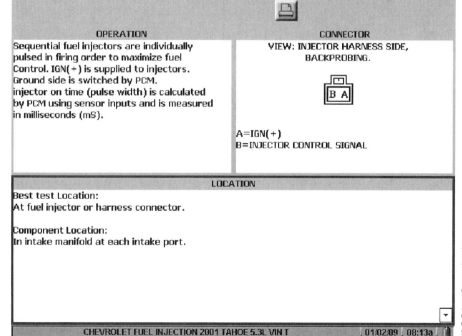

5. Press Yes to activate the Location box. Notice the thick blue banner at the top of the box, indicating it is activated.

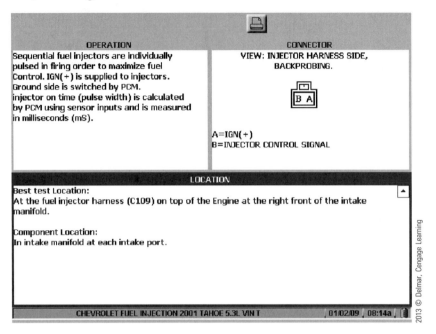

6. With the Location box activated, press Down to scroll through the information. You will be at the very bottom when the arrow moves to the upper right corner.

7. When finished reviewing the information about the Injector, press No to deactivate the Location box.

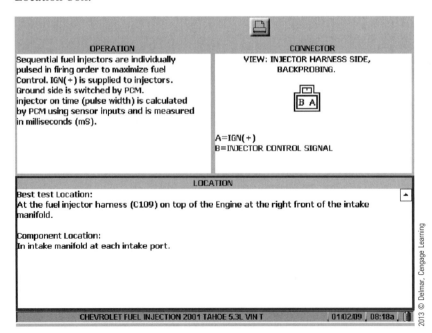

8. Press No again to un-highlight all of the information boxes and return to the Printer icon at the top of the screen, which should now be highlighted.

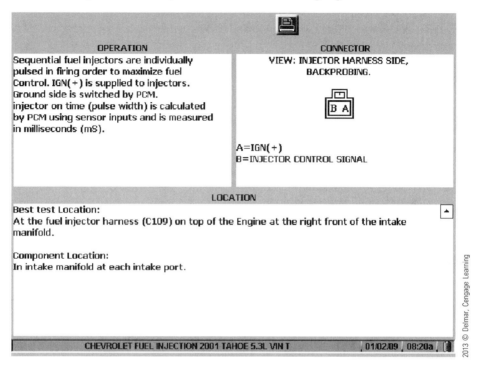

9. Press No one more time to exit and return to the Injector menu.
10. Scroll Down to Tests and press Yes.

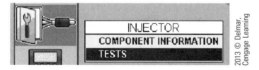

This component has five different preprogrammed tests available. The tests are listed in the order in which they should be performed. Each test takes you further down the diagnostic procedure to accurately condemn the component if all the tests fail. The test specifications are included with each test, so just connect, read, evaluate, and decide. Feel free to explore and look at each of the tests. For our demonstration, we'll be using the Current Ramp Test. As before, we will not be connecting to the vehicle. This test, when actually connecting to a vehicle, will require the use of the Low Amp Current Probe. This may be an unfamiliar device to some technicians, but this should not be a discouraging factor when you are using the Component Test Meter because all necessary information for performing the selected test will be available. Because of the amount of information provided, navigation can be come confusing.

It is important to remember that in a live diagnostic setting you would not need to view all of the screens available, but only the ones that you need to accomplish the job efficiently. Likewise, it is good to know that help, if you need it, is just a click away.

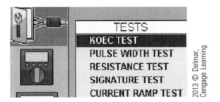

11. Scroll down to Current Ramp Test and press Yes.

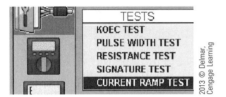

Your screen will look similar to this:

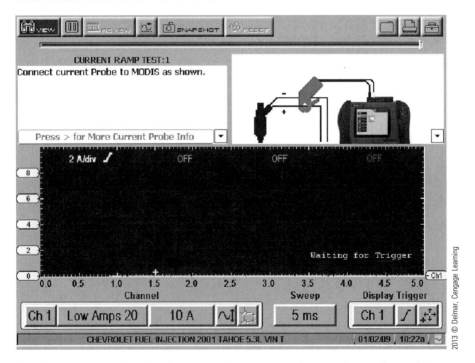

One important navigational concept that you need to understand on this screen is how to switch between the top and bottom toolbars and the middle informational boxes. The next set of steps will walk you through this concept.

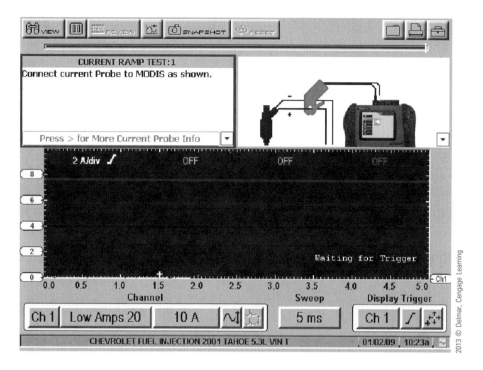

12. Press Down to highlight the Current Ramp Test: 1 box.
13. Press down again to highlight the Ch 1 button in the lower left corner.

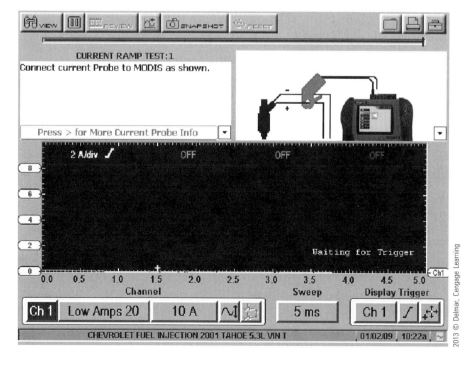

14. If you try to return to the Current Ramp Test: 1 box by pressing Up, you will notice that this will NOT work; pressing Up has no effect.

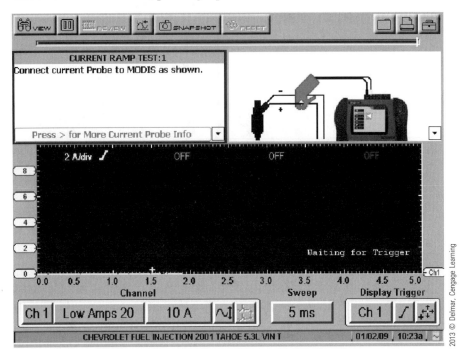

15. The quickest way to get back to the Current Ramp Test: 1 box is by pressing the No button. Press NO and see that the Current Ramp Test: 1 box becomes highlighted.

That navigational point should help as we continue to use the Component Test Meter and navigate through its features. Let's continue on with the Fuel Injector example demonstration. Read carefully because the next set of instructions will jump around a little, showing how to navigate *all* of the available information from the Component Test meter.

For this example, assume the technician who is performing the Current Ramp Test on the Fuel Injector has never done this before and may not even fully understand or appreciate the concept of current ramping.

16. Scroll until the Current Ramp Test: 1 box is highlighted.

17. Press Yes to activate this box and again look for the thick blue banner at the top of the box indicating that it has been activated.

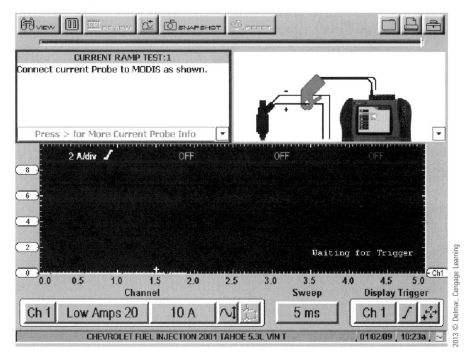

18. The instructions indicate to connect the Current Probe as shown in the box to the right. Assume you don't know what a current probe is or what it looks like. What can you do?

19. Press Right to get more Current Probe Info.

    NOTICE: After pressing Right to gain access to this extra Current Probe Info, there are now two *different* information boxes on the screen that can *each* can be activated and scrolled for even more information.

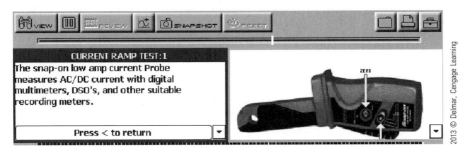

20. With the Current Ramp Test: 1 box still active, but now explaining the Current Probe in detail, press Down to scroll *all* of the information. Do not stop scrolling until the scroll arrow disappears from the lower right corner.

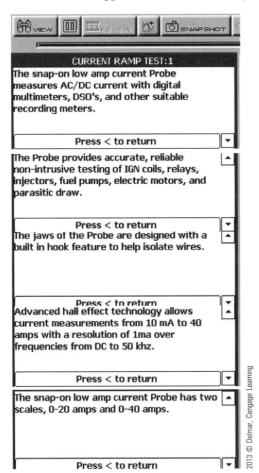

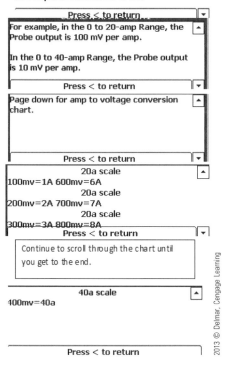

21. To get a better view of the Current Probe picture in the box to the right, you will need to: Press No to deactivate the Current Ramp Test: 1 box, press Right to highlight the Current Probe picture box, and finally press Yes to activate and enlarge the picture box. The final screen will look like this:

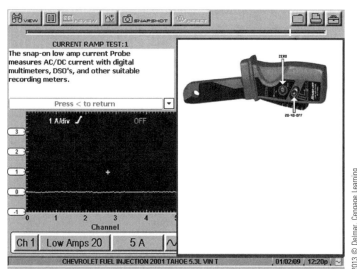

368 Chapter 12 Component Testing

Now that we learned all we could about the Current Probe itself, let's get back to the Current Ramp test procedure.

22. Press no to deactivate the picture box. It will shrink to normal size.
23. Press Left to move over to the Current Ramp Test: 1 box.
24. Press Yes to activate the Current Ramp Test: 1 box. Watch for the thick blue banner at the top of the box.
25. Press Left "to return" to the Current Ramp test procedure, which is where we started from.

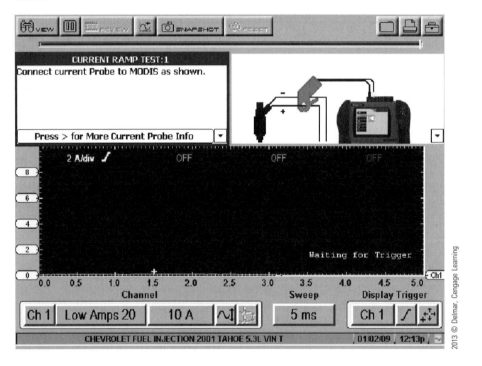

Quick reminder: It is not necessary to go through the detailed information about the Current Probe if you are already familiar with it, but be sure new technicians know this information is readily available to them in case they get stuck.

26. With the Current Ramp Test: 1 box still activated, Press Down to scroll to the next step.

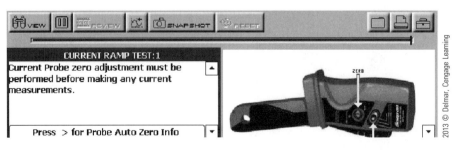

The instructions indicate that the Current Probe will have to be zeroed. What do you do if you're not sure how to accomplish this step? Just as before, press Right to go through a tutorial on how to zero the Current Probe. We are not going to list the instructions again because the procedure will be the same as what we just did, but here are the screenshots that you would see if you needed the information. Be sure to scroll all the way to the bottom.

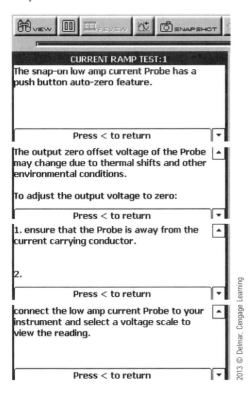

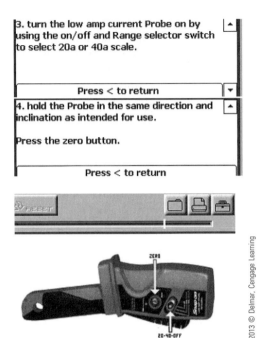

27. With the Current Ramp Test: 1 box still activated, Press Down to scroll to the next step. This step again reminds us to zero the Current Probe before clamping the jaw around any wire.

28. With the Current Ramp Test: 1 box still activated, Press Down to scroll to the next step. This step explains and shows that the Current Probe needs to be clamped around either the positive or negative side of the injector, but *not* both. Since amperage will be the same throughout this circuit, it doesn't matter what side we chose as long as we don't clamp around both wires. After you perform this step on a live vehicle, your waveform should show on the scope part of the screen. Since this demonstration is not actually connected, there will be no pattern to analyze.

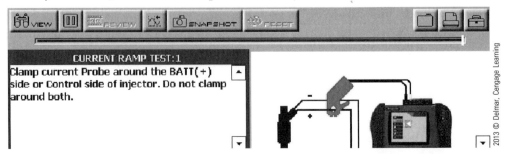

What if the fuel injectors are buried under the intake manifold or are obstructed by other engine components? How can we connect to the fuel injector wires? If you are able to connect using the first method you can scroll through this step, otherwise here is another option for connecting to the fuel injectors.

29. With the Current Ramp Test: 1 box still activated, Press Down to scroll to the next step.

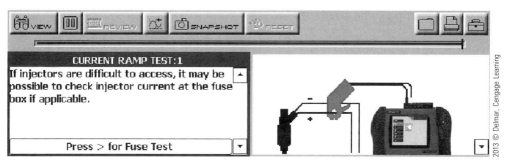

The instructions indicate that it may be easier to connect the Current Probe at the fuse box instead of at the individual injector. What do you do if you're not sure how to accomplish this step? Just as before, press Right to go through a tutorial on how to perform the Fuse Test. We are not going to list the instructions again because the procedure will be the same as what we did earlier in this section, but here are the screenshots that you would see if you needed the information. Be sure to scroll all the way to the bottom to access all of the available information.

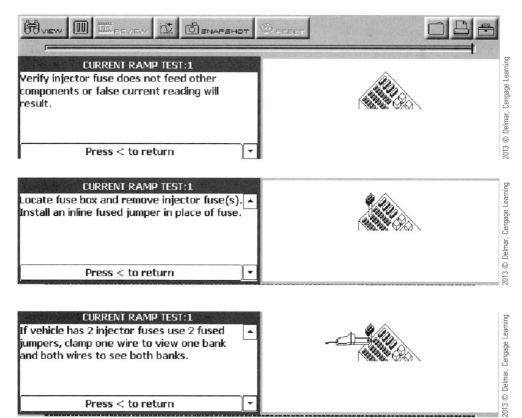

At this point, one of the two methods described above will have guided you through the hook-up procedure. If this were being done on a live vehicle, there would be a waveform on the scope screen. Now we need to know what the waveform should look like so that we can compare it to the waveform coming from the vehicle.

30. With the Current Ramp Test: 1 box still activated, Press Down to scroll to the next step.

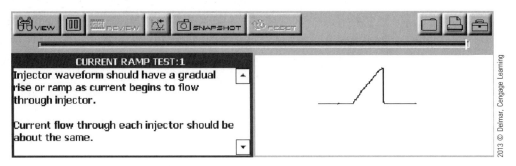

Continue to read and follow the directions on the screen and look at each fuel injector. Good waveforms are given, as well as some common problems that should be looked for in the waveform pattern.

31. With the Current Ramp Test: 1 box still activated, Press Down to scroll to the next step.

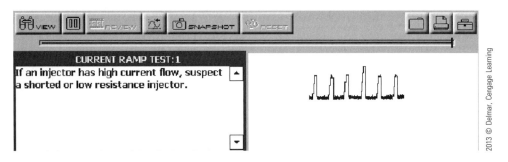

32. With the Current Ramp Test: 1 box still activated, Press Down to scroll to the next step.

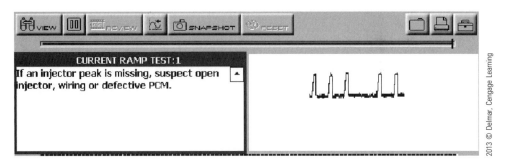

33. This brings us to the end of the procedure because there is no arrow in the lower right corner, so no more information can be accessed.

There may be a time when the specific component test you need to perform is not found in the CTM. With over 2 million specific tests, and that database continually growing each software upgrade, the odds of not finding the test are small. But if it should happen, do not give up on the CTM yet. There is a way to access some generic component testing. This will obviously not include all of the wiring diagrams and other VIN specific information but will help you to set up the appropriate lab scope or meter configuration to perform the test and give some general guidelines.

1. From the Component Testing menu, select New Vehicle ID and press Yes.
2. Scroll and select US Imports.

3. Scroll and select "Generic (All Others)" and press Yes.

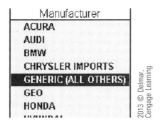

4. Then select the System Type you are working on and Press Yes. In this example, Fuel Injection is selected.

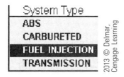

5. The Year will default to "Generic," so press Yes to continue.

6. Press Yes to accepts the changes.

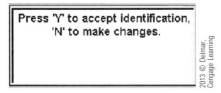

7. The list of Component Tests looks similar to before, but all of the components are now generic and not VIN specific. This option can be very helpful when testing an odd-ball

component. Pick the component that is closest to the test you need to perform, and the scope or meter will automatically be configured. The information windows will now provide some generic connection instructions and component information.

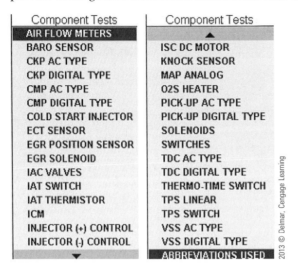

Component Test Meter Summary

Hopefully, from the two demonstrations we just did you can see the value of the information available through the Component Test Meter. This is an excellent way for newer technicians to learn to use the Lab Scope because they will be guided through connecting to the component and analyzing the waveform, using known good and bad patterns as their reference point. Locations and electrical diagrams are also available to speed up the process for veteran technicians. And all technicians benefit from the preset scope parameters—little has to be manipulated to get the waveform on the screen. Ultimately, this saves time, which can lead to greater profits for the service center.

Online Info

This feature is found on the MODIS and is used to access the internet. It will require a wireless CF card.

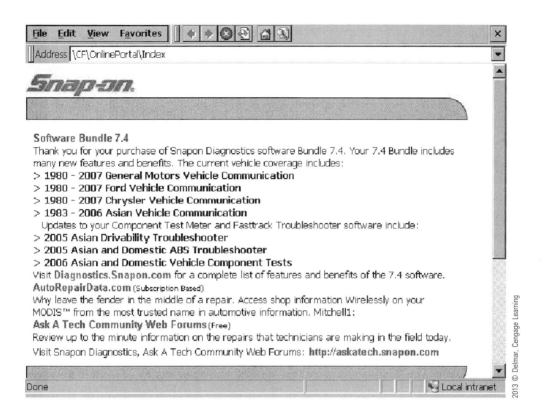

Power User Tests

This feature explains some of the advanced testing techniques of the Power Graphing Meter and Lab Scope. Some of these tests will require the purchase of additional accessories, such as Pressure Transducers. The tests are available with or without help. Tests without help are closely related to a Preset, as all of the parameters of the scope are set, but no additional information or procedures are provided. Tests with help are just like tests found in the Component Test Meter. All of the scope parameters are set, but there will also be information boxes that include specifications and procedures. Let's take a look at some of the Power User Test. This is an important part of learning and understanding how to use 90+ percent of the diagnostic tool.

1. Scroll and select Power User Tests.

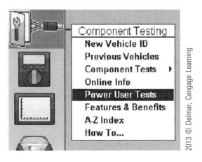

2. Select Power User Tests (with help).

3. Select Current Probe Tests.

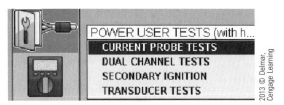

4. Keep pressing Yes to select the type of current probe you are using.
5. Here is a list of the Current Tests that you can perform.

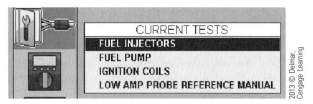

6. Keep pressing No until you back out to the Power User Test Menu.
7. Scroll and select Dual Channel Tests.
8. Here is a list of the Dual Channel Tests that you can perform.

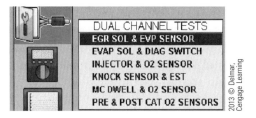

9. Keep pressing No until you back out to the Power User Test Menu.
10. Scroll and select Secondary Ignition Tests.
11. Here is a list of the Secondary Ignition Systems that can be tested. These are all set for Single Cylinder testing that can be accomplished with the single cylinder test lead and can be expanded with additional accessories.

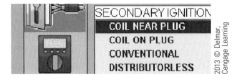

12. Keep pressing No until you back out to the Power User Test Menu.

13. Scroll and select Transducer Tests.
14. Here is a list of the Transducer Tests that you can perform with the optional Pressure Transducers.

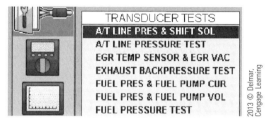

To make any of the Power User tests more convenient to perform it is beneficial to build or bundle adapters that can make the connections quicker and easier. This is especially true with the pressure transducers. Having a technician be able to see pressure changes over time in a lab scope format is a great diagnostic advantage, but this cannot be hindered by a disadvantage with the practical connections of the transducer itself. Here are some tips to make connecting transducers quicker and easier and will greatly expand the usefulness of the lab scope and the diagnostic capabilities of the technician.

Fluid Pressures (Fuel, Transmission, Power Steering, ABS, etc)

Connecting the transducer to these fluid systems can be complicated, but utilizing the same pressure testing kits you already have in your tool box will make this much easier. I'm going to use fuel pressure as the example. Most technicians will have a "master" fuel injection testing kit that comes with a gauge and a variety of adapters to accommodate most vehicles that enter the shop. Using this same kit the technical can adapted it to use with the pressure transducer by simply removing the gauge and screwing in the pressure transducer. This can be seen in the picture below and is also described more in the Lab Scope section of this book.

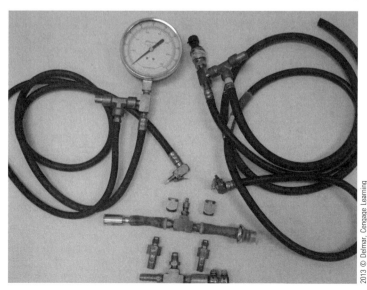

To make this a complete separate testing system I went to the Snap-On website and ordered "spare parts" through their Parts Catalog with included the adapter hose, bleed off valve, and the brass T-connector. This was cheap investment that allowed greater and more convenient use of my more expensive investments such as the lab scope itself and the transducers. This same idea can be applied to any "master" fluid pressure kit such as automatic transmissions, power steering, ABS pressure, engine oil, etc. Simply replace the analogue gauge with the proper psi transducer to see and record the pressure changes in real time on the lab scope.

Compression Tests (Air/Gas/Vacuum Pressure)

Just as we adapted the transducers for fluid pressures we can also make them easier to use when measuring air or gas pressures. One of the simplest is adaptuing them for compression testing. By installing the transducer into a pneumatic quick coupler (some ones commonly used on most air tools in the shop) that is compatible with the compression adapters you probably already have in your toolbox, you once again increase the usefulness of your higher priced investments (lab scope, transducers), with minimal cost. The picture below shows a simple compression set.

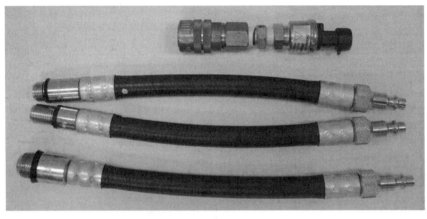

A common thread size for oxygen sensors is M18-1.5mm, which is also available as a compression adapter, so buying this size adapter and will allow you an easy way to connect to the exhaust system to check backpressure. Simply remove the oxygen sensor, thread in the adapter, and connect the transducer via the quick coupler. Just as with the fluid gauge adapters most technicians have a vacuum gauge with various adapters/connectors. Once again replace the gauge with the proper psi transducer and you'll have a quick and easy way to monitor and record manifold vacuum changes using the labs scope features. Use these tips to increase the diagnostic capabilities of the tools you already own with little additional expense.

Features and Benefits

The Features and Benefits selection explains in detail the diagnostic tool in general, optional accessories, general use of the tool, and warranty information. It is similar to the user manual and the other informational papers that come with the tool in booklet or electronic form. The electronic version, saved right in the tool, can't be lost or misplaced and thus is always accessible right when you need it. This is an excellent section to review soon after you purchase the tool because it provides a hands-on tutorial that shows how to perform some basic measurements using the Component Test Meter. To fully interact with the built-in tutorial you will need to purchase the demonstration board (Demo Board). See **Figure 12-1**.

Figure 12-1 Waveform Demonstration Tool Part # EESX306A

1. If you are in another menu area, press No until you return to the Main Component Testing Menu.
2. Scroll and select Features and Benefits.

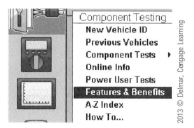

3. Most of the menu options are self explanatory. Please take a few minutes to browse the options to familiarize yourself with the information the tool can provide you under this feature.

4. Scroll and select Accessories.

5. Here is a list of the accessories that are available for the diagnostic tool. This may be helpful when ordering repair parts or are confronted with a diagnostic problem that requires another piece of equipment to complete the diagnostic testing procedure.

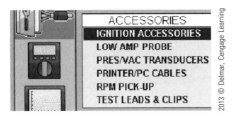

A-Z Index

The A-Z is a quick way to find answers to your questions about how to use the diagnostic tool itself, but it also contains other general automotive terminology. You can find out how to use the tool to perform a parasitic draw test, how to set trigger levels, and how to set up and connect to different ignition systems. Along with this tool-specific information, you can also find out information about the function of oxygen sensors, potentiometers, and how to read fault codes. There are standard to metric conversion charts for the various units of measure and definitions for different functions and terms used with the Lab Scope and other meters. The A-Z Index is one more feature that enables the technician to be more independent and to remain at the vehicle diagnosing the problem and not having to go elsewhere to search for needed information.

You are diagnosing an O2 sensor problem and the trouble codes are indicating it might be a malfunction in the heater circuit. You are not exactly sure how a heated O2 sensor functions. You remember hearing about them in school/learning seminar but didn't pay close enough attention, and now you need a quick reminder of the basics in order to perform this diagnosis. What can you do?

1. If you are in another menu area, press No until you return to the Main Component Testing Menu.
2. Scroll and select A-Z Index.

3. Scroll and select "H" for Heated O2 Sensor.

4. Scroll and select Heated O2 Sensor.

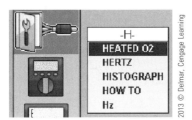

5. Select General Info.

6. Read and review the information as needed. Be sure to notice the scroll arrow in the bottom information box. You'll want to see all the information that is available.

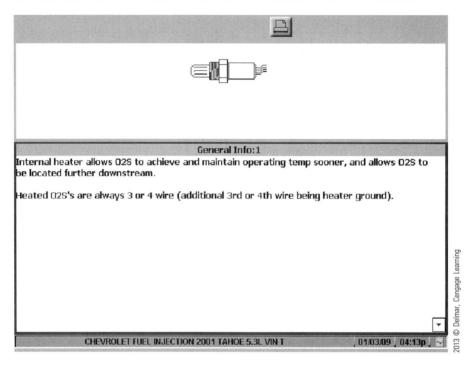

7. Press Down twice to highlight the bottom information box.
8. Press Yes to activate that box and be sure to look for the thick blue border at the top indicating that it is active.

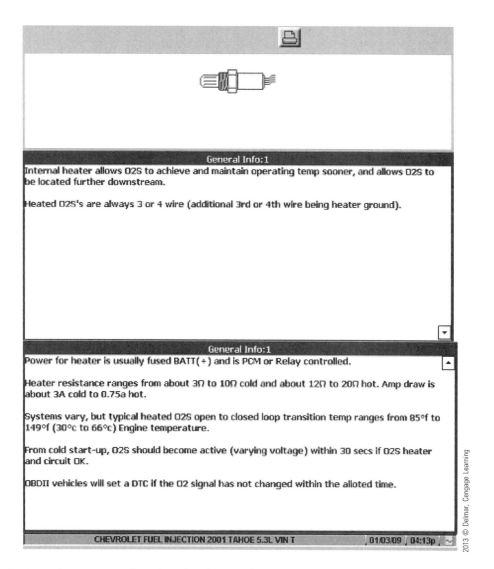

9. Press down to scroll within that box and view all of the available information.

How To ...

This menu section is filled with mini tutorials and tips that cover basic automotive principles and incorporate ways of using features found in the diagnostic tool. It is an excellent way for students and new technicians to learn how to use the tool while also reviewing and learning many important automotive and electrical concepts. The No-Start Basics option is worth noting. It is based on a diagnostic flow chart but opens in the Component Test Meter and prompts the technician to perform various tests on a vehicle that will not start. As the technician performs the tests and answers the questions, the tool moves through the diagnostic flow chart, systematically focusing in on the cause of the no-start condition. All of the features in this section are very helpful in learning how to use the tool in diagnostic situations and building that foundational automotive knowledge needed to diagnosis and repair today's complex vehicles. Let's look at a few examples in this section.

10. If you are in another menu area, press No until you return to the Main Component Testing Menu.

11. Scroll and select How To....

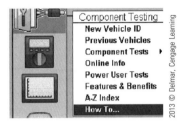

We are not going to go in depth into each of the tutorials but will show screenshots of each section so that you can see what is available. Here is the main How To... menu that will be referred to in the following steps.

12. Here are the 10-Minute Electronic Class lessons.

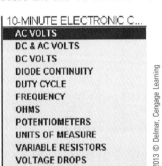

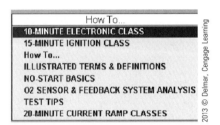

13. Selecting the 15-Minute Ignition class will bring you directly into the tutorial.

14. Here are the How To... lessons.

15. Here are the Illustrated Terms and Definitions.

16. No-Start Basics is broken down into different systems.

    Remember, this feature works like a diagnostic flow chart and uses the diagnostic tool to help determine the cause of a no-start condition.

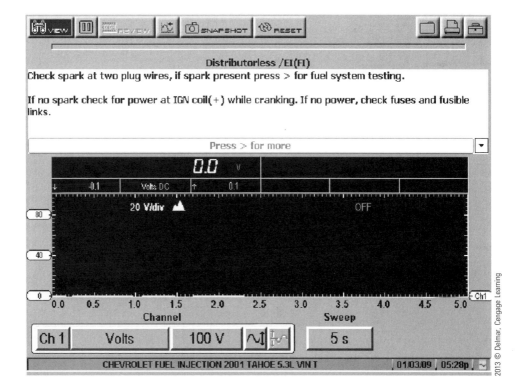

17. Here are the O2 Sensor and Feedback System Analysis lessons.

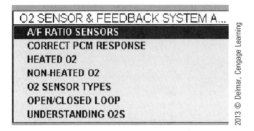

18. Here are the Test Tips.

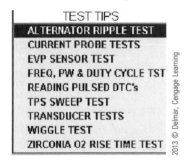

19. Here are the 20-minute Current Ramp Classes.

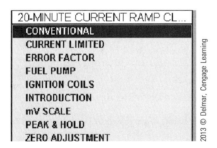

Component Testing Summary

The Component Testing menu of the diagnostic tool is full of valuable information. It is the library aspect of the tool. The information in this library was developed by technicians in the field and used by them to perform daily diagnostic chores. The heart of this section is the Component Test Meter (CTM), which allows anyone with basic automotive knowledge to use a Lab Scope and Power Graphing Meter to diagnosis specific components by viewing their waveform signature or looking for glitches in the waveform signature. Remember, when using a Lab Scope or Graphing Meter to view electrical flow, you are looking at front-door data. This is where the cause of the problem will be found. With over 2 million vehicle-specific component tests in the CTM, there is never a reason to guess if a component is bad. The tests for each component are listed in the order in which they should be performed. Each test takes you further down the diagnostic procedure to accurately condemn the component if all the tests fail. The wiring diagrams needed for connection and test specifications are included with each test, so just connect, read, evaluate, and decide. This is how you can confidently condemn a component and know that it needs to be replaced. Your diagnostic testing will ensure that you are confident of your repairs and save time and money by avoiding unneeded repairs.

Review Questions

1. Technician A states that the Component Test Meter (CTM) information is developed by technicians currently working in the automotive repair industry. Technician B states that the CTM database goes back to the early 1980s. Who is correct?

    a. Tech A only
    b. Tech B only
    c. Both Tech A and Tech B
    d. Neither Tech A nor Tech B

2. Technician A states that the tests found in the CTM are generic and not specific to any vehicle. Technician B states that there are over 2 million component tests in the CTM database. Who is correct?

    a. Tech A only
    b. Tech B only
    c. Both Tech A and Tech B
    d. Neither Tech A nor Tech B

3. Technician A states that the CTM will always bring you to the best tool for the job, which could be the Digital Meter, Power Graphing Meter, or Lab Scope. Technician B states that the CTM Previous Vehicles menu saves the last 25 vehicles that have been identified. Who is correct?

    a. Tech A only
    b. Tech B only
    c. Both Tech A and Tech B
    d. Neither Tech A nor Tech B

4. Two technicians are discussing Component Information option. Technician A states that this option contains wiring information to help in connecting to the component. Technician B states that both the component location and the best test location will be included with the information. Who is correct?

    a. Tech A only
    b. Tech B only
    c. Both Tech A and Tech B
    d. Neither Tech A nor Tech B

5. Technician A states that Tech Notes are included for every component. Technician B states that Tech Notes identify helpful tips or component service information that can make diagnosing the problem less frustrating. Who is correct?

   a. Tech A only
   b. Tech B only
   c. Both Tech A and Tech B
   d. Neither Tech A nor Tech B

6. Two technicians are discussing the screenshot below. Technician A states that this is found under the A-Z Index menu option. Technician B states that pressing Yes will take you to a functional test of the component. Who is correct?

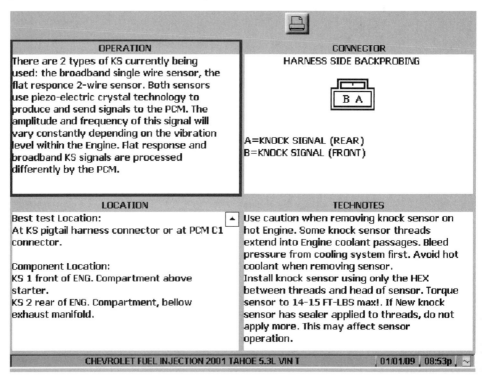

   a. Tech A only
   b. Tech B only
   c. Both Tech A and Tech B
   d. Neither Tech A nor Tech B

7. Two technicians are discussing the screenshot below. Technician A states that pressing Yes will activate the dialogue box, and this will be indicated by a thick blue banner at the top of the box. Technician B states that pressing the Down arrow button now will allow the user to view more information. Who is correct?

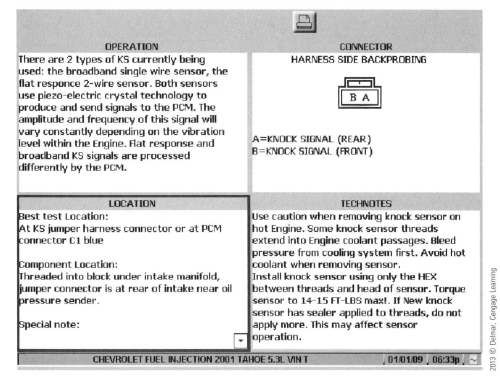

a. Tech A only
b. Tech B only
c. Both Tech A and Tech B
d. Neither Tech A nor Tech B

8. Two technicians are discussing the screenshot below. Technician A states that these tests are listed in a specific diagnostic order and should be performed in that order. Technician B states that pressing Yes now will bring the user into the best diagnostic application to perform this test and will include directions to help connect the tool to the component. Who is correct?

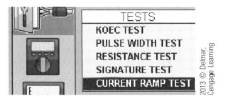

a. Tech A only
b. Tech B only
c. Both Tech A and Tech B
d. Neither Tech A nor Tech B

9. Two technicians are discussing the screen hot below. Technician A states that there is no option to remove the information windows and view a full screen of the diagnostic tool. Technician B states that Pressing Yes now will highlight an option on the lower toolbar. Who is correct?

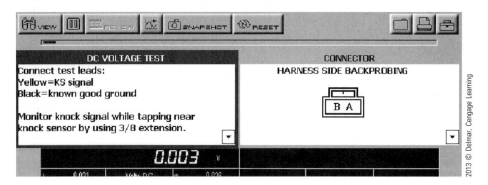

   a. Tech A only
   b. Tech B only
   c. Both Tech A and Tech B
   d. Neither Tech A nor Tech B

10. Technician A states that some small, self-contained automotive diagnostic lessons are accessed in the Features & Benefits menu option. Technician B states that Power User Tests is a great way to explore some of the other testing possibilities of the diagnostic tool that may require some additional accessories. Who is correct?

   a. Tech A only
   b. Tech B only
   c. Both Tech A and Tech B
   d. Neither Tech A nor Tech B

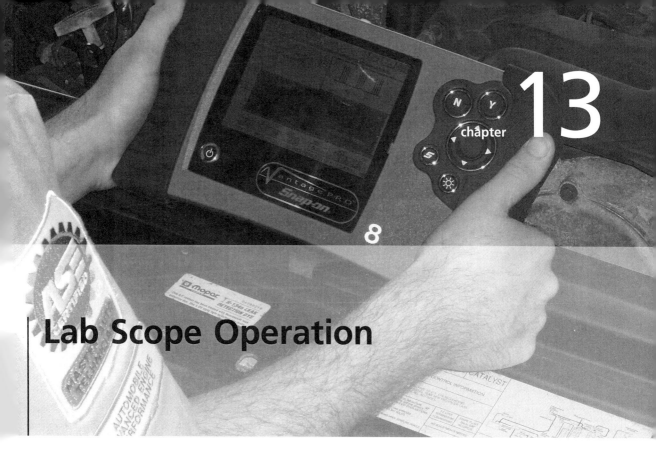

Lab Scope Operation

Lab Scope Section Outline

1. Overview and Connections
2. Display Configurations
3. Channel Configurations
4. Trigger Configuration
5. Presets
6. Interpreting and Saving Lab Scope Data
7. Ignition Scope Configuration
8. Reviewing Ignition Scope Movies and Snapshots
9. Finding Glitches using a Lab Scope
10. Summary

Upon completion of the Lab Scope module, you will be able to:

- Explain the necessity of Lab Scopes in the diagnostic procedure of late model vehicles.
- Efficiently set the possible viewing options and parameters in order to get the most information from the desired waveform.
- Quickly configure each required channel to give you the best waveform possible to make a diagnostic decision.
- Configure and save lab scope Presets that will allow quicker diagnosis of similar problem in the future.
- Easily save Movies and Snapshots so data can be saved and analyzed later.
- Quickly find electrical glitches by using the zoom features of the lab scope.

REMINDER: The Lab Scope section will provide the information for both the MODIS and Vantage Pro lab scopes. The software running these two diagnostic tools is virtually identical. Some of the icons on the Vantage Pro may look a bit different on the screen due to a resolution difference, but the functions are the exactly same. The reader will more clearly understand the information presented in this section if following along using the diagnostic tool. Using the tool in conjunction with the book builds the hands-on knowledge required to pass the Snap-On Certification Exams.

Need for Lab Scopes

One of the reasons that electrical problems are so frustrating to many technicians is the physical fact that we can't see electricity. The lab scope helps to solve this problem. The lab scope is going to show us a picture of electrical values over a specified time frame. With this we can actually see what is happening in an electrical circuit and then make more accurate diagnostic decisions. Almost all technicians have used a digital multimeter (DMM) in the past to test an electrical circuit. For many years, a DMM was a satisfactory tool to diagnose most electrical problems. The problem arose when vehicle control computers started to perform calculations faster than the DMM was sampling the electrical circuit. If there is a problem, or glitch, in one of the input components sending a signal to the computer, and this glitch is happening faster then the sampling rate of the DMM, the DMM would not be able to detect and display the incorrect value. If the DMM got lucky and did catch the glitch and display it, the technician would have a hard time seeing the result on the digital screen because the incorrect value would appear as a flicker and then return to normal. Literally a blink of an eye would cause this glitch to go unnoticed. In the 1980s and early 1990s, a 500 millisecond (ms) (1000 milliseconds = 1 second so 500 ms = ½ second) glitch might not have caused a drivability problem at all. In today's vehicles a 500 microsecond (μs) (1 million microseconds = 1 second) glitch can cause a drivability issue. Most DMM's will sample and display about four times a second or 250 ms, but a fourth-generation Ford Electronic Engine Control (EEC-IV) computer makes about 625 calculations per second and an EEC-V about 1.5 million calculations per second. Any electrical glitch faster than 250 ms could certainly cause a drivability concern in a vehicle with these control systems but would not be able to be accurately diagnosed with a DMM. That is why lab scopes are so important when diagnosing vehicles today. As with all computer technologies, we can expect faster speeds as time goes on, which will only make lab scopes more significant in diagnostic work.

Lab Scope Basics

The advantage of a lab scope is its ability to draw a picture of a value over a specified amount of time. This is much easier for the human eye to distinguish, and it is the only way for the human eye to see high-speed problems that occur in electrical circuits. When viewing a single channel or waveform, we call this a signal's signature. If we know what a good signature looks like for a specific component, and then compare this to the waveform on the lab scope, we can make diagnostic decision about the quality of that specific component. When viewing multiple channels, we are usually looking for relationships between the different components being tested. Cause and effect can more easily be perceived using multiple channels, and this makes finding the root cause of a problem or the first defective part more of a science and less of a guessing game. There are some fundamental aspects of any lab scope that you'll need to know in order to make these diagnostic decisions and for the rest of this section to make sense. **See Figure 13-1.**

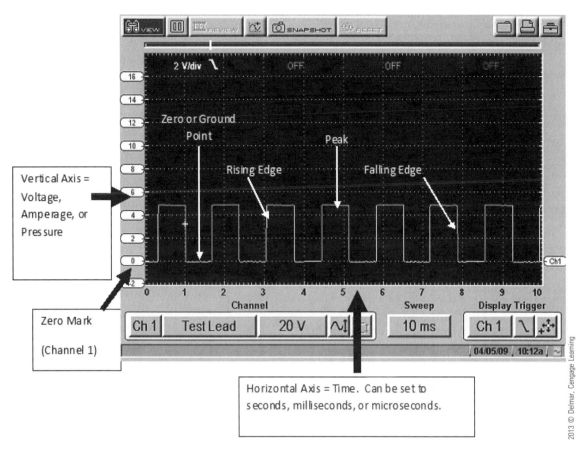

Figure 13-1

Vertical Axis: This is where the value of what is being measured will be counted. There will be a zero point that will not always be on the bottom of the screen, especially when you are using multiple channels. This is normally voltage but may include other values, including amps and pressure. The user can decide what scale he or she wants to use. On a MODIS, the screen is divided into ten equal divisions so the chosen scale is overlaid on these divisions. Assume the user decides on a 20-volt scale, then zero will be the start, and each division will be worth 2 volts (20 volt scale over 10 divisions so 20 volts/10 divisions = 2 volts per division). The Vantage Pro functions the same, but due to screen size only has five divisions instead of ten.

Horizontal Axis: This axis represents time, which is also referred to as sweep. The start or zero for time is always the far left of the screen. The user will decide how much time he or she wants to look at and will choose that time scale or sweep. Assume the user chooses a 10 ms sweep: The zero will be the far left of the screen and 10 ms will be the far right. The screen is divided into ten equal sections, so, in this case, each section is 1ms (10 ms / 10 divisions = 1 ms per division)

Ground (Zero) Point: This is where the value being measured is zero. Usually, since this measurement is voltage, this is referred to as the ground point. This point does not have to be at the bottom of the screen. It can be moved by the user, and we'll discuss how to do this later in the lab-scope section. The key point is to be able to clearly identify the mark or icon on the screen representing the zero reference point regardless of the labscope you are using.

Raising Edge: The edge of the waveform that is produced as the value rises from zero is called the raising edge. This will become more important when we start to learn how to set trigger levels.

Peak: This is when the waveform reaches its maximum value.

Falling Edge: The edge of the waveform that is produced as the value falls towards zero is called the falling edge. This will become more important when we start to learn how to set trigger levels.

Labe Scope Channels and Connections

Although the color coding makes connecting the test leads to the different channels an intuitive process, the ground connection requires a little explanation. As you will notice on the channel one yellow/black test lead, there are two ground connections. The purpose of the double-ended ground connection is to allow it to be piggybacked by other ground connections from other test leads such as the channel two green/black test lead and the Amp Current Probe which also has its own ground connection. The double-ended ground connection is called the flying ground as it hangs off the side of the piece of equipment and all other ground connections are connected to it. Only the single-capped ground connection is physically connected to the diagnostic tool. **See Figure 13-2.** The reason behind this is to protect the diagnostic tool from damage. If all the grounds were stacked on top of the tool, this would create a leverage point such that, if the tool were to fall, it could hit this stack, resulting in severe damage. **See Figure 13-3.**

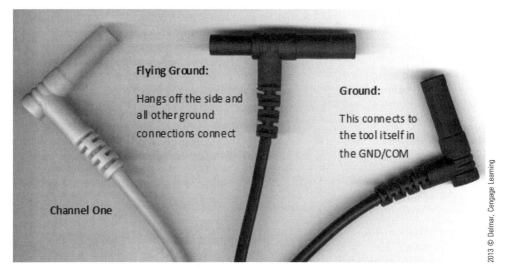

Figure 13-2 There should be supporting text as part of the picture, can add caption: The Yellow/Black test lead has two ground connections; one for connecting to the scope and the other, double ended one, to hang off the side and allow for additional ground connections from other accessories

It is also important to understand the difference between the channels so the full potential of the lab scope can be used. The general rule is that channel one is the fastest channel and should be used whenever you are testing a single component. When multiple channels are

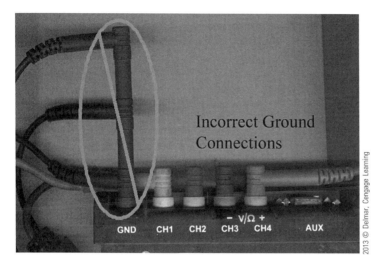

Figure 13-3 Stacking the ground connections in this manner is not advised as it could damage the lab scope module during a fall.

being used, the fastest reacting component should be connected to channel one and the next fastest reacting on channel two. Channels one and two are the best to use when looking for high-speed glitches. See Figure 13-4.

MODIS™ LAB SCOPE SPECIFICATIONS BY SWEEP RATE

Sweep[1]	Channels	Data points per screen	Buffer storage/Ch	Max # Screens	Total time[2]	Sample rate[3]	Peak Detect[4]
50 µs	Ch 1 only	300	524.288	1747	87.3 ms	6.0 MHz	N
100 µs	Ch 1,2 only	300	261,1'	870	87.0 ms	3.0 MHz	N
`0 µs	Ch 1,2,3,4	300	131.040	436	87.2 ms	1.5 MHz	N
500 µs	Ch 1,2,3,4	500	131.070	262	131 ms	1.0 MHz	N
1 ms	Ch 1,2,3,4	500	131.040	262	262 ms	500 KHz	Y
2 ms	Ch 1,2,3,4	500	131.040	262	524 ms	250 KHZ	Y
5 ms	Ch 1,2,3,4	500	131.040	262	1.3 S	100 KHz	Y
10 ms	Ch 1,2,3,4	500	131.040	262	2.6 S	50 KHz	Y
` ms	Ch 1,2,3,4	500	131.070	262	5.2 S	25 KHz	Y
50 ms	Ch 1,2,3,4	500	131.070	262	13.1 S	10 KHz	Y
100 ms	Ch 1,2,3,4	500	131.070	262	26.2 S	5 KHz	Y
`0 ms	Ch 1,2,3,4	500	131.070	262	52.4 S	2.5 KHz	Y
500 ms	Ch 1,2,3,4	500	131.070	262	2.2 M	1.0 KHz	Y
1 s	Ch 1,2,3,4	500	131.070	262	4.3 M	500 Hz	Y
2 s	Ch 1,2,3,4	500	131.070	262	8.7 M	250 Hz	Y
5 s	Ch 1,2,3,4	500	131.070	262	21.8 M	100 Hz	Y
10 s	Ch 1,2,3,4	500	131.070	262	43.7 M	50 Hz	Y
` s	Ch 1,2,3,4	500	131.070	262	87.3 M	25 Hz	Y

` = 20, so (`0 µs = 200 µs), (`ms = 20 ms), and so on.

[1] 50 and 100 µs sweeps are only available in Bundle 5.4 and later software.
[2] Total time is equal to the sweep times the number of frames.
[3] Actual sample rate for sweeps 50-`0 µs. Effective sample rate for sweeps 500 µs and longer. The effective sample rate is based on the number of sample points stored to the data buffer memory over the selected time sweep. On all sweeps 500 µs and longer, the ADC samples at 1.5 MHz per channel regardless of sweep. The number of sample points is greater than the number of points needed to complete a screen. Only enough points to complete a screen are selected to be stored to the data buffer. This results in the effective sample rate being lower than the actual sample rate of 1.5MHz.
[4] When Peak Detect is On, all samples are evaluated. The points stored to the buffer are intelligently selected to capture fast events that might be missed at slower effective sample rates. Peak Detect will capture fast changes at an effective sample rate of 1.5MHz.

diagnostics.snapon.com

Figure 13-4 Modis™ Lab Scope Specifications by Sweep Rate

Lab Scope Display Configurations

It is best if the reader follows along using the diagnostic tool. There is no need to have the scope connected to a vehicle or anther electrical circuit at this point.

1. From the main menu scroll down to the lab scope option, then select "Lab Scope", and finally select "4 Ch Lab Scope".

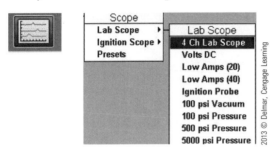

Notice that there are already some pre-configured lab scopes to choose from depending on what you are testing. The pre-configurations are self explanatory based on their names and may require additional equipment. The Low Amps configurations will require the Amp Current Probe, the ignition configuration will require ignition test leads and adapters, and the vacuum and pressure configurations will require the appropriate pressure transducer along with the split lead adapter.

2. Your tool may have a different number of channels turned on or off but at this time that will not matter, so ignore this difference.
3. With the "View" button highlighted, press Yes.
4. Scroll down to "RPM" and press Yes. This will place a check mark next RPM and engine RPM will then be displayed at the top of the screen. For this option to work, you will need to connect the Inductive RPM Pickup to the AUX port on the diagnostic tool and then clip it around a spark plug wire.

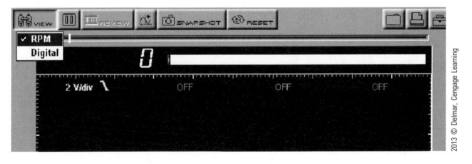

5. Scroll down to "Digital" and press Yes. This will now digitally display the average value of the signal being sampled as well as the min and max for the signal. This will be displayed for all the channels that are turned on.

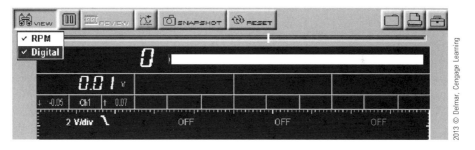

6. Scroll up to RPM and Press Yes to turn Off the RPM display.
7. Press No to exit the View menu options.
8. To reset the captured Minimum and Maximum values scroll over to "Reset" and press Yes. A dialogue box will appear telling you that the values are being reset.

9. Scroll over to the right to the end icon, the tool box, and then press Yes. This option provides a shortcut to some of the options we have already talked about in the utility menus of the various tools. There are also some important lab scope options that will greatly improve your ability to gather data from the scope.

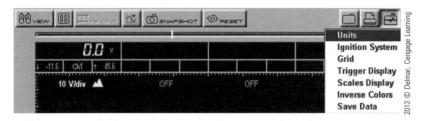

10. Scroll down to "Units" and press Yes. This is a shortcut to the utilities menu. This allows the user to switch units on the fly without having to back all the way out of the scope.

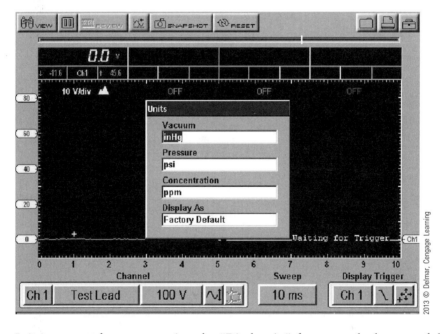

It is important for you to review the "Display As" feature at the bottom of the Units dialogue box. This will have a direct impact on how information is displayed on the lab scope. This box allows the user to change how the units are displayed in the lab scope. When adjusting voltage in the lab scope using either the Factory Default or Full Scale option, the voltage reading will be that of the total scale. Since, for the MODIS, there are always 10 divisions on the lab scope screen, the total voltage chosen by the user will be overlaid on these 10 divisions, and if the user picks the 100 volt scale, each division will be worth 10 volts and the lab scope will read 10 volts per division. If the Display As setting is changed to Units/Division, then the voltage adjustment in the lab scope is set to increments of volts per division, not to the total voltage scale. As an example, to select the 100 volt scale, you have to choose the 10 volts per division option. This gives the user flexibility to change the tool to his or her thinking style. If one usually thinks of units in per division increments, then switch this to the Units/Division setting. If one thinks of units as a total scale, then leave it at either Factory Default or the Full Scale Option. The following screenshots will illustrate this example. You do not need to use the tool to follow along for this illustration. Remember the Vantage Pro will function the same, but only has five total divisions instead of ten.

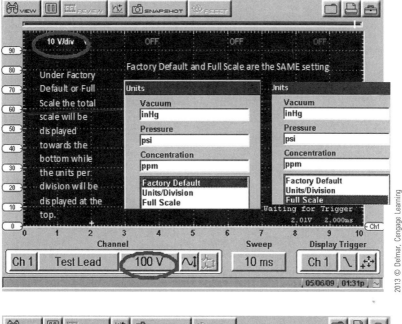

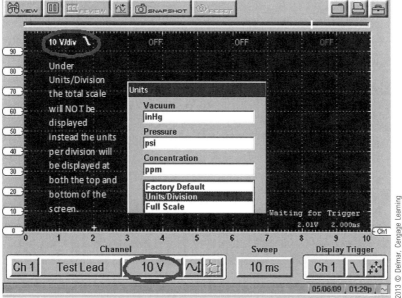

11. Back out and then scroll down to "Ignition System" and press Yes. This is another shortcut to the utilities menu. This allows the user to configure the Ignition System information without having to back all the way out of the scope.

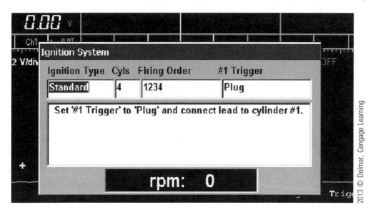

12. Back out and then scroll down to "Grid" and press Yes. When this option is selected, a check mark will appear next to the word Grid. When the Grid is turned on, there are dotted lines extending from each of the large 10 division marks on both the horizontal axis and vertical axis. This makes it easier to visually gather data from the scope.

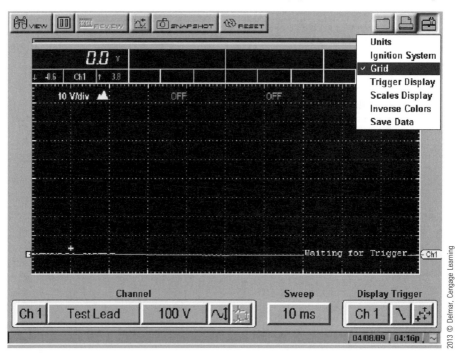

13. Scroll down to "Trigger Display" and press Yes. When this option is selected, a check mark will appear next to the words Trigger Display. In the lower right corner will appear the exact numeric location of the trigger cursor. This will be very important if you need to make sure a particular waveform is triggered at a specific value or time. In this example the value is in volts and time is in milliseconds. Think of this as GPS

coordinates of the trigger cursor. You could estimate the cursor location by using the Grid, but if a more exact trigger is needed, use the trigger display and read the values.

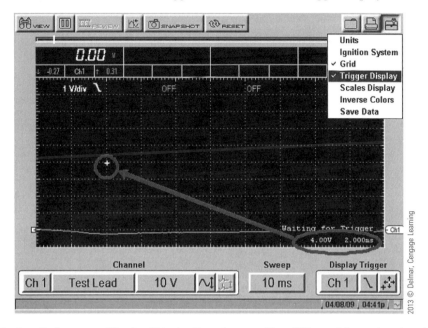

14. Scroll down to "Scales Display" and press Yes. When this option is selected, a check mark will appear next to the words Scales Display. With this option activated, numbers will be displayed on both the vertical axis and the horizontal axis, thus labeling the ten large divisions of each. This option, along with the Grid display, makes visually gathering data from the scope easy: The trigger cursor can clearly be seen at 4 volts and 2 milliseconds, just as the trigger display in the lower right corner is confirming.

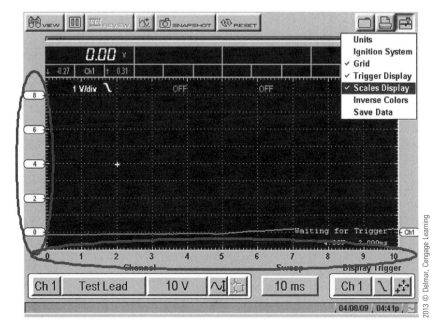

15. Scroll down to "Inverse Colors" and then press Yes. When this option is selected, a check mark will appear next to the words Inverse Colors. This option will make the background a white/grey color instead of the black. This is mainly used when printing out lab scope screenshots, as the background will print white, thereby saving you a lot of black printer ink. To help conserve black ink, this option is automatically chosen when you print a lab scope screen, so with software 7.4 and newer, you will *not* have to manually select this when printing. When you have finished, press Yes to return to the normal black background.

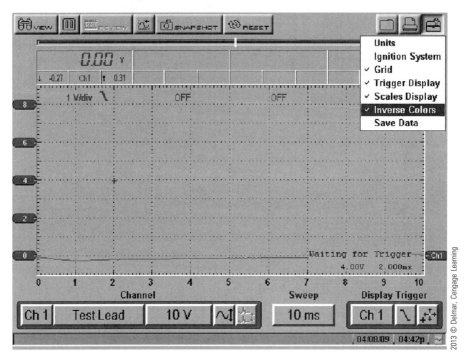

16. Scroll down to "Save Data" and press Yes. This is another shortcut to the utilities menu. This allows the user to configure Save Data information without having to back all the way out of the scope. For more information, please refer back to the Utilities and Tool Setup options detailed in your tool's specific section.

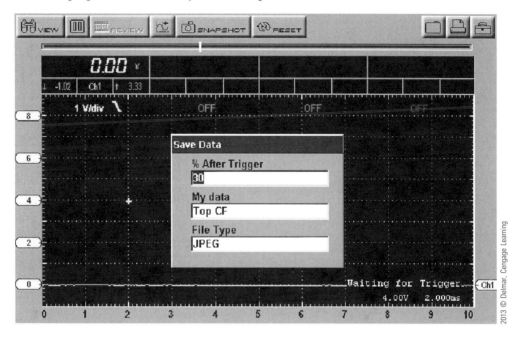

17. Scroll Down and over to "Sweep" and look at the various Sweep (time) setting options. Sweep is independent from the channels and thus is constant for all the traces on the scope. Sweep is the amount of time being displayed on the screen. It always starts on the left side of the screen and ends on the right side of the screen. It is the last option that we'll look at in this section that changes the overall view on the lab scope. The remainder of the options discussed in the next section are channel specific so they may change the view for that channel but not the entire lab scope. The example screenshots show a common fuel injector signature at different time sweeps. As the time sweep gets larger, the resolution of the individual patterns goes down but the number of patterns visible to the user increases. The user can select from an extremely fast 50 microsecond (50 millionths of a second) to a very slow, in lab scope terms,

20 second overview. The Ignition Scope times at the bottom are grayed out and not available unless the ignition probe is selected as the input lead, which we'll discuss later in this chapter.

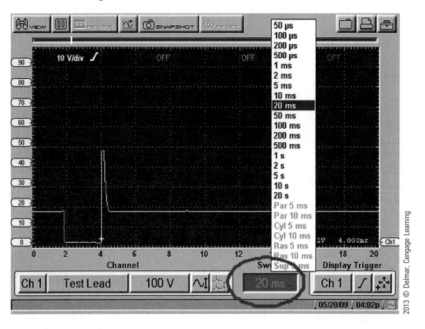

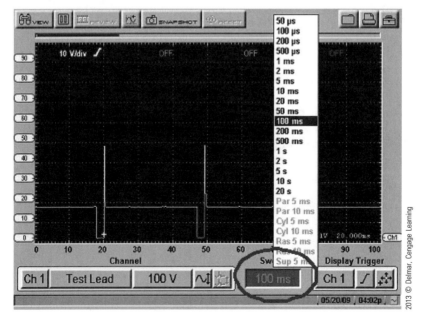

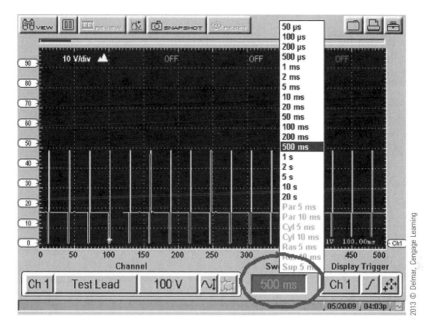

18. As a quick review and to get us on the same page before we continue to the next section, scroll back to View and turn off Digital (remove the check mark) by pressing Yes.

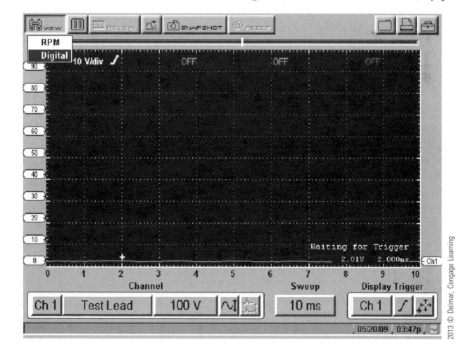

Lab Scope Channel Configurations

Each channel on the lab scope can be individually configured to show the best possible picture of the incoming signal.

1. Scroll down to Ch 1 and press Yes. At this point, you will see a dialogue box that lists channels one through four with an arrow pointing to the right after each channel. The first box is a quick way to select which channel is actively being controlled on the scope screen, as well as which channel's measurement scale is being displayed on the left side of the screen. The color coding on the physical module corresponds to the colors on the screen. Channel 1 is Yellow, Channel 2 is Green, Channel 3 is Blue and Channel 4 is Red. In the example shown, Channel 1 is active and currently being displayed—as indicated by the yellow signal line and the yellow scale measurements on the left side of the screen. Each channel is labeled on the far right side of the screen. This identification box also serves as a zero reference for the measurement. This is very important to remember as we turn on and activate multiple channels. Knowing and quickly identifying each channel's zero reference makes gathering data from the waveforms more efficient and accurate.

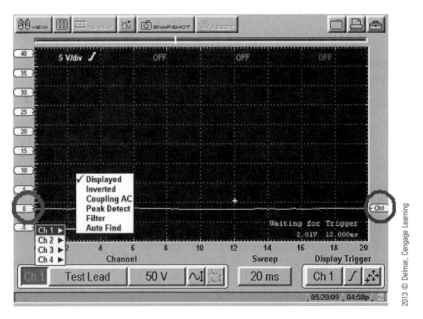

2. Scroll down to Ch 2 and then scroll over to the right and highlight "Displayed." Press Yes to display Channel 2. Pressing Yes on any of the Displayed commands for any of the channels will toggle on and off that particular channel. At this point, you will see the green signal line appear, as well as the green measurement scale on the left side of the screen. Each channel also has information displayed at the top of the screen. By quickly referencing this information, you can tell which channels are active or turned off as well as what voltage scale (other measurement) the channel is set to as the measurement per division will be displayed, as well as icons indicating trigger slope and the other features we will be discussing next.

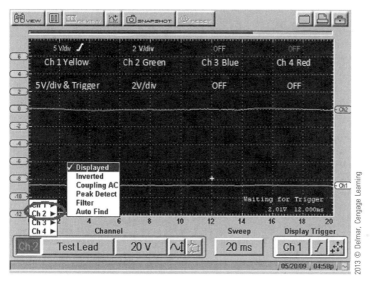

The following examples use Channel 1 as a base reference and Channel 2 as the variable. Each channel can be tweaked by using the various features to give you the best picture possible of the incoming signal. Understanding each feature will allow you to more quickly identify the proper application for each one and thus get the required waveform on the screen quickly and easily. You will not see the waveforms on your screen because you are not connected to a live signal. Each of the following examples was captured by connecting Channel 1 and Channel 2 to the same signal output while having the entire lab scope set exactly the same *except* for the feature in question. This way, you can literally see what the feature is doing to the signal pattern and learn the advantage of each.

3. Scroll down to "Inverted" and press Yes. This is a toggle on and off feature that is active when indicated by a check mark to its left. Also notice the inverted arrow next to the Green Channel 2 display information near the top of the screen. This icon can be quickly used to see what channels are inverted. The example shows a common fuel injector waveform on Channel 1 and then the same signal on Channel 2, but inverted this time. Use this feature if the physical connection to the component was done backwards. Instead of having to physically change the connection, use this feature to flip the waveform to the correct position. This is most commonly done with the secondary ignition adapter and amp clamp. These probes may have to be put in tight places that are difficult to access. In these instances, instead of trying to physically move the probe, just remember to use the built-in software that will allow quick and easy access to the waveform that you wish to see.

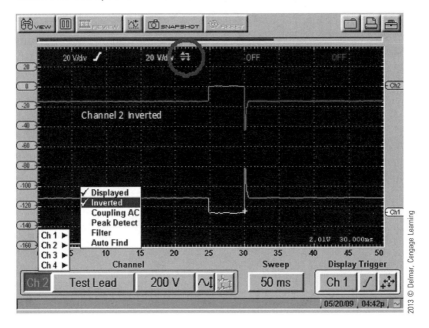

4. Press Yes to remove the check mark and turn off the Inverted feature.
5. Scroll down to Coupling AC and press Yes to activate this feature. Again, a check mark will appear indicating this feature is turned on, as well as a small AC wave at the top of the screen next to the channel that has it activated. In this example, it is on channel 2. AC Couple blocks the DC voltage portions of the input signal so that small amounts of AC voltage can be seen. This is an excellent feature to use when viewing alternator AC ripple or fuel pump amps. The example shows the output of an alternator. Both Channel 1 and 2 are connected to the same output of the alternator. Channel 1 is set to a 20 volt scale and is reading the expected 14+ volts DC. When AC Couple is activated on Channel 2, the DC portion (14+ volts) is automatically subtracted and what remains is only the AC voltage, which is just a few millivolts.

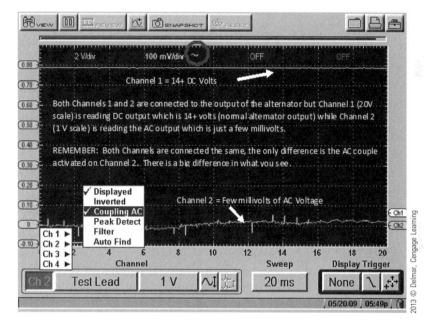

6. Press Yes again to remove the check mark and turn off the AC Couple feature.
7. Scroll Down to Peak Detect and Press Yes to activate this feature. Again, a check mark will appear indicating this feature is turned on and well as a small mountain icon next to the channel that has it activated. Peak Detect is used to capture high-speed events in both positive and negative directions. Don't let the name "Peak" Detect fool you. It will not only catch fast spikes but also fast-occurring dropouts. Under normal conditions, with Peak Detect off, the lab scope will collect enough data to plot a waveform on the screen. This is standard for many lab scopes. The Peak Detect feature overrides this normal operation and causes the lab scope to sample more points than necessary to plot the waveform, but this then allows the lab scope to catch high-speed events or glitches. With Peak Detect on, the lab scope is giving you the most accurate picture possible for the given sweep setting. Although this feature is not on by default, we recommend that you remember to use it whenever you are trying to catch a high-speed event.

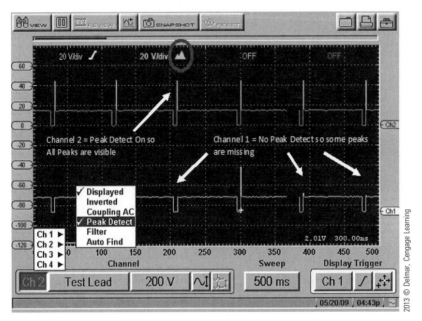

8. Press Yes again to remove the check mark and turn Peak Detect off.
9. Scroll Down to Filter and press Yes. A check mark will appear next to the feature, indicating that it is turned on. But notice this time, on the MODIS, that no small icon appears next to the channel near the top of the screen. If using a Vantage Pro you will see an icon that resembles the Peak Detect (mountain peak) icon, but has line cutting off its top. This feature cannot be used at the same time as Peak Detect, as it forces the lab scope to do the exact opposite of Peak Detect, which is to sample more slowly. Filter is used when the voltage scale is less than 2 volts, and RFI (Radio Frequency Interference) or stray voltage can interfere with a small voltage signal. Filter will smooth out voltage

spikes but still provides good signal integrity. The example shows both Channels 1 and 2 again connected to the same test point, but Channel 1 has a lot of noise or interference on it, which makes it hard to interpret. Channel 2 has Filter turned on, so this same signal is now much cleaner and easier to read. This feature is commonly used when testing an O2 sensor on a one- or two-volt scale. Depending on your testing conditions, any time the voltage scale is set to under one or two volts, Filter is an important feature to remember. It will make viewing the waveform much easier. Because the Low Amps Probe actually sends a low-voltage signal to the lab scope, which is then interpreted to an Amps reading, the Filter option can sharpen up any amp patterns, especially fuel pump velocity waveforms.

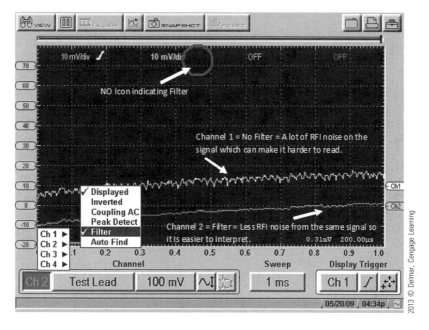

10. Press Yes again to remove the check mark and turn off Filter.
11. Scroll down to Auto Find and press Yes. This function will automatically analyze the incoming signal and set the scale, trace position, and trigger level to get the waveform on the screen. When Auto Find is used on a channel also designated as the trigger, the trigger will be automatically set half way between the minimum and maximum values. Auto Find will leave the sweep set to the current value as this is independent of all the channels. Once you have your desired sweep set then use Auto Find to quickly fit each channel on the screen. This may not always give you the best picture, but it will make it easier to tweak the waveform into a better view. As stated earlier this function is

channel independent so if you have four signals coming into the lab scope; set your desired sweep, then use Auto Find on each channel to get all the waveforms on the screen, and then start to manipulate each waveform to the desired view.

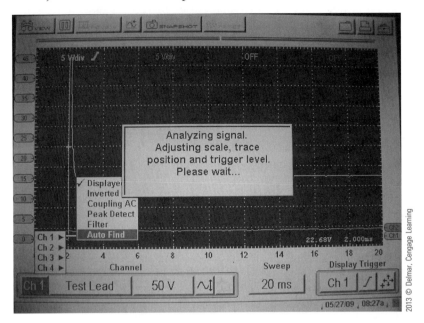

12. Scroll over to the right and select Test Lead and press Yes. This will open a dialogue box that will allow you to select the type of testing probe being used. This selection will allow the lab scope to automatically configure itself to give the best information possible. An example of this is when Test Lead is selected the scale is set to voltage and is displayed as volts per division but when switched to either the Low Amps 20 or 40 probe the scale is changed to amperage and expressed as amps per division. Other selections include the Ignition Probe and one of four pressure Transducer settings. This will help to configure the display to allow for the best waveform view and most accurate information gathering.

Lab Scope Channel Configurations 413

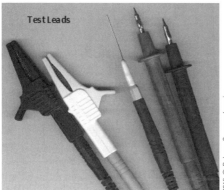

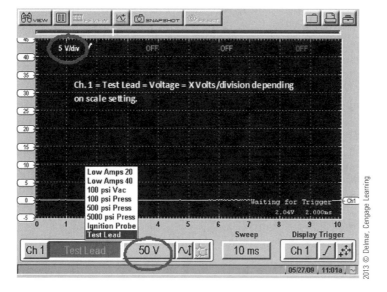

414 Chapter 13 Lab Scope Operation

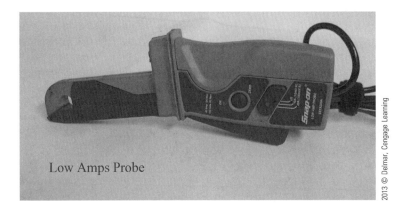

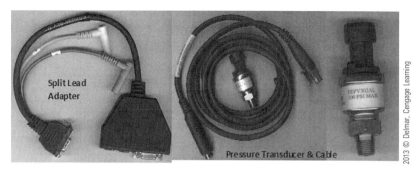

For more information on using pressure transducers, please see the 'Power User Tests' section in the previous Component Testing chapter.

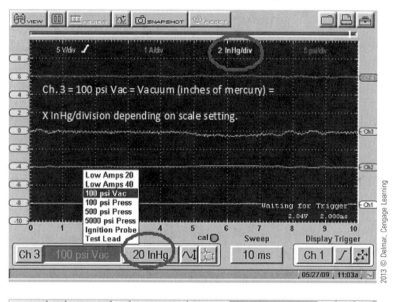

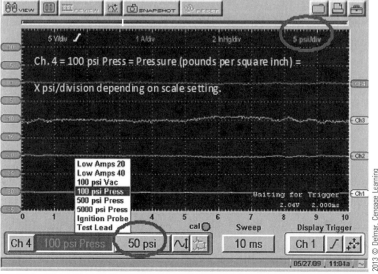

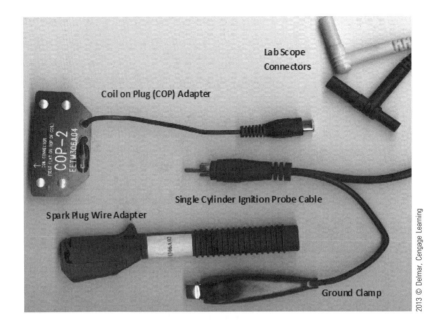

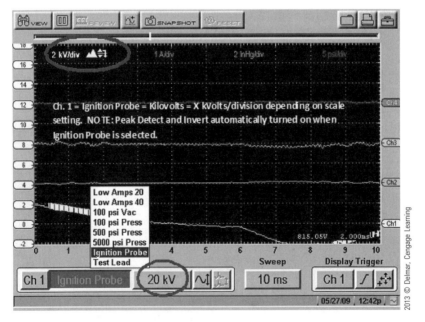

It is important to select the type of testing probe that you are going to use *before* you select the scale and other channel settings because the type of test probe may affect these configurations, causing you to have to go back and change these settings a second time.

13. Scroll over to the right and select the Scale button and press Yes. This will open a dialogue box of the available scale settings. As stated before, the scale options will change depending on the selected test probe.

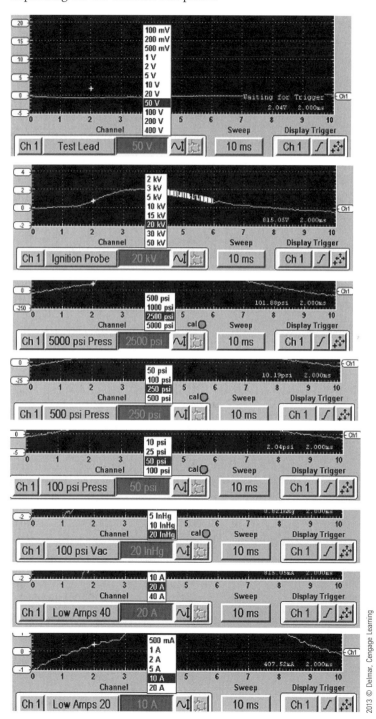

418 Chapter 13 Lab Scope Operation

14. Scroll over to the right and select the Zero Offset button and activate it by pressing Yes. With the Zero Offset button activated, use the up and down arrows to change the zero location of the vertical scale. It is helpful to place the zero on a large division mark for easier reading of values but this is not a requirement. This feature is very important when trying to fit multiple traces on the scope screen at one time. It also allows the user to stack specific waveforms while looking for relationships and to superimpose multiple waveforms if required.

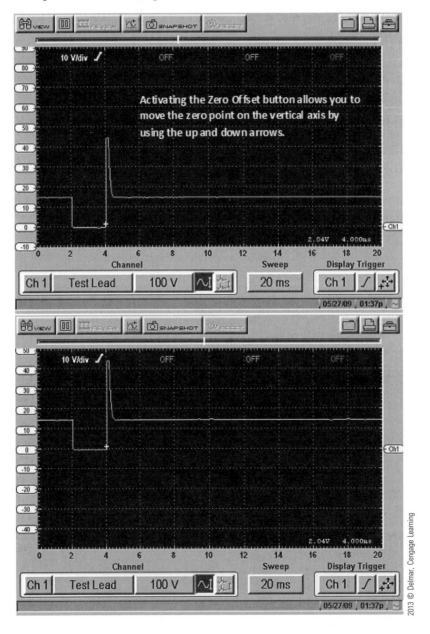

Zero Offset can be used to space out the individual channels and see each pattern separately, yet look for relationships between them.

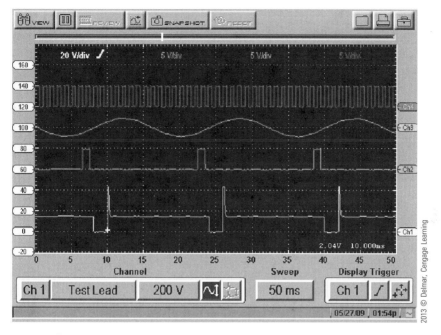

Zero Offset can also be used to superimpose multiple channels to help look for relationships. Use Zero Offset to move the patterns to any vertical location on the screen.

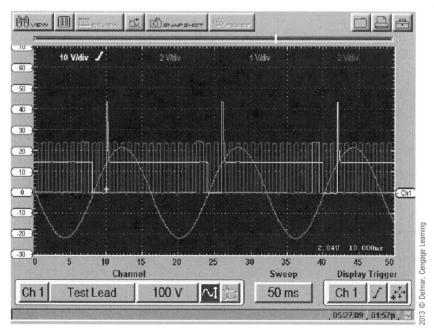

420 Chapter 13 Lab Scope Operation

15. The next button to the right is used only with the ignition scope and changes the vertical distance between the Raster secondary ignition patterns. When you are using the regular lab scope, this button will be grayed out and inactive. The following screenshots will demonstrate how this feature works and more details will be given in the ignition scope section. Notice how the distance between the top four waveforms changes as the Raster Offset button is activated by pressing Yes and then using the up and down arrows.

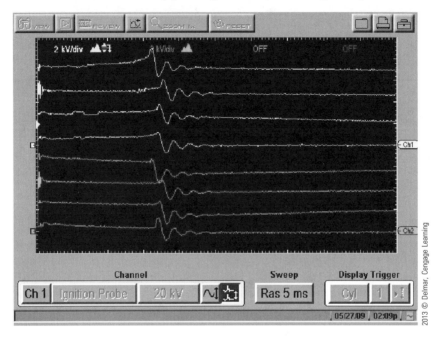

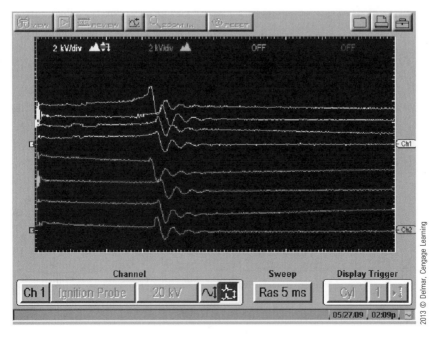

Trigger Configuration

A trigger point can be set to start and update the waveform. This is also useful to help stabilize the waveform on the screen, as well as to synchronize multiple waveforms on the screen. You can either set the trigger to "None," after which the lab scope will display data as fast as it is received, or you can set one specific trigger point. Only one trigger can be set. Precise trigger values for both scale and sweep can be set to help diagnose problems. To view a clear signature, set the trigger level to a known good value that will cross the trigger point. This will produce a clear and stable waveform. If you have an idea of a bad or problematic signal value that will only be reached when a problem occurs, then set the trigger for that value so you will not have a waveform to view until the problem occurs. This will have the lab scope only show a waveform when the problem occurs. The trigger can be adjusted at any time and has a separate control that is not dependent on what channel is currently displayed or activated. Be sure to pay attention to the colored border around the trigger area if you are triggering from a channel, or ensure that either "Cyl" (cylinder) or "None" is chosen as the trigger.

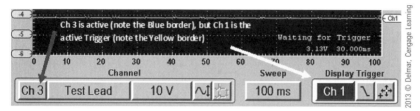

1. Scroll over to the right and select "Ch 1" under the Display Trigger area and press Yes. This will open a dialogue box that displays the various trigger-setting options. Only activated channels will be available as triggers. If a channel is turned off or not displayed, then that channel will be grayed out and inactive in this menu. Any one of the four channels can be selected as the trigger. You can quickly determine what channel is being used as the trigger and what slope the trigger is set for. The Trigger Slope will be discussed in detail shortly, but it is important to understand where to look to identify the trigger channel and the trigger slope. Each one of those channels has a sub option of either being set to Auto or Normal, which we'll look at next.

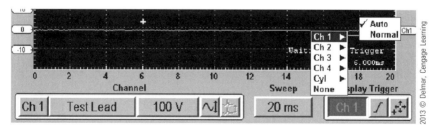

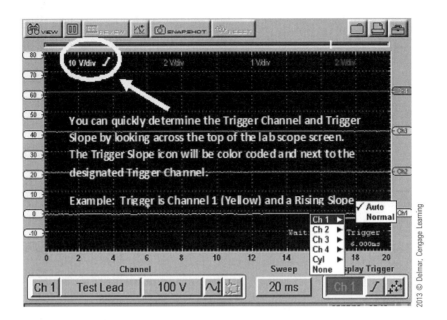

2. Scroll over to the right and highlight Auto. Auto is either an activated or deactivated feature designated by the check mark next to it. Either Auto or Normal has to be selected so pressing Yes when highlighted will move the check mark and activate that feature. When Auto is selected, the screen is automatically updated when the signal crosses the trigger point. The advantage of using Auto is that even if the signal does not hit the trigger point, the screen will automatically update after a short period of time, flashing the waveform signal. This lets you know that the signal is being received and that the trigger needs to be changed for optimal viewing. For general diagnosis and waveform viewing, this is a helpful feature to have selected as it is more forgiving of poorly set trigger levels.

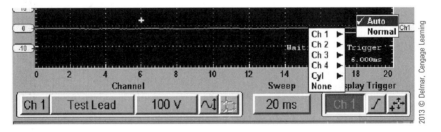

3. Scroll down and select Normal by pressing Yes. With Normal activated, there will be no screen update unless the signal crosses the trigger point, including *no* flashing waveform after a short period of time. If Normal is selected, your connections and test leads are all correct, but if your trigger is set wrong you will *not* see any waveform. The advantage of this feature is its ability to catch intermittent glitches. Assume a 4.5-5 volt reference signal is suspected of an intermittent glitch. If you set the trigger to a dropout value of 3 volts or below *only* when the intermittent problem occurs will any waveform be displayed.

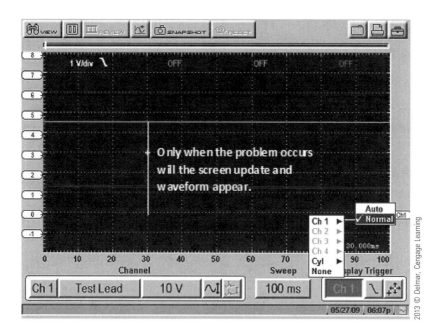

4. Scroll down and select "Cyl" (cylinder). This setting triggers the lab scope from the signal coming from the inductive RPM attachment. When the lab scope receives the spark plug firing signal, the waveform is then displayed on the screen. As with the Cylinder triggers, all waveforms are synchronized from this point. This is commonly used with the Ignition scope to synchronize multiple sparks with the engine's firing order, as the RPM pick-up would be placed on the number one cylinder. This can also be used to synchronize any other event that occurs in unison with the firing order, such as fuel injector operation on a SFI system.

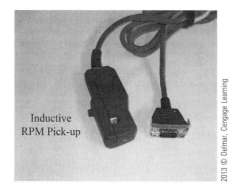

Inductive RPM Pick-up

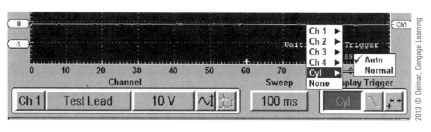

5. Scroll over and select "Auto." The Auto feature on the Cylinder trigger is very similar to the Auto feature on the Channel triggers. Either Auto or Normal can be selected, and this is done by highlighting your choice and pressing Yes, With Auto activated, the screen will automatically update and display the incoming signal even if no trigger signal is received from the RPM pick-up. This option can be more user friendly and allow the waveform to be more easily seen.

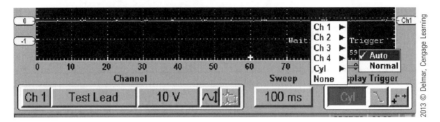

6. Scroll down to "Normal." Again, the Normal feature of the Cylinder trigger is similar to the Normal of the Channel trigger. Either Auto or Normal can be selected, and this is done by highlighting your choice and pressing Yes, With Normal activated the screen will *not* automatically update and display the incoming signal when no trigger signal is received from the RPM pick-up. No trigger signal means no screen update and thus no waveform.

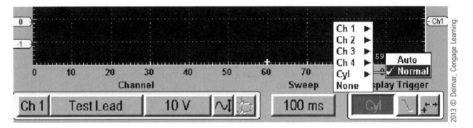

7. Scroll down and select "None." With None selected, the lab scope will update the screen and display data as fast as it is received. This is commonly used to view oxygen sensor oscillations or other patterns that are not directly tied to other signals. Many times, with None selected, the waveform will roll across the screen and may seem unstable.

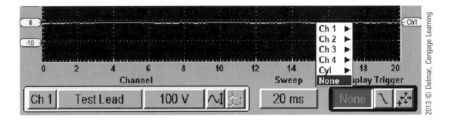

8. Scroll back up and select Ch 1 as your trigger. Make sure that Auto is activated.

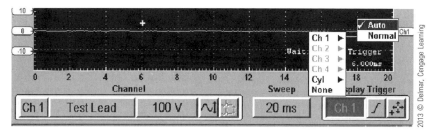

9. Scroll over to the right and select Trigger Slope icon. The example is that of a common fuel-injector signature. Currently, the Trigger Slope is set to the Rising edge or Upward slope. You can see the Trigger Cursor set for 2.04 volts on the Rising edge. Depending on the signal you are viewing or the glitch you are attempting to capture, it may be beneficial to trigger from either the raising or falling edge.

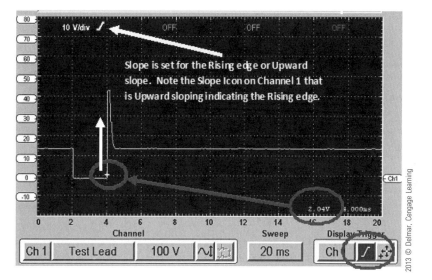

10. Press Yes to change the Trigger Slope from a Upward slope or Rising Edge to a Downward Slope or Falling Edge. The trigger voltage will stay the same, 2.04, but the Trigger Cursor will now move to the Falling edge. This may make it easier to catch a dropout glitch or provide a better view to see the problem. Whether you trigger from the Raising or Falling edge really depends on the specific circumstance. Pressing Yes will continue to change the Trigger Slope back and forth from Rising to Falling edge.

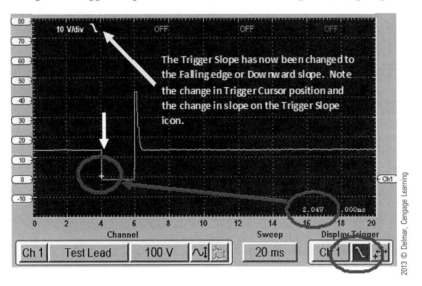

11. Scroll over to the right and select the Trigger Position Cursor button and activate it by pressing Yes.

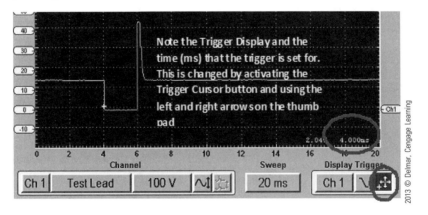

12. With the Trigger Position button activated, use the right and left arrows to move the Trigger time to your desired value. When you do this, the waveform will also move left and right across the screen. This allows you to position the waveform anywhere on the horizontal axis.

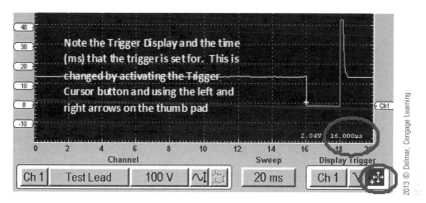

13. With the Trigger Position button still activated, use the up and down arrows on the thumb pad to change the trigger scale value. In the example, this value is measured in volts. If the voltage adjustment is too coarse, and you can't move to the exact voltage you wish, fix this by temporarily changing the voltage scale to a smaller scale. This will provide a finer adjustment on the Trigger Display, and when you change the voltage scale back to your desired viewing scale, the triggering voltage will remain the same. Remember to use this in conjunction with the Trigger Slope to choose between the Rising and Falling edge of the waveform. This will allow you to trigger at very specific scale values.

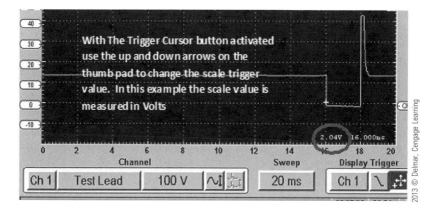

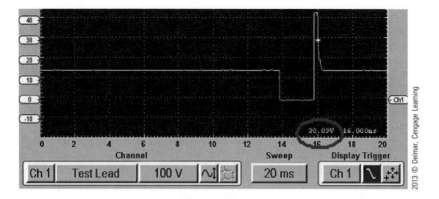

14. Assume you wish to trigger the fuel injector waveform as close to three (3.00) volts as possible. The current scale setting of 100 volts makes this impossible because, when you move the Trigger Cursor, it will jump between 2.82 and 3.13 volts—too large of a scale for this fine of an adjustment.

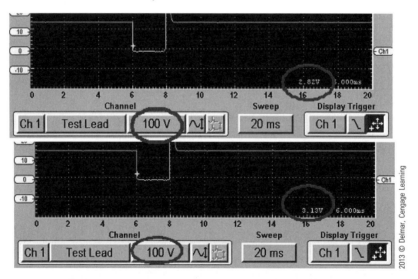

15. To fix this problem, temporarily adjust the scale down to a lower value such as a 5-volt scale. Then go back and activate the Trigger Position button and more finely adjust your trigger value. When you have finished, return the scale to the original value for best viewing and notice that the trigger value will remain unchanged.

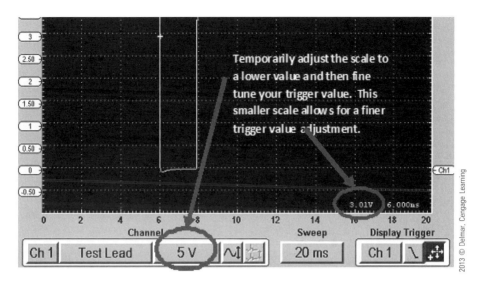

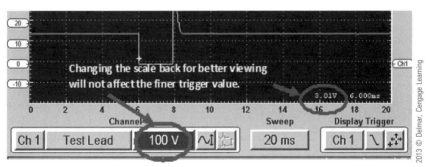

16. Scroll over and change the trigger to Cylinder (Cyl). Notice that when the trigger is set to Cylinder that the Trigger Position button icon changes and the trigger cursor is at the very bottom of the lab scope screen and can only be moved horizontally, thus changing the trigger time but not the trigger scale value.

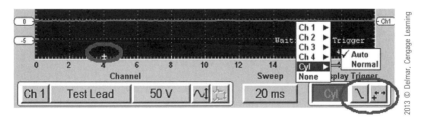

17. Scroll over to the Trigger Position button and activate it by pressing Yes. Notice that in the example the trigger position is set at 4.000 milliseconds.

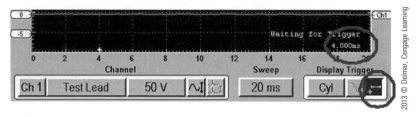

18. Using the left and right arrows on the thumb pad, move the trigger cursor along the bottom of the screen to change the time at which the trigger is set. Notice that the up and down arrows will not function because only the time and not the trigger value can be changed.

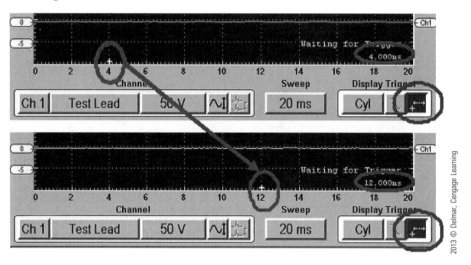

Lab Scope Presets

As you can see from the previous sections, there are many configurations and features that can be modified to change the function of the lab scope. Some features, like Sweep and Trigger, are the same for all the channels, but other features such as Inverted, Peak Detect, Couple AC, and Filter can be set differently for each individual channel. Other viewing options such as RPM, Digital, Grid, and Scales give more or less information on the screen depending on your unique diagnostic situation. It can be time consuming to set and then have to reset all of these features. This is where Presets come into play. The diagnostic tool comes with some factory installed Presets that you can use, and there is an easy way for you to develop and save your own. Presets save the current configuration of the lab scope so that it can be recalled and used at a future time.

1. To view some factory Presets, scroll down to the Lab Scope icon and then over to the right. Then scroll down until Preset is highlighted and press Yes.

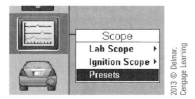

2. This automatically selects the internal drive and brings up all of the saved Presets. If you have saved your own Presets, they too will be found here along with the factory Presets.

   It is important to notice that all Presets have the file type of LS(C). This stands for Lab Scope Preset.

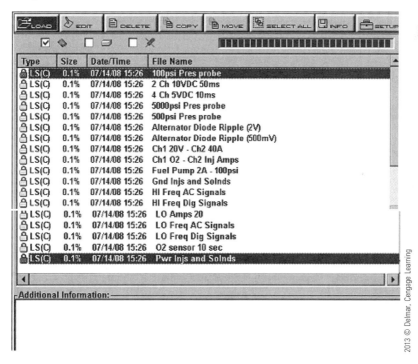

3. Scroll down to see all of the available Presets.

   None of the Presets will tell you how or where to connect the test leads as occurred in the Component Test Meter (CTM), but simply configure the various features and settings to a preset condition on the Lab Scope in order to more quickly view the waveform. If you do not need the extra help with the test lead connections and information, and don't want to waste the time having to do all of the tedious configurations, then Presets are the option you want to use. Presets are helpful without getting in the way.

4. To make or save your own Presets, which we recommend you do whenever you take the time to configure the labscope, start by selecting the 4 Channel Lab Scope and pressing Yes.

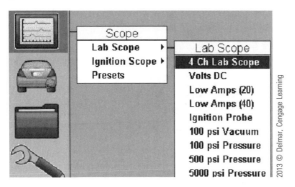

5. As we described earlier, go through and make all of the detailed adjustments, activate needed features, and configure the required viewing options until the Lab Scope is set up perfectly to view the waveforms that are needed.
6. When this is completed, scroll over to the Save Folder and press Yes.
7. Then scroll down to Save Preset and press Yes.
8. Press Yes to Continue.

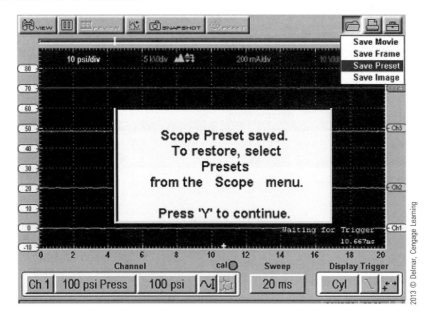

9. The Save Dialogue box will appear. Pressing Yes right now will save the Preset using a default file name with a User designation and a date and time stamp. If you are in the middle of a job, this is probably the best and easiest thing to do. The File Name and details can be edited later. To edit this information to make future use easier and more likely, open the Preset as described in the previous directions and then scroll over to the right and highlight Edit to do the following steps. If you have a USB keyboard available and have a minute to change the file name, this can also be done right away.

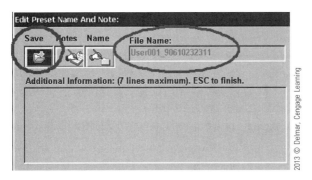

10. With the USB keyboard plugged in, scroll over to the right, highlight the Notes icon and press Yes. A typing cursor will appear in the Additional Information box, and you are free to type in any information you would like to help identify the Preset you just created. We suggest that you identify each channel, plus the year, make, model, and engine size of vehicle. The Sweep and Trigger are the same for all the channels, so this can just be listed on Channel one. This is a fictitious example illustrating the range of options that can be chosen. The extra minute it will take to save this the first time will save many minutes each time it is used, so the Lab Scope does not have to be reconfigured.

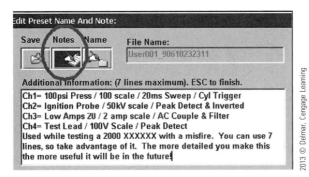

11. When you have finished typing additional information, press ESC on the keyboard to exit the typing box.
12. Scroll over to the right and highlight the Name box and press Yes.
13. Use the USB keyboard to delete the old file name and type in a new one that will easily and quickly identify this Preset to you in the future. Possible examples might be to label it by the component(s) being tested, the vehicle it is being connected to, or whatever makes it easiest for you.

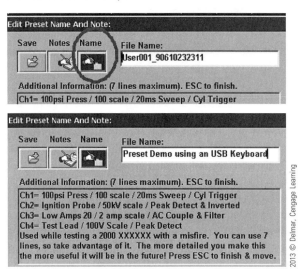

14. When you have finished typing the new file name, press ESC on the keyboard to exit the file name box.
15. Scroll over to the left until Save is highlighted and Press Yes.

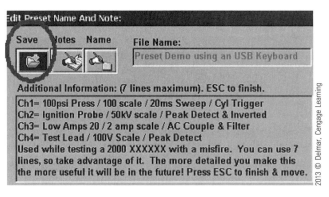

16. You will now see that the file name has changed and the additional information you added is visible when selecting presets.

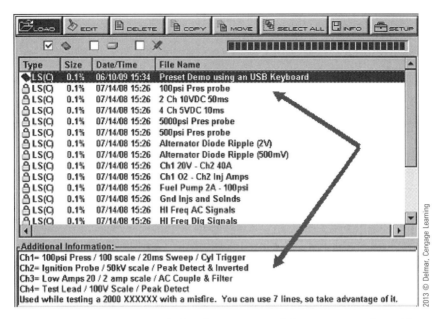

Presets are a great way to maximize the amount of time you spend working on the vehicle and to minimize the amount of time you spend on configuring the tool. Remember to check out the factory Presets that come with the tool and use them to your advantage. If you spend a good chunk of time setting up a lab scope configuration, then it is well worth the extra minute or two to save it and then have it on file and ready to go for the next time.

Interpreting and Saving Lab Scope Data

With the vast amount of information that can be gathered using a lab scope, it is important to be able to interpret this data and save it for later reference. Since most lab scope data is displayed as a pattern or picture, the first step in interpreting that picture is to extract actual numerical data or values from the measurements we are making with the lab scope. After this, we can save and archive this information for future reference or easily send it electronically to a colleague for a second opinion or post it to an Internet forum for even more different opinions. All of the diagnostic tools make this very easy and convenient to do. For more detailed information about saving data, please refer to the Data Management section of this book.

1. From the 4 Channel Lab Scope, scroll over to the right until the Folder Icon is highlighted and press Yes.

2. With "Save Movie" option highlighted, press Yes.

3. A dialogue box will appear and allow you to input information to identify the file you are about to save.

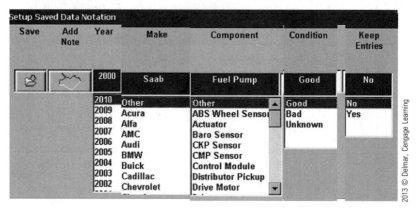

Here is a breakdown of the features found in this box.

- **SAVE:** When the "Save" button on the far left is highlighted, pressing Yes will save the file. If the additional information boxes are not used, then the file will be saved using a default file name that includes a date and time stamp. This is the quickest way to save a file that you do not need to identify right away. When you have finished inputting information and are ready to save the file, highlight this box and press Yes.
- **ADD NOTE:** If a USB keyboard is plugged into the top, large USB port, then this button becomes active and can be highlighted, and when Yes is pressed you will be able to add 7 lines of additional information describing the file about to be saved. This is only available after a UBS keyboard has been plugged into the top, large USB port.
- **YEAR:** When this box is highlighted, press Yes to open a drop-down menu in which the year of the vehicle being tested can be selected. Do this by scrolling down to the appropriate year and pressing Yes.
- **MAKE:** When this box is highlighted, press Yes to open a drop-down menu in which the make of the vehicle being tested can be selected. Do this by scrolling down to the appropriate make and pressing Yes.
- **COMPONENT:** When this box is highlighted, press Yes to open a drop-down menu in which the component being tested can be selected. Do this by scrolling down to the appropriate component and pressing Yes.
- **CONDITION:** When this box is highlighted, press Yes to open a drop-down menu in which the condition of the component being tested can be selected. Do this by scrolling down to the appropriate condition and pressing Yes.
- **KEEP ENTRIES:** When this box is highlighted, press Yes to open a drop-down menu to select either Yes or No to keep Entries. This feature allows you the option of keeping the last selected vehicle information in the tool's memory to make saving additional information from the same vehicle quicker and easier. If you are planning to save multiple movies, frames, or snapshots from the same vehicle select Yes and the next time you save a file the vehicle's information will already be selected. Selecting No will default all the information boxes to the generic settings each time a file is saved.

4. A Movie will save all of the information collected in the memory buffer up until the time you decided to save the Movie. The number of frames saved in a Movie is dependent on

the Sweep at which the lab scope was set when the movie was taken. If you look at The Lab Scope Specification by Sweep Rate chart near the beginning of this chapter, you can see how many frames (screens) are able to be saved for each channel. The breakdown is simple: A 50 microsecond sweep can save 1747 frames, 100 microsecond sweep saves 870 frames, '0 microsecond saves 436 frames, and the rest of the sweep settings are able to save 262 frames. Remember, this is all the data in the memory buffer that was recorded before a Movie was saved, with no data after the Movie was saved. A Movie file is an active file that can be treated and manipulated just as if it were a live lab scope signal. Different channels can be displayed, frames can be reviewed, viewing options can be turned on or off, offsets can be moved, and the cursors can be activated. The only limitation is that no new data can be collected because this is a saved file and the tool is not connected to a vehicle.

5. Scroll down to Save Frame. Saving a frame will save only the current frame of data. Only one frame is saved. A frame is also an active file. It can be treated just as you would a Movie file, except there is only one frame to review. All of the other features are still available.

** Older versions of software may show different wording for these features, but they work the same. If you see "Save All Pages," that is the same feature as "Save Movie." And "Save Page" is the same feature as "Save Frame."

6. Save Preset was discussed earlier in this section. After completely configuring all of the lab scope features, options, and views, this can be used to save all of those configuration settings so that they can be used in the future without having to do all of the set-up work again.

7. Scroll down to Save Image. When this is highlighted, pressing Yes will save a screenshot, that is, a picture of the entire screen just as it is at that second. This is not an active file type, so no manipulations can be made of it when viewing it on the tool. It is simply a frozen picture of the screen. Use the "Save Data" options discussed earlier in this section to change the file type if you wish. You have a choice between a bitmap (bmp) or jpeg file type.

8. Press No to exit out of the save menu box.

9. Scroll over to the left until Snapshot is highlighted and press Yes. A small option box will appear with "Manual" highlighted.

10. Press Yes to chose "Manual." Wait a couple of seconds, and then the Snapshot box will start to flash blue. Once blinking, the Snapshot is armed and ready to record. The next time you press Yes, a snapshot will be taken.

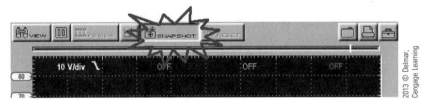

11. Press Yes to record a Snapshot. A series of yellow dialogue boxes will appear near the center of the screen. Allow the tools to finish collecting all the data.

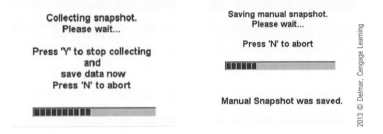

It takes longer to capture and save a Snapshot than a Movie because a Snapshot includes data collected after the triggering event (pressing yes). As explained in the Data Management chapter, a Snapshot is slightly different from a Movie; it is made up of both past data stored in the buffer and new data that is continually being fed into the buffer. The moment the user activates or takes the Snapshot is the trigger. The % after trigger option, which is found in the "Save Data" option found under the Tool Box icon to the far right, allows the user to change how much old information is taken from the buffer and how much new information is collected. The default setting is typically 30% after trigger. Assuming a buffer of 262 frames of data (common for most sweep scales), this means that once the user triggers a Snapshot, the tool takes 183 frames (70%) of the data from the buffer, which was automatically recorded before the trigger, and then continues to capture 79 frames (30%) of data after the trigger. These 262 frames are bundled together and stored as a Snapshot file with the triggering event located between the beginning and the end of the entire Snapshot file and at the user specified percent. If the user wants to see more of the effect of the triggering event, then the percent after trigger should be increased; if the user wants to see more of the cause of the triggering event, then the percent after trigger should be decreased. When viewing a Snapshot file, treat it just as you would a Movie file, as there are many active features that can be changed. But remember, a Snapshot contains both cause-and-effect data.

12. Scroll over and highlight the Freeze button and press Yes. This will stop the tool from recording new data into the memory buffer and allow you to review the data that is already saved to the memory buffer. The example screenshot has a simple fuel injector signature waveform to illustrate the functions.

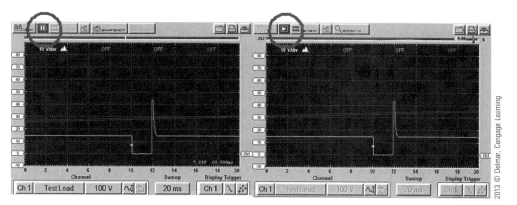

13. Scroll over to the Review button and activate by pressing Yes. The Review has slightly different functions depending on what Zoom scale you are looking at. Currently, the Zoom is set to 1x.

14. Pressing the Up and Down arrows while on 1x Zoom will move you left and right in fractions of a frame. This will shift and slide the waveform to the left and right. You will have to look carefully, as these are very small movements. To move faster, press and hold down either the up or down arrows and watch the waveform move accordingly.

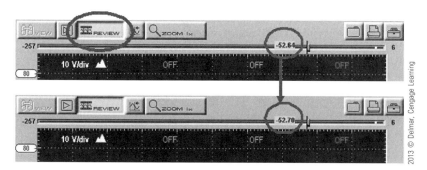

15. While still on 1x Zoom, you can use the Left and Right arrows to move one whole frame at a time.

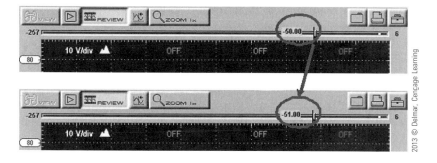

16. Press No to deactivate Review.
17. Scroll over to Zoom and press Yes.
18. Then scroll down to 16x and press Yes again.

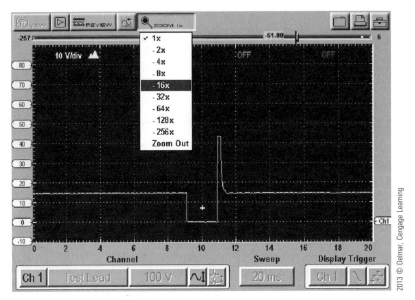

19. There are a couple of options for Reviewing when Zoomed out. You can just use the Up and down arrows on the thumb pad to move one frame left and right. This is a fine or small adjustment. To review more quickly you can use the Left and Right arrows on the thumb pad, which will move you in frame chunks of 10 or more. This is the coarse, or large, adjustment. The amount of frames moved with each push of the right or left arrow will depend on the Zoom level.

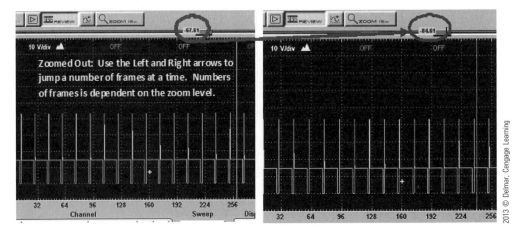

20. Once you have Zoomed out at a level greater than 1x, you can also scroll over to Review and activate that button, just as we did when at 1x zoom. Again, you can use the Up and down arrows on the thumb pad to move one frame left and right. This is a

fine or small adjustment. To review more quickly you can use the Left and Right arrows on the thumb pad which will move you in frame chunks of 10 or more. This is the coarse, or large, adjustment. The amount of frames moved with each push of the right or left arrow will depend on the Zoom level.

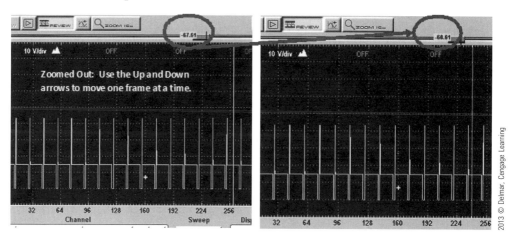

21. Press No to deactivate the Review button, if used.
22. Scroll over to Zoom press Yes and select 1x Zoom.
23. Press No to deactivate the Zoom button and scroll over to the Cursor button and press Yes. Two cursor lines appear, along with a dialogue box in the upper right corner of the screen that reads out values. This allows you to read the exact value anywhere along the waveform.

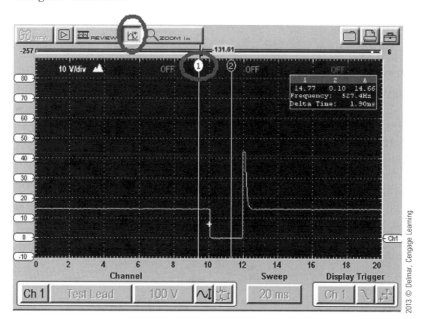

24. Cursor 1 is highlighted. So, using the left and right arrow buttons on the thumb pad, move the cursor to a point of interest and read the value.

25. When you have finished, press Yes. That will highlight Cursor 2. Next, again use the left and right arrows to move Cursor 2 to a point of interest, and again read the values. In the example, the voltage value at Cursor 1 is 14.66 volts, and the voltage value at Cursor 2 is 43.42 volts. The third column is labeled with a small triangle at the top. This is the delta symbol or "change." This functions as a built-in calculator. The difference between the incoming voltage and the high voltage spike is 28.76 volts. Notice that at the bottom of the information box is a Delta Time. This calculates the amount of time between Cursor 1 and Cursor 2. In this example, the pulse width or "on time" is 1.92 ms.

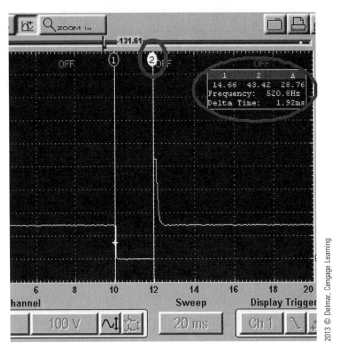

Here is an example of the cursors on a four channel trace. The information box looks more complex, but it is used just as the last single trace example. Notice that the numbers are color coded to the channels on the Lab Scope module. The exact value of where each cursor hits the waveform is displayed in the information box. The Delta (triangle) column still calculates the difference between Cursor 1 and Cursor 2 for each channel, and the amount of time between the two cursors is displayed at the bottom of the box and is labeled Delta Time. Use the cursors to get any value needed from the waveforms, as well as to measure time and frequency. The cursors can be used at any Zoom level, so even while you are zoomed out you can still gather precise data. The cursor function is a very powerful tool that should not be overlooked. Give the picture on the screen real meaning by attaching accurate and precise numerical data to it and quickly calculate differences using the Delta column.

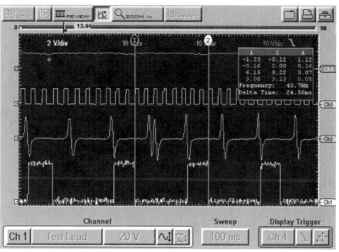

26. When you have finished using the Cursors, press no to exit. A yellow dialogue box will appear and ask if you want to leave the cursors on for reference or to turn them off. If you leave them on, they will remain on the screen but will not be moveable. Turning them off will make them disappear.

Press 'Y' to leave cursors on for reference, 'N' or down-arrow to turn cursors off.

Cursors are an excellent way of quickly gathering accurate data from the lab scope. They help make the lab scope an even more power diagnostic tool.

Ignition Scope Configuration

The lab scope is capable of displaying secondary ignition patterns to help diagnose hard-to-find ignition system issues. There is a separate selection for the ignition scope, but this feature can be manually configured through the 4 Channel Lab Scope selection as well. Most of the configurations and features that are part of the lab scope have already been explained in the previous Lab Scope section. If at any time there is a feature or button whose function you don't know, and it is not explained in the Ignition Scope section, please refer back to the Lab Scope section—its function will be described there. The MODIS and the Vantage Pro have the same Ignition Scope capabilities. The MODIS kit comes with the single cylinder ignition adapter that will allow ignition patterns to be viewed when the probe is connected to a spark plug wire. This ignition adapter has to be purchased separately for the Vantage Pro. There are also many other ignition scope accessories that can be purchased to increase the capabilities of the ignition scope of both the MODIS and Vantage Pro. It is the goal of this textbook to make sure you are aware of 90% or more of the diagnostic tools' capabilities, so optional test leads and accessories will be listed as the various Ignition Scope functions are explained.

1. From the Main Screen, scroll down to the Lab Scope Icon, then over to the right dialogue box, then down to Ignition Scope, and finally over to "Parade."

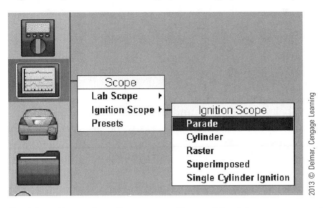

The Ignition Scope has five preconfigured ignition patterns that can be used. Next we'll look at each one and explain its purpose.

2. With Parade highlighted, press Yes. When multiple cylinders are connected to the Ignition Scope, Parade can be used to view all of the firings at the same time. When properly configured, the cylinder number will be displayed along the horizontal axis, making Parade excellent for comparing firing voltages.

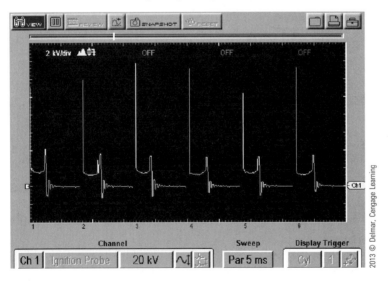

3. Press No to return to the Ignition Scope selection box.
4. Scroll down and highlight Cylinder and press Yes.

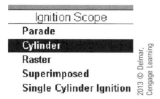

Cylinder is also used when multiple cylinders are connected to the Ignition Scope. Cylinder allows you select and view one cylinder at a time. The scope is triggered from the inductive pick-up attached to the number one cylinder and also defaults to an ignition scope sweep setting. With this setting, the trigger sensitivity can easily be adjusted, and the amount of burn time can easily be calculated.

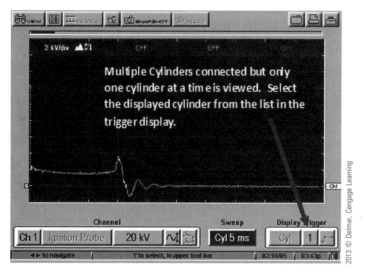

5. Press No to return to the Ignition Scope selection box.
6. Scroll down and highlight Raster and press Yes.

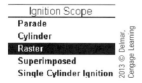

Raster is also used when multiple cylinders are connected to the Ignition scope. In Raster view, all the cylinders are displayed on the screen. Cylinder number one is at the bottom of the screen and the rest are displayed in firing order, moving towards the top of the screen. The Raster view is best used to compare burn time measurements for all the cylinders so that you can easily see which cylinder is not acting normally. The space between each cylinder pattern can be changed by highlighting and activating the Raster Offset button and using the up and down arrow buttons to change the distance.

Chapter 13 Lab Scope Operation

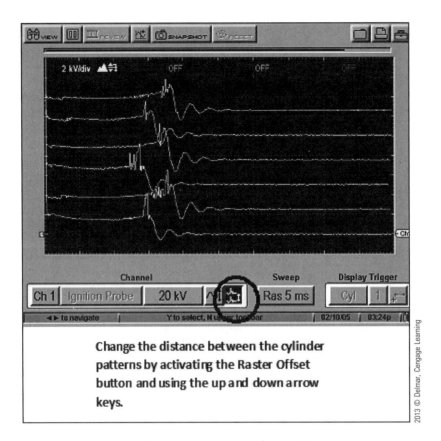

Change the distance between the cylinder patterns by activating the Raster Offset button and using the up and down arrow keys.

7. Press No to return to the Ignition Scope selection box.

8. Scroll down and highlight Superimposed and press Yes. Again, Superimposed is used when multiple cylinders are connected to the Ignition Scope. Superimposed stacks all cylinders on top of each other so that they can be compared to each other.

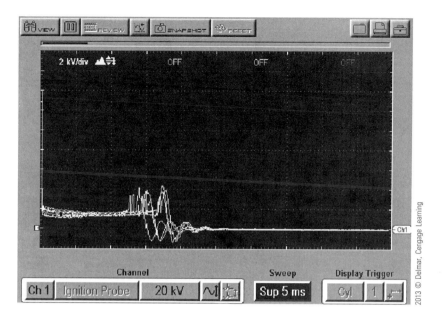

9. Press No to return to the Ignition Scope selection box.

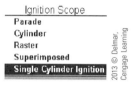

10. Scroll down and highlight Single Cylinder Ignition. This option is used when only one cylinder can be connected to the scope at a time. This view is most similar to the normal 4 Channel Lab Scope view. The trigger is not set to cylinder number one using the inductive pick-up; it is set using channel one and a voltage value, usually around 2 kilovolts. By default, Peak Detect and the Inverted features are activated for channel one. One common problem found when viewing ignition pattern using the Single Cylinder Ignition feature is not seeing the pattern due to an opposite firing cylinder. If there is no pattern after the proper connections have been made, try to fix this by turning off the Inverted feature. What commonly happens is that the pattern is firing down towards the bottom of the screen and not hitting the trigger point. If you turn off the Inverted feature, this will flip the pattern, causing it to hit the trigger and thus display the

ignition pattern. Refer back to the normal lab scope configurations and trigger setting options for more information on configuring this screen.

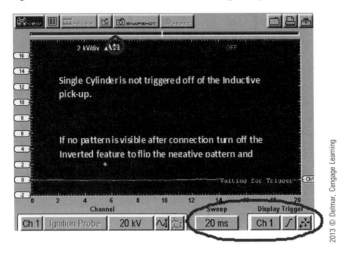

Single Cylinder Ignition

To view a single ignition pattern, you need to select the Single Cylinder Ignition option under the Ignition Scope, as discussed previously. The connection for this option comes standard with the MODIS and can be purchased separately for the Vantage Pro. The Single Cylinder Ignition Probe cable, officially called the Secondary Coil Adapter, connects to the Lab Scope just as any other test lead, and the other end has a separate grounding clip along with a RCA-type connector that looks similar to an audio cable. This RCA-type connector allows you to use various coil adapters. The standard Secondary Clip-on Wire Adapter (Spark Plug Wire Adapter) also comes with the MODIS and has to be purchased separately with the Vantage Pro. This is the traditional way of connecting to an ignition system via the spark plug wires to view the secondary ignition pattern. If the vehicle you are working on has Coil on Plug (COP) or Coil in Cap (CIC), then additional adapters will be needed. The picture illustrates one such adapter, the COP-2 adapter. This would connect to the RCA end of the ignition test lead and again allow you to view one secondary pattern at a time. There are many COP and CIC adapters to choose from, designated by manufacturer and year.

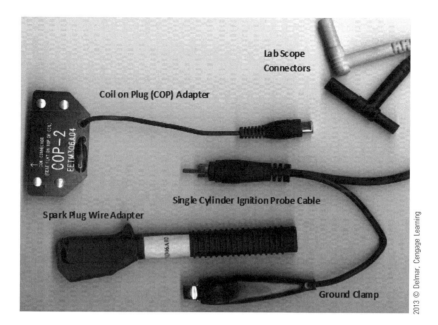

Viewing Multiple Ignition Scope Patterns

To view multiple ignition scope patterns at the same time, you will need additional equipment. The SIA 2000 Ignition scope adapter will allow you to view multiple patterns. The SIA 2000 comes with a set of standard cables that clip onto the secondary ignition wires. With this set, up to eight cylinders can be viewed. If you are trying to view multiple patterns on a direct ignition system, you can also use the SIA 2000, along with multiple COP adapters. The SIA 2000 also comes with a cable that has multiple RCA jack ends that will allow you to use multiple COP adapters. The following section will explain how to connect and use the SIA 2000 ignition scope adapter.

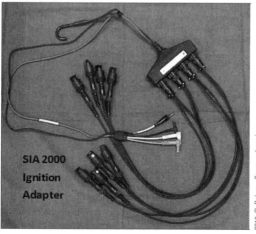

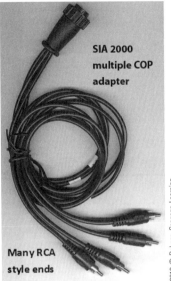

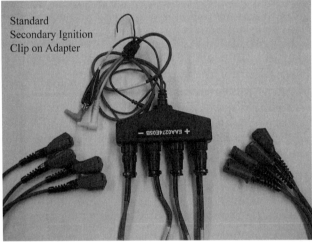

1. Connect the SIA 2000 to the lab scope as shown. Use the color-coded cables to connect to channels one, two, and ground. With multiple cylinders being viewed, it is important to connect the inductive RPM pick-up so cylinder number one can be identified, connected to, and thus identify the remainder of the cylinders based on the firing order.

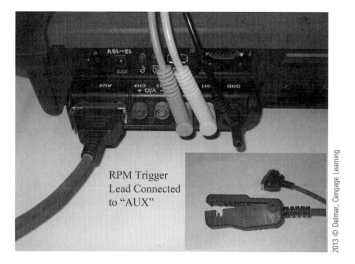

RPM Trigger Lead Connected to "AUX"

2. Connect the ground lead clip to a good engine ground.

3. The next step is to determine each cylinder's polarity, so after the final connections the patterns will be firing upwards on the scope screen. The SIA 2000's user manual comes with a reference section that is very helpful in finding cylinder polarity. Because

of aftermarket coils and ever-changing ignition systems, it is best to test for the polarity manually using your SIA 2000 adapter.

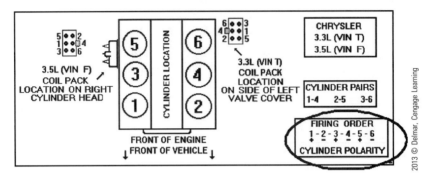

4. To find Cylinder Polarity, choose "Single Cylinder Ignition" from the main menu of the Lab Scope.

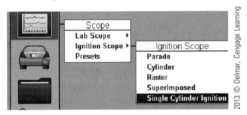

5. You need to find cylinder polarity because the SIA 2000 has positive and negative test leads, and connecting them to corresponding positive and negative cylinders will give us the best ignition pattern to view. Here is how the cables are defined on the SIA 2000 adapter. All the negative cables are grouped together, and all the positive cables are grouped together.

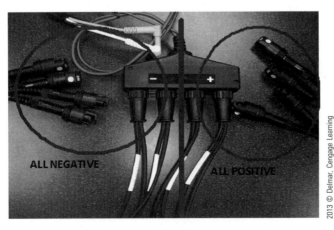

6. Connect one Negative test lead to any plug wire.
7. Be sure that the ignition scope channel one Inverted option is active.
8. Set the channel one trigger level to a positive value and an upward slope (usually 1kv or higher).

9. Your goal is to have all the patterns fire upwards after we determine the cylinder polarity, and this will happen if we connect the correct polarity test lead to the corresponding cylinders. A negative test lead is currently being used, and channel one is Inverted. This is important for you to understand because it will help you decide whether the cylinder is a positive or negative firing cylinder. With this configuration, all negative firing cylinders will fire upwards, and all positive firing cylinders will fire downwards (or not at all, due to the positive trigger).

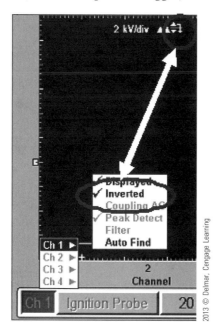

With a Negative Test lead and Inverted channel:
- Negative Cylinders will fire Upwards
- Positive cylinders will fire Downwards (or not at all due to the positive trigger)

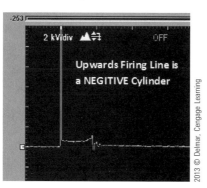

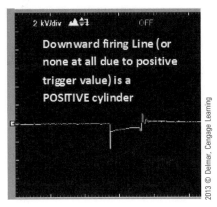

10. Start the engine and observe the pattern on the first cylinder. Remember, negative is upwards and positive is downwards or nothing at all. Record the cylinder's polarity.
11. Move the test lead to the next cylinder and repeat until all the cylinders have been identified as positive or negative.
12. With all the cylinder polarities recorded, you are now ready to use the ignition scope.
13. Connect the RPM Trigger Lead to Cylinder Number One.

14. Connect a negative test lead to all negative firing cylinders.
15. Connect a positive test lead to all positive firing cylinders.

16. From the Lab Scope Main Menu, select "Parade."
17. Scroll over to the far right and open the Toolbox by pressing Yes.
18. Then scroll down and select Ignition System by pressing Yes.

19. Begin to input the data for the ignition system you are working on.

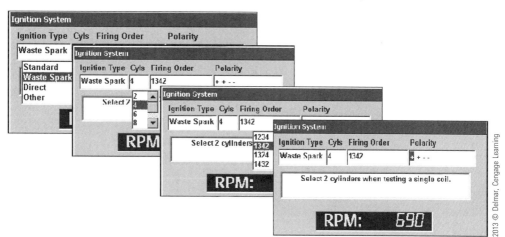

20. Change the cylinder polarity by pressing Yes. When changing the polarity for the cylinders, only half will have to be changed because the companion cylinders are automatically changed.

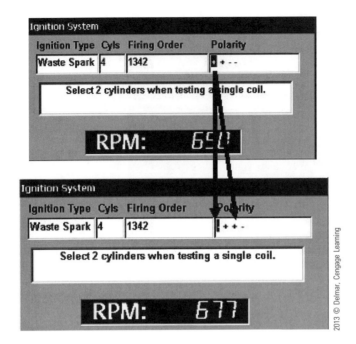

21. After this information is completed and all physical connections have been made, multiple patterns will be displayed. In this Waste Spark example, firing events are on channel one (yellow), and waste spark events are on channel two (green). Note the cylinder designations along the bottom. The first number is the firing cylinder, and the second is the waste spark cylinder.

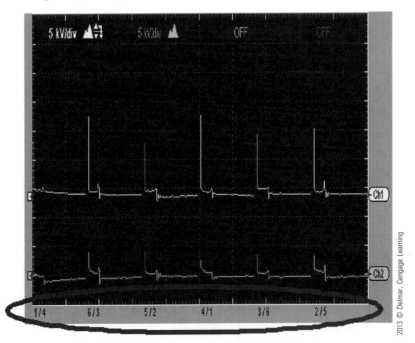

22. With the ignition scope gathering live data, you can change between the different ignition scope views. To do this, scroll down to Sweep and press Yes.
23. This will bring up the options available. Parade (Par), Cylinder (Cyl), Raster (Ras), and Superimposed (Sup), are all available in both 5 and 10 millisecond sweeps. All of these options were discussed previously in this section.

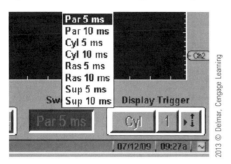

24. One other helpful ignition scope view is the digital view. This is located under the View option.
25. Scroll to the View button and press Yes.
26. Scroll down to Digital KV and press Yes.

27. The Digital KV View quickly shows the Firing KV, Spark KV, and Burn Time for each cylinder. The cylinder numbers are listed on the left side of the screen. It also records the minimum and maximum of each value and displays this. This is an excellent way to quickly see if all the cylinders are producing similar values and to identify problem cylinders. This viewing option can only be used on a live display and will not be available from a saved Movie or Snapshot file.

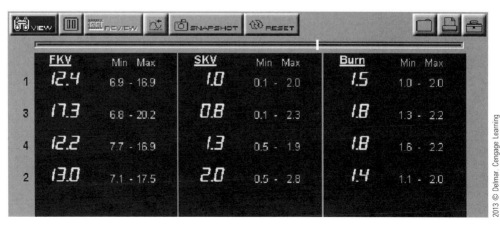

Reviewing Ignition Scope Movies and Snapshots

Saving an Ignition scope Movie or Snapshot can be very helpful in trying to find ignition system problems. In order to diagnose problems efficiently, it is important for you to know what information can be accessed from a saved Movie or Snapshot file. One important detail is that the Digital KV view is not an option with a saved file. This can only be viewed from a live display. Here is an example of an Ignition Scope Movie from a 1998 Mercury Tracer with a 2.0L engine. If you wish to follow along ,you can choose the IS(M) Demo 8 file preloaded on your lab scope. The patterns will be different, but the procedure for reviewing the information is the same.

1. Select Data Management from the Main Menu.
2. Load the file you wish to review. (IS(M) Demo 8 if you do not have another Ignition scope Movie to view.)
3. Refer back to the Data Management section if these directions are unclear to you.
4. The example movie was recorded in Parade 5ms. Remember to use the Review button at the top to scroll through the saved movie. This was discussed in more detail earlier in this section.

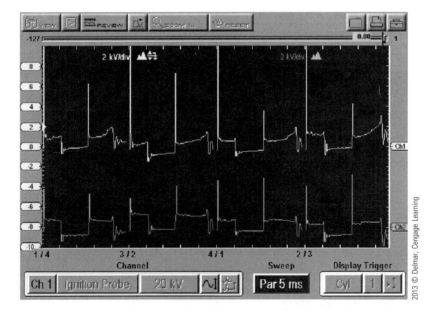

5. Even though this Movie was saved in Parade view, all the other views are still available.
6. Scroll to Sweep and press Yes.
7. Scroll down to "Cyl 5ms." This displays each cylinder individually so it can be diagnosed more closely.

Reviewing Ignition Scope Movies and Snapshots 459

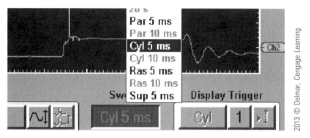

Cylinder allows you select and view one cylinder at a time.

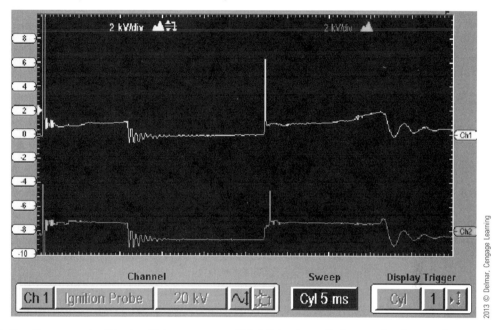

8. Scroll over to the Display Trigger area and highlight the cylinder number box and press Yes.

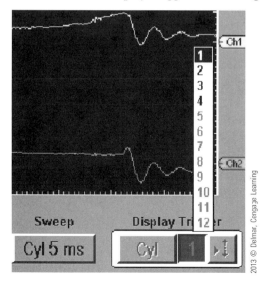

9. Scroll down to the cylinder you wish view and press Yes.
10. Here, cylinder 2 is being viewed.

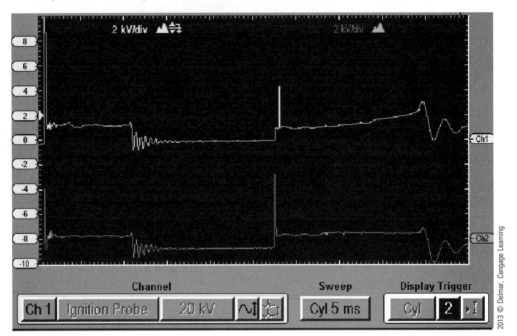

11. Scroll to Sweep and press Yes.
12. Scroll down to "Ras 5ms."

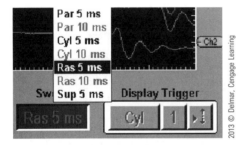

In Raster view, all the cylinders are displayed on the screen. Cylinder number one is at the bottom of the screen, and the rest are displayed in firing order moving towards the top of the screen. The Raster view is best used to compare burn time measurements for all the cylinders to easily see which cylinder is not acting normally.

Reviewing Ignition Scope Movies and Snapshots 461

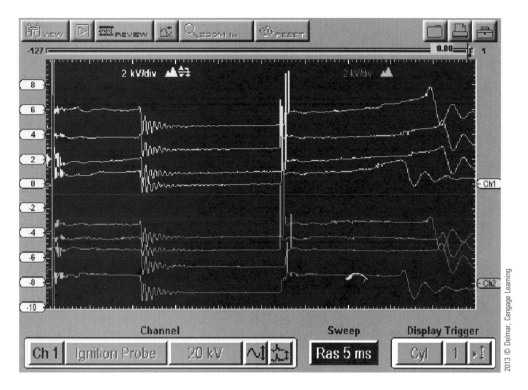

13. Scroll to Sweep and press Yes.
14. Scroll down to "Sup 5ms."

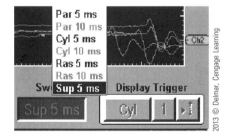

Superimposed stacks all cylinders on top of each other so they can be compared.

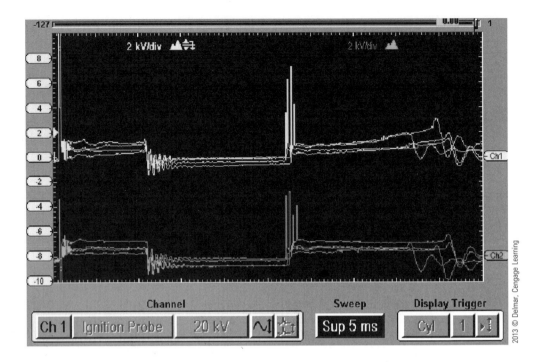

15. Once an Ignition Scope Movie or Snapshot is saved, no matter what view was used during the save, all of the information is available. Use Raster and Superimposed to quickly find cylinders that stick out, and then go to Cylinder to look at each pattern individually to determine the exact cause of the problem.

Finding Glitches Using a Lab Scope

This is a summary that combines many features already discussed about the lab scope, but puts them into a systematic order that helps you to more efficiently find glitches or problems in lab scope patterns. Depending on the pattern and problem, a technician may use one or more of the following steps

Steps to Finding a Glitch Using a Lab Scope

Step 1: Capture/Freeze the Data

It is usually difficult to find and analyze the problem with the lab scope actively collecting data. This step may be as simple as pressing the Freeze/Pause button, or it may be more sophisticated, such as saving a Movie or Snapshot.

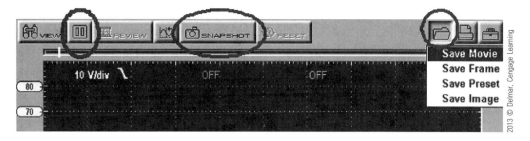

Step 2: Zoom Out

It can be difficult to see the problem when looking a small amount of the pattern or time. Stepping back and getting a bird's-eye view of the pattern will make abnormalities easier to see. Zooming will become more effective after some practice and experience. Zooming out too far sacrifices too much detail, and the pattern becomes worthless. Not zooming out enough will hinder the diagnostic process. An intermediate place to start is 16x.

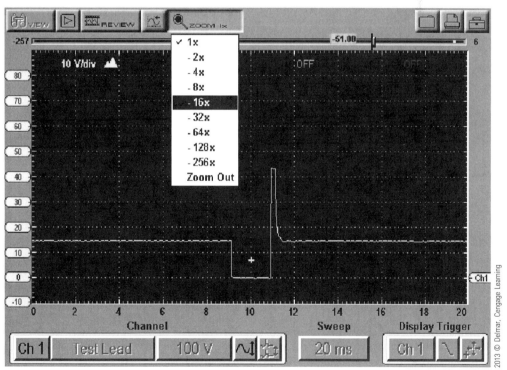

Step 3: Review the Pattern

Reviewing the pattern is usually done with the Review button activated, but it can automatically be done with the zoom button activated. Remember to use the Right and Left arrows to quickly scroll through the data, and use the Up (moves right) and Down (moves left) to look at each individual data point. Right and Left are a quick (coarse) review, whereas Up and Down is a slower (fine) review.

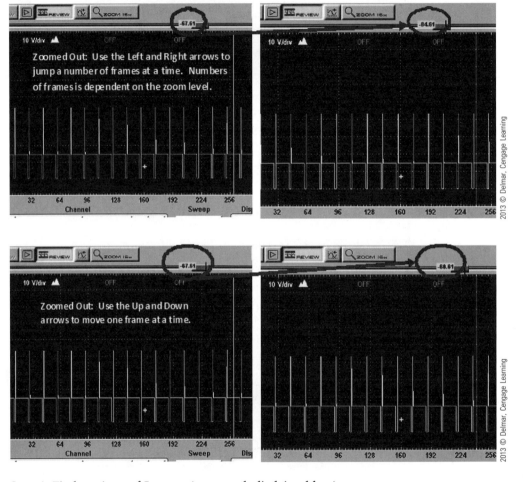

Step 4: Find an Area of Interest (suspected glitch/problem)

While you are zoomed out, there will be a vertical white line that moves across the screen while reviewing the patterns. This white vertical line is the Zoom line. Using the Up and Down arrow buttons move this white line as close to the suspect area as possible. It may not be possible to get it centered on the area which is all right; just move it as close as possible.

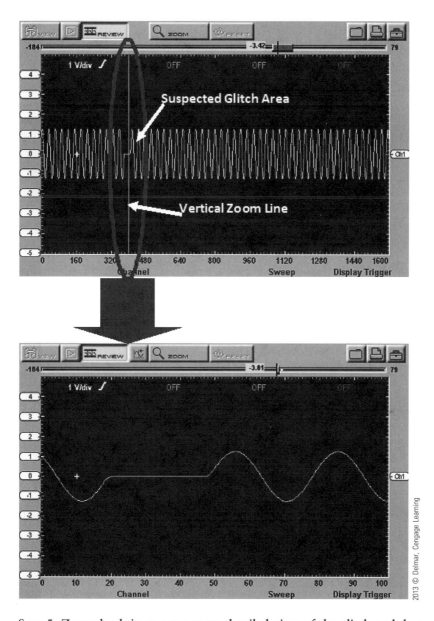

Step 5: Zoom back in to get a more detailed view of the glitch and determine the cause.

This is more important when viewing multiple patterns because the pattern that glitched first is the root cause. If components are going to be replaced, probably only the component that glitched first needs to be replaced. The other odd patterns from other components may be due to their compensating for the problem with the first component. Zooming out helps find the problem area, but zooming back in helps determine the actual cause of the problem. To move the glitch to the center of the screen first, activate the review button and then use only the Up and Down arrow buttons to move the pattern left and right. Holding down the Up and Down arrow buttons will slide the pattern across the screen. Using the Left and Right arrow buttons moves the pattern too much and will not allow for detailed placement of the glitch on the screen.

Glitch Example #1: Bad Fuel Injector

Step 1: Capture/Freeze the Data

Currently, no glitch can be seen, pattern looks normal/OK. Proceed to step two.

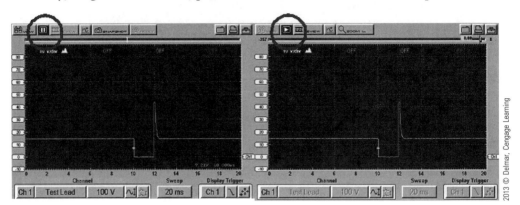

Step 2: Zoom Out

Choose something like 16x, which is a good start because it is in the middle. No glitch can be seen yet, so proceed to step three.

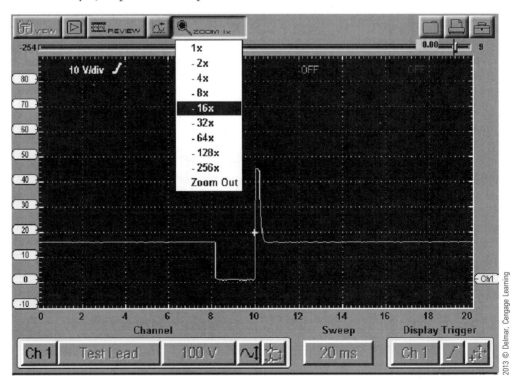

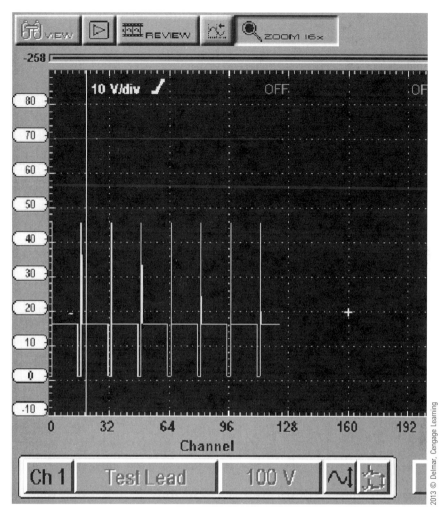

Step 3: Review the Pattern

Either activate the Review button, or with Zoom still activated just use the Up and Down arrow buttons to slowly review the pattern and look for a suspicious area that may need a closer look. Don't get frustrated; keep looking. If you know there was a problem, then the Lab Scope captured it, and the problem is in there waiting to be discovered.

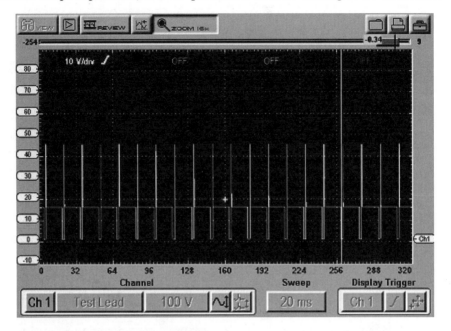

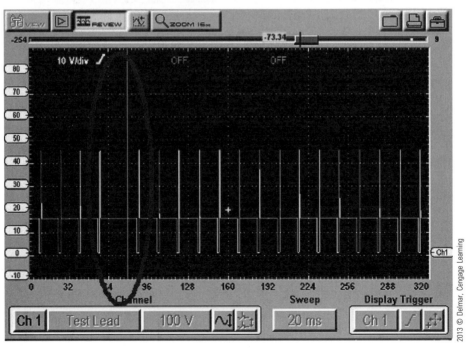

Step 4: Find an Area of Interest (suspected glitch/problem)

This took some time to review, but after starting at frame 8.34 and scrolling all the way to frame 73.34, finally a problem was spotted. In this case, the problem is a missing injector pulse. You can see the blank area that is inconsistent with the pattern to either side of it. At this point, we move the vertical white zoom line as close to the area as we can to mark it. If we needed to view the details of this glitch, we would proceed to Step Five and zoom back in. But since this glitch is a missing pulse, there really isn't much detail to view, and the problem is clearly understood as an intermittent non-firing injector. At this point, the glitch is captured, and there is no need to continue to the next step.

Glitch Example #2: Bad Wheel Speed Sensor

Step 1: Capture/Freeze the Data

In this case, we'll save a Movie and review it through the Data Management section.

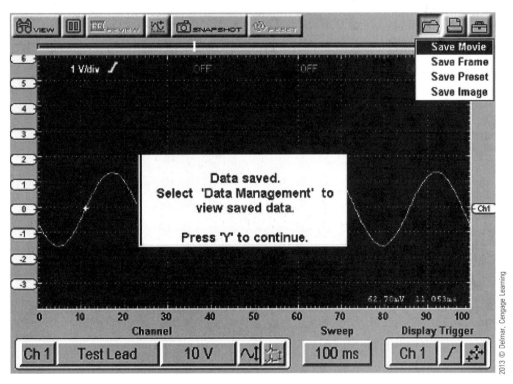

Step 2: Zoom Out

Pick something like 16x, which is a good start because it is in the middle. No glitch can be seen yet; proceed to step three.

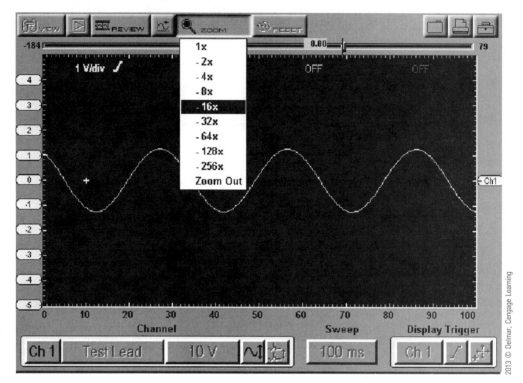

Step 3: Review the Pattern

Either activate the Review button, or—with Zoom still activated—just use the Up and Down arrow buttons to slowly review the pattern and look for a suspicious area that may need a closer look. Don't get frustrated; keep looking. If you know there was a problem, then the Lab Scope captured it, and the problem is in there waiting to be discovered.

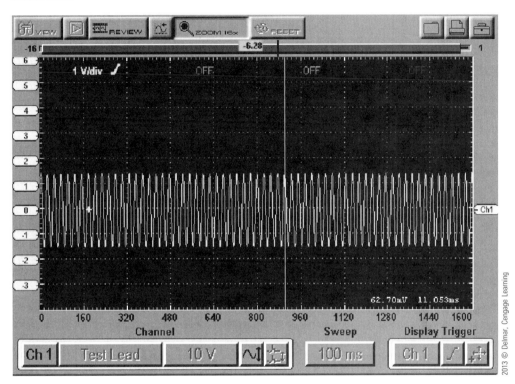

Step 4: Find an Area of Interest (suspected glitch/problem)

Again, be patient and don't get frustrated. If the problem occurred during the test drive or while you were monitoring the vehicle, then the problem is stored somewhere in the memory and it just has to be found. After finding the suspected glitch area, place the white vertical zoom line as close to the area as you can.

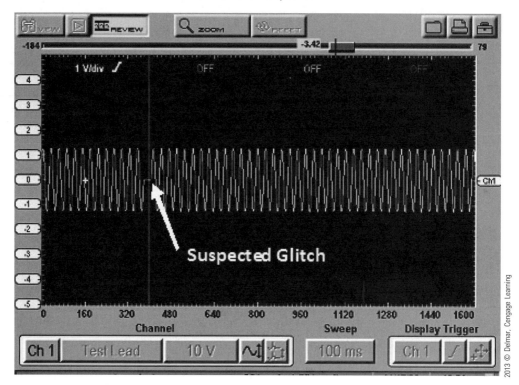

Suspected Glitch

Step 5: Zoom back in to get a more detailed view of the glitch and determine the cause.

Zooming back in will allow you to view more details of the pattern, or to decide what problem happened first if viewing a multiple trace scope pattern.

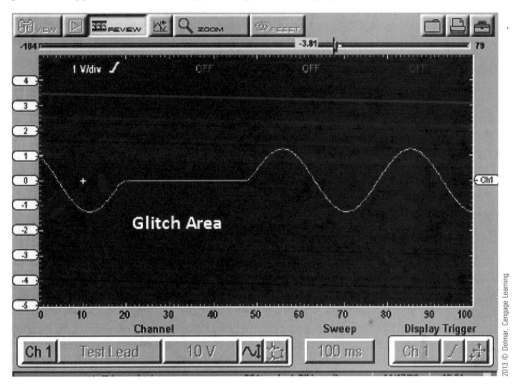

Summary

Lab scopes are a necessary diagnostic tool when working on today's automobiles because they plot a measured value over a specified amount of time. The speed at which electrical signals are being transmitted, received, and processed by computer-controlled vehicles is too fast for a common digital multimeter to accurately measure, especially when looking for a quick intermittent glitch. The lab scope allows you to see a picture of the electrical signal and thus makes it easier to find electrical problems. Snap-On's lab scopes are capable of measuring much more than just common voltage readings. Amperage, pressure, vacuum, and high secondary ignition voltages can also be measured with additional accessories. Each channel on the lab scope can be configured to give the best possible view of the signal it is sampling. The Ignition Scope can be used to view secondary ignition from a single cylinder and—with additional accessories—multiple cylinders at the same time, along with direct or coil over plug (COP) ignition systems. Lab scope data can be saved either by saving a Movie or Snapshot file, and these can be reviewed at a later time to find the glitch or diagnose the problem. Using the zoom and review features is the most efficient way of quickly finding a glitch and diagnosing the root cause. Lab scopes can seem complex and intimidating at first, but after you grasp the fundamentals of what any lab scope does and then the specifics of the scope you are using, this tool will help you solve complex vehicle problems.

Review Questions

1. Technician A says that the value being measured is always on the vertical axis of the lab scope screen. Technician B says the amount of time displayed on the lab scope screen is referred to as the sweep. Who is correct?
   a. Tech A only
   b. Tech B only
   c. Both Tech A and Tech B
   d. Neither Tech A nor Tech B

2. Technician A says all the channels of the lab scope are capable of sampling at the same maximum rate. Technician B says the AUX port is used for connecting the Inductive RPM Pickup and Split Lead Adapter accessories. Who is correct?
   a. Tech A only
   b. Tech B only
   c. Both Tech A and Tech B
   d. Neither Tech A nor Tech B

3. Technician A says Peak Detect will only capture fast acting high value events. Technician B says Filter should be used when viewing small value scales such as 2 volts or less. Who is correct?
   a. Tech A only
   b. Tech B only
   c. Both Tech A and Tech B
   d. Neither Tech A nor Tech B

4. Two technicians are discussing the screenshot below. Technician A says Channel 2 is set to a 5 volt scale. Technician B says Channel 1 is set to trigger on the rising edge and at 10 milliseconds. Who is correct?

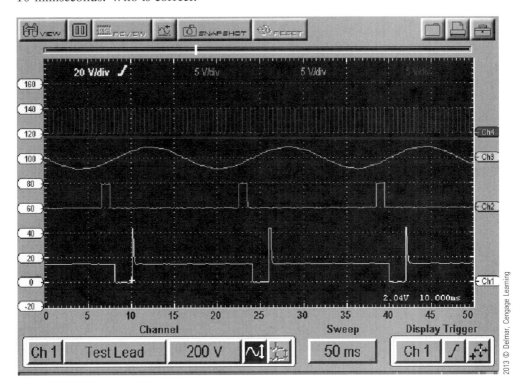

a. Tech A only
b. Tech B only
c. Both Tech A and Tech B
d. Neither Tech A nor Tech B

5. Two technicians are discussing the screenshot below. Technician A says Peak Detect is activated on Channel 4. Technician B says the Low Amps Probe is being used on Channel 3. Who is correct?

a. Tech A only
b. Tech B only
c. Both Tech A and Tech B
d. Neither Tech A nor Tech B

6. Two technicians are discussing the screenshot below. Technician A says that to remove the Grid and Scales from this screen, the display options are accessed through the Toolbox icon to the far upper right of the screen. Technician B says the Zero Offset button is currently highlighted in blue. Who is correct?

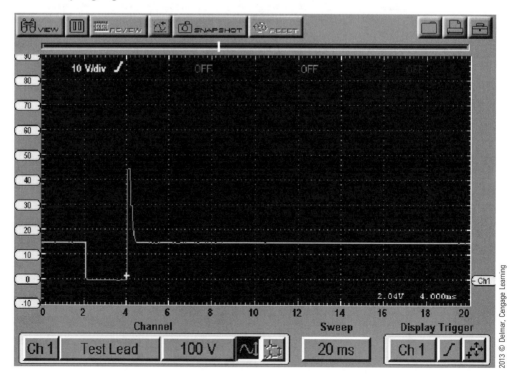

a. Tech A only
b. Tech B only
c. Both Tech A and Tech B
d. Neither Tech A nor Tech B

7. Two technicians are discussing the screenshot below. Technician A says Channel 1 could be Filtered. Technician B says that in order to decrease the number injector pulses on the screen, the sweep should be changed to 1 second. Who is correct?

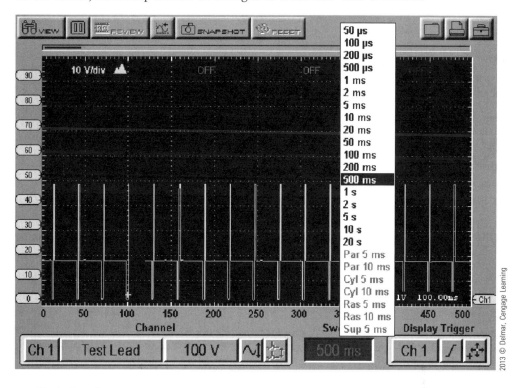

a. Tech A only
b. Tech B only
c. Both Tech A and Tech B
d. Neither Tech A nor Tech B

8. Two technicians are discussing the screenshot below. Technician A says Channel 2 is set as the trigger. Technician B says that in order to shorten (vertically) the Channel 3 sine wave, the Channel 3 scale should be decreased. Who is correct?

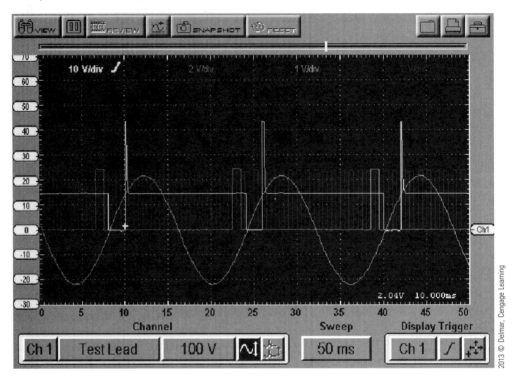

a. Tech A only
b. Tech B only
c. Both Tech A and Tech B
d. Neither Tech A nor Tech B

9. Two technicians are discussing the screenshot below. Technician A says pressing Yes now will Zoom out and allow more time to be displayed. Technician B says the current view is set to 2x zoom. Who is correct?

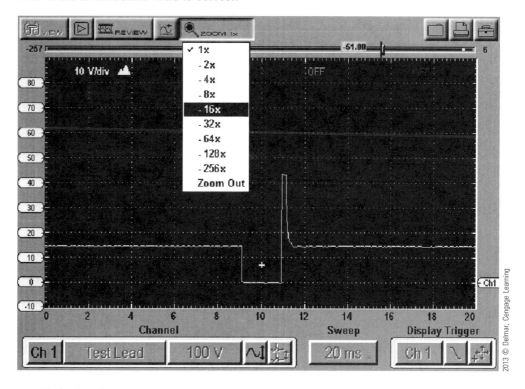

a. Tech A only
b. Tech B only
c. Both Tech A and Tech B
d. Neither Tech A nor Tech B

10. Two technicians are discussing the screenshot below. Technician A says that when Review is activated, the Right and Left Buttons will scroll the data multiple frames at a time, while the Up and Down arrows will scroll the data at one frame at a time. Technician B says that the white vertical bar (circled in blue) on the screen represents the zoom line where the pattern would be centered if zoomed back to a lower level. Who is correct?

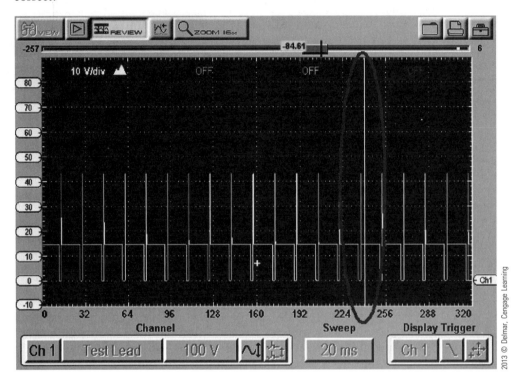

   a. Tech A only
   b. Tech B only
   c. Both Tech A and Tech B
   d. Neither Tech A nor Tech B

11. Two technicians are discussing the screenshot below. Technician A says the firing events are on Channel 1 (yellow). Technician B says this ignition pattern is triggered off Channel 1. Who is correct?

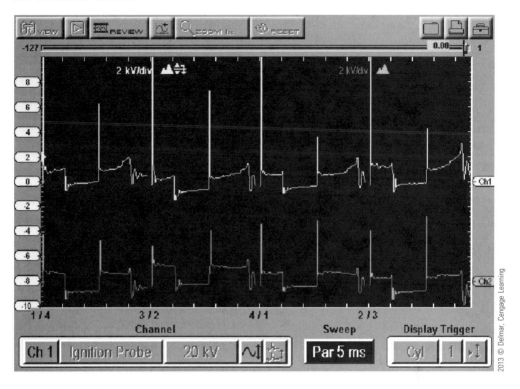

a. Tech A only
b. Tech B only
c. Both Tech A and Tech B
d. Neither Tech A nor Tech B

12. Two technicians are discussing the screenshot below. Technician A says scrolling over to the Sweep button (blue circle) and pressing Yes will allow the technician to switch between Parade, Cylinder, Raster, and Superimposed views. Technician B says Cylinder 2 is currently being viewed independently from the other cylinders. Who is correct?

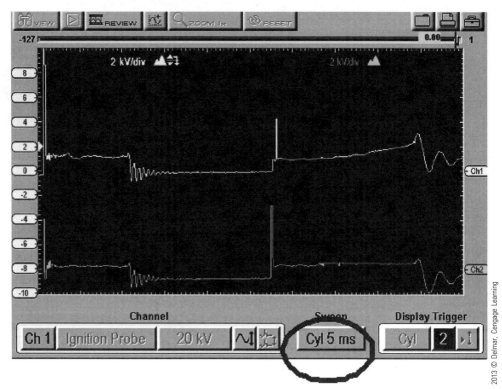

   a. Tech A only
   b. Tech B only
   c. Both Tech A and Tech B
   d. Neither Tech A nor Tech B

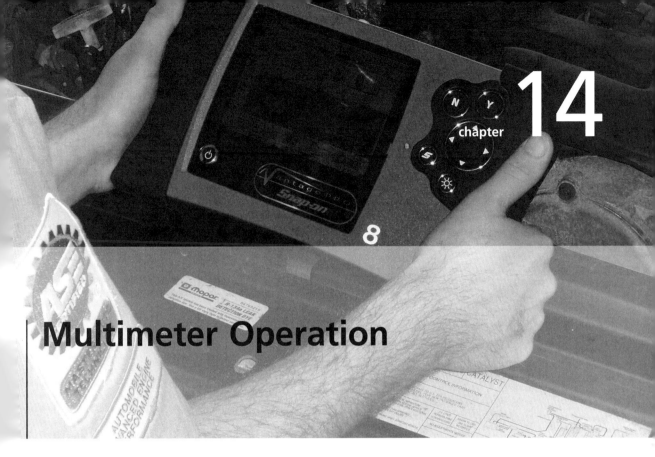

Multimeter Operation

chapter 14

Upon completion of the Multimeter module, you will be able to:

- Explain the difference between the Digital and Graphing meters.
- Compare and Contrast the advantages and disadvantages of each type of meter.
- Efficiently navigate the controls and features within each meter.
- Explain situations where the Graphing Meter will find a high glitch more easily than a lab scope.
- Collect, capture, and review measurements from the Graphing meter.

REMINDER: The Multimeter section will provide the information for both the MODIS and Vantage Pro Multimeters. The software running these two diagnostic tools is virtually identical. Some of the icons on the Vantage Pro may look a bit different on the screen due to a resolution difference, but the functions are the exactly same. You will more clearly understand the information presented in this section if you follow along using the diagnostic tool. Using the tool in conjunction with the book builds the hands-on knowledge required to pass the Snap-On Certification Exams.

Task: Your instructor may have you take Screenshots (Save Images) of the following screens as proof you are navigating the tool, making the connections, and capturing readings. The easiest way to do this is to program the S-button (Brightness/Contrast button for MODIS) to "Save Image." To gather most of the measurements, we'll use the Diagnostic Waveform Demonstrator board, part number EESX306A. See Figure 14-1. If this board is not available, other components can be used. Substitute as needed so you can actually use the meter while following along.

Figure 14-1 Diagnostic Waveform/Signal Generator board with a built in glitch to allow hands-on practice using lab scope and meter functions.

Multimeter Overview

Even though the MODIS and Vantage Pro are best known as lab scopes, both have a variety of electrical test meters built into them. There are times when these meters are simpler to use and represent the better tool for the job., mainly because a lab scope does have its limitations. A lab scope simply plots a voltage over time and nothing more, so it cannot display a resistance measurement in ohms or a frequency measurement. These are calculated values that are displayed by a meter, not a scope. There are times when a meter is still your best or only option for taking a reading. The Multimeters are broken into two groups: the Digital Meter and the Power Graphing Meter (PGM). The Digital Meter is similar to any common handheld digital multimeter (DMM) that you have used. It is capable of taking a variety of readings, including AC and DC Voltage, Resistance, Continuity, Diode testing, and Amps. Amps can be measured directly using the Vantage Pro by connecting it in series while the MODIS requires the use of a low amps probe. The digital meter is good for quickly reading steady constant signals. Digital meters typically display or update a measurement about four times a second, and the samples that are measured are usually averaged. This is why finding high-speed glitches is difficult using a meter that samples and averages the readings before displaying them. The odd value, or glitch, is intermittent, so this abnormal value is averaged away and looks closer to normal. A numeric digital display that is being updated four times a second can result in a missed reading by the operator. Even blinking could cause you to miss the displayed bad value. In response to these limitations, the Power Graphing Meter (PGM) was developed. The primary benefit of the PGM is clearly its graphing capabilities, which allow the measurement to be viewed over time, so that the human eye can more easily see the changes. The second advantage of the PGM is that it does not average the samples. It instead samples and plots individual values as they change, always keeping track of the minimum and maximum. This type of sampling works well with repetitive signals, such as duty cycles, pulse widths, and frequencies. The PGM is looking for changes, so as soon as a high-speed glitch changes the repetitive signal, the PGM displays that change, and the glitch appears on the meter. The PGM is capable of displaying a 300-microsecond glitch 100% of the time. This type of glitch would be impossible to detect using an averaging meter and difficult to find with a lab scope. This concept can be confusing, so let's look at an example. If you have read the Lab Scope section of this book, you are familiar with the process of finding a high-speed glitch. We first gather data with the lab scope and then freeze (pause) or capture the data in a Movie or Snapshot file so we can review it. To make the review process quicker, we zoom out and scroll through the data looking for the problem so that we can then zoom back in and analyze the problem area. Let's put the PGM up against the Lab Scope.

This example is generated from the waveform board. We will perform an exercise like this later in this section.

The suspected problem is a high-speed glitch in a frequency signal. We'll try the lab scope first.

I enter my lab scope with my demo board connected and manipulate the settings to get the waveform on the screen. As expected, it is a square wave pattern, but there is no sign of the glitch. From here we follow our systematic diagnostic procedure as described in the Lab Scope sections.

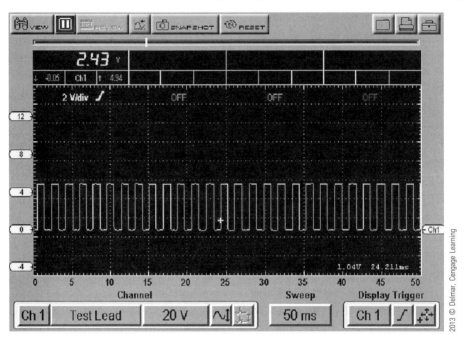

The next step is to Freeze the data and Zoom out. There is still no sign of the glitch.

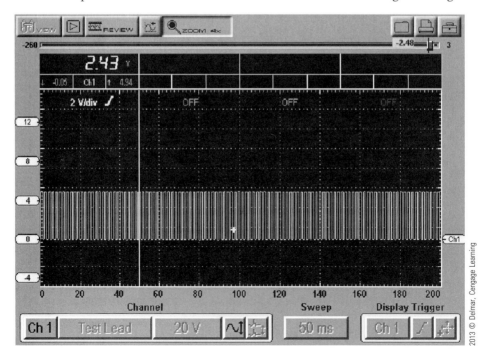

Next we review the data and look for the glitch. From the last screenshot you could see we started on frame number 2.48. According to the memory buffer bar going across the top, I had to review to frame number 13.48 to find the glitch. Using the Lab Scope, we confirmed the problem and found the glitch. Let's see if the PGM can make this any easier.

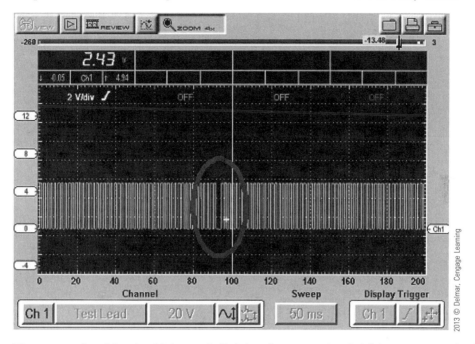

The suspected problem is a high-speed glitch in a frequency signal. A lab scope can only plot voltage over time, but the PGM can calculate values such as frequency and is preset to measure frequency. From the start there should be less setup and tool manipulation to get the pattern on the screen.

First, I have to open up the PGM and choose the frequency meter.

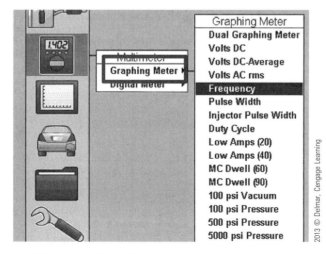

I still have the demo board connected, and as soon as I enter the PGM, here is what I see on the screen. The high-speed glitch is easily visible.

The PGM does not average the samples but instead is always looking for changes in the signal. When a change is found, it is displayed.

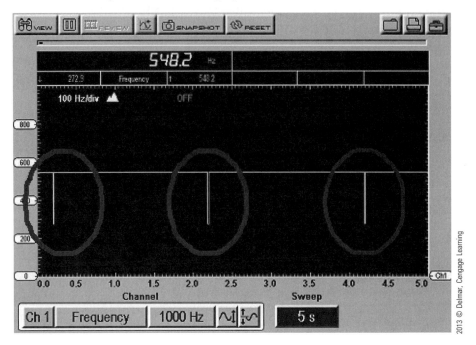

The PGM's lowest sweep rate is one second, so you'll never be able to get the zoomed-in view found in a lab scope with much smaller sweep rates. However, using the PGM to first determine if there is a problem with the signal and then switching to the lab scope to see more details about that problem is a good diagnostic strategy. Had I used the PGM first in my example, I would have then gone to the lab scope knowing there was a problem and that I was in the right place and looking at the right component. Having this information ahead of time will give any technician a confidence boost when using the lab scope. It can be very frustrating reviewing large lab scope files searching for the problem, when in the back of your mind you are not one hundred percent sure that the problem even exists on this component. Use the PGM to quickly check for abnormalities in a signal, and then use the lab scope to uncover the details regarding those abnormalities.

Waveform Demonstrator Board

Before you use the different meters, you will need an explanation of the Waveform Demonstrator board. (From this point forward, it will be referred to as the demo board.) The demo board is a compact signal generator that allows us to practice and train with a variety of signals using the lab scope and meters when not in a shop setting. The demo board is set to automatically time out and turn off after about 15 minutes to conserve on battery power, so if the signal is not showing up on the screen, be sure to check that the demo board is on.

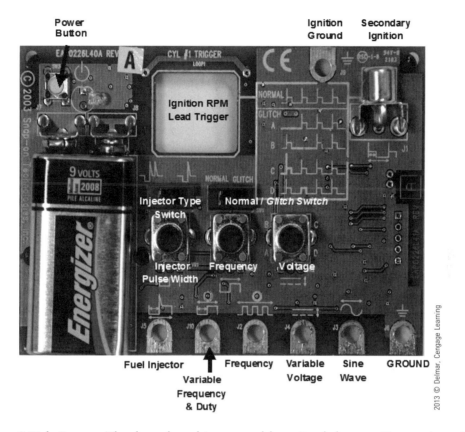

9 Volt Battery: The demo board is powered by a 9-volt battery. Use caution when installing the battery, so as not to break off the connector posts.

Power Button: Press and hold it down for a couple of seconds until the LED lights up.

Ignition Ground: This is where the single cylinder ignition test lead grounding clamp will be connected.

Secondary Ignition: The RCA-style (looks like an audio/video connector) ignition test lead will plug into this connector to produce sample secondary ignition patterns.

Injector Type Switch: Switched to the left, the board will produce a double peak and hold type injector signal and switched to the right it will produce the standard single peak fuel injector signal.

Normal/Glitch Switch: To produce good signals, have the switch set to Normal, and if you wish to program in a high-speed glitch, switch it to the right. The glitch is commonly used in training situations so that technicians can practice using the tool and finding the glitch.

Injector Pulse Width Dial: The far left dial controls the on time of the simulated injector. Note the notch in the dial: We use this as an indicator of dial position. If the notch is straight up, that would be considered the 12 o'clock position.

Frequency Dial: The middle dial can change the frequency of the signals coming from the frequency connectors.

Voltage Dial: The far right dial controls the voltage coming from the variable voltage connector.

Fuel Injector Connector: Connect here to view a fuel injector signal. This connector is controlled by the fuel injector switch and the pulse width dial.

Variable Frequency and Duty Connector: This connector is used to explain frequency and duty cycle or on time. Both the injector pulse width and frequency dials control their respective parts of this signal.

Frequency Connector: This connector produces a frequency signal with a fixed duty cycle. The frequency of the signal can be controlled by the middle frequency dial.

Variable Voltage Connector: From this connector comes a DC voltage signal that is able to be controlled by the far right dial.

Sine Wave Connector: An AC sine wave is produced from this connector. The frequency of the sine wave is controlled by the middle frequency dial.

Ground: This is the common ground for all the other connectors.

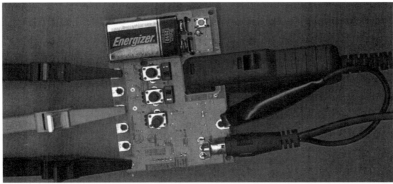

Example of various test leads connected to the Demonstration Waveform Board

Digital Meter Navigation

The digital meter is an on-screen version of a common handheld digital multimeter commonly referred to as a DMM. The benefits of using the diagnostic tool's DMM rather than a common handheld are many. First is the convenience factor of having multiple tools in one. Most diagnostic situations will require the use of a lab scope, but some preliminary checks may need to be done using a meter. Rather than having another tool out and in the way along with another set of test leads, it is easier to initially use the meter that is on the lab scope tool. Second, the display is much easier to read because it is larger and in color when compared to a normal DMM display. Finally, the Digital Meter also continually updates and displays the minimum and maximum values and these can be rest by the user at any time. This can be helpful for confirming an abnormal reading that may have been too fast to see on the screen or was inadvertently missed while displayed.

Task: Your instructor may have you take Screenshots (Save Images) of the following screens as proof you are navigating the tool, making the connections, and capturing readings. The easiest way to do this is to program the S-button (Brightness/Contrast button for MODIS) to "Save Image." After collecting the screenshots, you may be asked to put those pictures into a PowerPoint or Word file and print that out to turn in for credit.

1. Scroll to highlight the Meter icon, press Right to activate the Multimeter menu, and then scroll to highlight Digital Meter.

2. Press Yes (or the Right arrow) to activate the Digital Meter box and then scroll to Volts DC and press Yes.

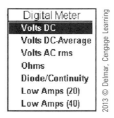

3. Connect the Yellow/Black test lead to the diagnostic tool using Channel 1 (Yellow) and Ground (Black).

4. Using the alligator clip test leads connect to the 9-volt battery used to power the demo board.

5. From this screen, we have the current voltage reading (large arrow), as well as the minimum and maximum readings (small circles).

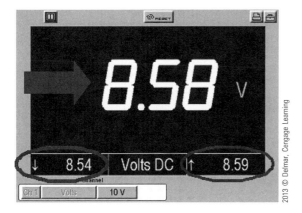

6. Press Yes to Pause the screen and make reading a fluctuating number easier. Notice the Pause button turns to a Play button after it is pressed.

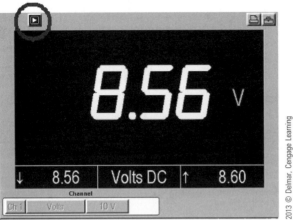

7. Press Yes to start sampling data again.
8. Scroll over to the Reset button and press Yes. This will reset the minimum and maximum values.

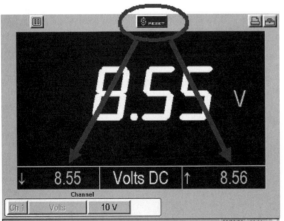

9. Scroll over to the right to highlight the Printer icon and press Yes.

   The only printing option available is the Screen print. After you connect to a printer, pressing Yes again will print out a picture of the screen. What you see is what you get. It is usually best to Pause the value before printing so that you are sure of the printing results.

10. Press No to exit the Printer icon options.
11. Scroll over to the right to highlight the Toolbox icon and press Yes.

    This opens the available shortcut options menu. Units and Saved Data are discussed under the Utilities menu for the specific tool you are using. Please refer to your tool's specific section in this book for more information.

    <u>Inverse Colors:</u> **Inverse Colors** will make the background white instead of black. When colors are inversed, a small check mark appears next to the feature label. This was originally designed for printing so as not to waste a large amount of black ink. Updated software has fixed this problem and now, no matter when you print, it will always print with a white background. Inverse Colors is more of a user display preference now.

    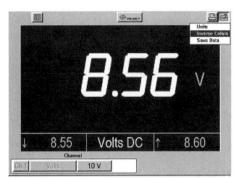 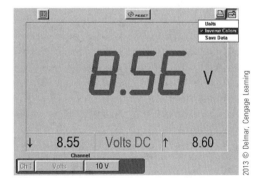

494 Chapter 14 Multimeter Operation

12. Scroll down to the Voltage Scale options and press Yes.

    This opens up the available voltage scales. Scroll to your desired setting and press Yes.

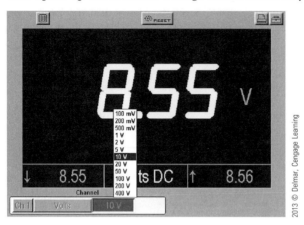

If you choose a voltage scale that is too low, the reading will be off the screen, and this will be indicated by many upward-pointing arrows.

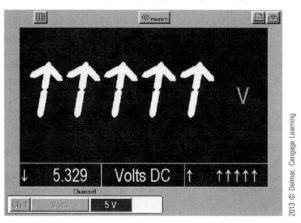

Increasing the voltage scale lowers the sensitivity of the meter, and you will eventually lose decimal places in the readings.

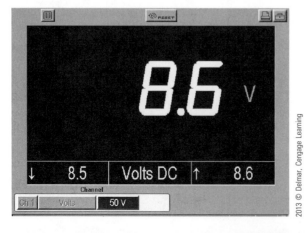

If Easy Scroll is enabled, this can be done without pressing the Yes button. Simply scroll and highlight the voltage scale button, and use the Up and Down arrow buttons to change the scale.

13. Keep pressing No until you return to the Digital Meter menu.
14. The other Voltage Settings are identical in function. The only difference between these settings is how the signal is being sampled.

   Volts DC: Measures direct current voltage using channel one and ground.

   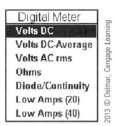

   Volts DC-Average: Measures DC voltage but averages the sample to act as a filter against radio frequency interference and other electrical noise. This reading is taken using channel one and ground.

   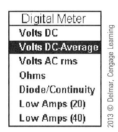

   Volts AC rms: Measures the effective AC voltage rather than the peak or average voltage. This is the best measurement of AC current when looking at the potential electrical work that can be done using it. This reading is taken using channel one and ground.

   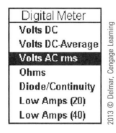

The next example is tool specific. Please read the procedure for the tool you are using.

Vantage Pro Specific

1. Scroll down to Ohms and press Yes.

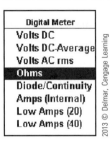

2. Connect the test leads (alligator clips) together but Press No and to *not* calibrate and use a previous calibration.

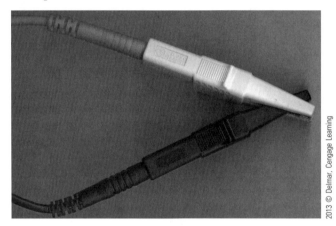

3. You initial screen should show a resistance near zero, and the numbers should be red in color.

4. Press No to exit.
5. Press Yes to return to the Ohms Meter, but this time, with the test leads connected, press Yes to calibrate them.
6. This time, the screen should read exactly zero, and the numbers should be green, as well as the calibration indicator towards the lower left of the screen.

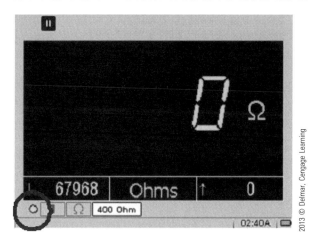

Calibration will always set the Ohms Meter to zero, so this way you can test resistance from a relative position if required. An example of this would be if you are testing a component that will change resistance as the temperature increases. The initial resistance may be 1000 ohms, but after calibration this first 1000 ohms would be zeroed out, so any change in resistance would now be displayed from the zero point.

NOTE: Vantage Pro users need to recognize that Ohms, Diode testing, and Continuity can only be performed using channel one and ground. This is why there is a volt and ohm symbol next to channel one and not channel two. The volt symbol references the voltage drop measurement used when checking continuity and testing diodes and, of course, the ohms symbol designates resistance testing. Be sure to use Channel 1 and Ground for these tests.

The common procedure for both tools will begin again after the following MODIS-specific section.

MODIS Specific

1. Scroll down to Ohms and press Yes.

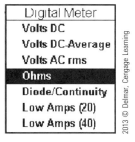

2. Connect the test leads (alligator clips) together but Press No and to *not* calibrate and use a previous calibration.

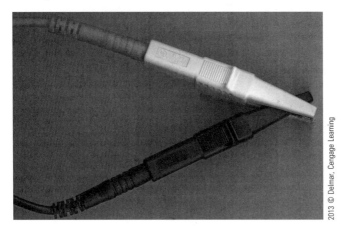

Remove leads from vehicle and connect together. Press 'Y' to calibrate, 'N' to use the previous calibration value.

3. What is the rationale for this reading? The meter is indicating an open circuit, but the test leads are clipped together. This is a common mistake when using the MODIS. To resolve this issue, look at the lab scope module where the tests leads plug into the MODIS. Notice anything? Channel 3 and Channel 4 are designated with a positive and negative symbol, as well as a volts and ohms symbol.

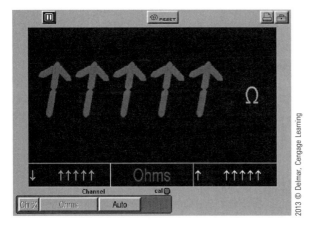

NOTE: MODIS users need to recognize that Ohms, Diode testing, and Continuity can only be performed using channels three and four. This is why there is a volt and ohm symbol next to these channels. The volt symbol references the voltage drop measurement used when checking continuity and testing diodes and, of course, the ohms symbol designates resistance testing. Be sure to use Channels 3 and 4 when performing these tests.

4. To fix this problem, move the test leads to channels three and four. The black ground should go to channel three—the negative channel—and the yellow test lead should go to channel four—the positive channel.

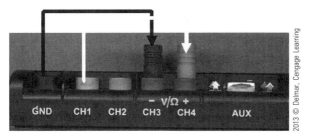

5. Now the meter should give us a more expected reading.

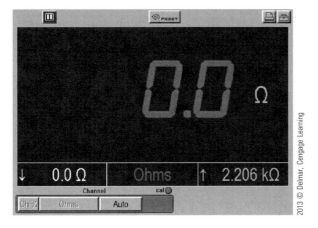

6. Press No to exit.
7. Press Yes to return to the Ohms Meter, but this time with the test leads connected, press Yes to calibrate them.

8. This time, the screen should read exactly zero, and the numbers should be green, as well as the calibration indicator towards the center of the screen.

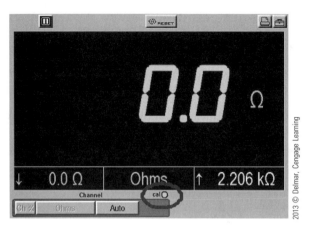

Calibration will always set the Ohms Meter to zero, so this way you can test resistance from a relative position if required. An example of this would be if you are testing a component that will change resistance as the temperature increases. The initial resistance may be 1000 ohms, but after calibration this first 1000 ohms would be zeroed out, so any change in resistance would now be displayed from the zero point.

MODIS and Vantage Pro

All the screen features and options discussed when using Volts DC are the same when testing resistance. The only slight difference is found under the scale button.

1. Scroll down to the scale button and press Yes.
2. Notice that in addition to the different ohm scales there is also an "Auto" option.

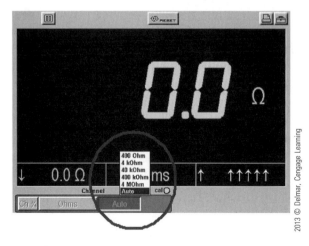

3. Make sure that this option is selected because it provides the easiest way to use the digital meter when measuring resistance.

Task: Resistance Testing Practice

Use the resistance testing capabilities of the tool to check each connector along the bottom of the demo board and find the resistance value of each. Do this by connecting the black test lead to the far right ground, and then move the yellow test lead to each connector, checking their resistance values. Be sure to remove the battery from the demo board before performing the tests.

Step: 1 Remove Battery

Step 2: Connect Black test lead to the far right ground.

Step 3: Move the Yellow test lead to each connector one at a time, record the resistance, and take a screenshot.

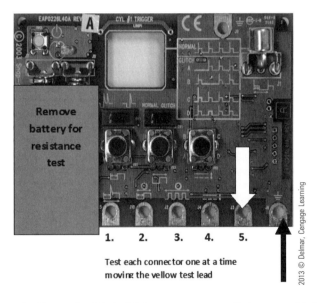

Test each connector one at a time moving the yellow test lead

1. From the Digital Meter menu, scroll down to highlight Diode/Continuity and press Yes.

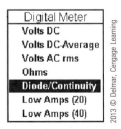

2. With the test leads connected together press Yes to calibrate.

Be sure that you have the test leads in the proper channels depending on the tool you are using. This test is essentially a voltage drop test, and that is what is actually measured on the meter.

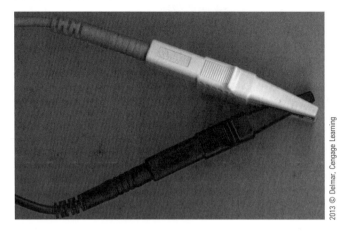

Remove leads from vehicle and connect together. Press 'Y' to calibrate, 'N' to use the previous calibration value.

3. After calibration, you should hear an audible alarm, indicating that continuity is being made.
4. As with the other screens we have discussed, this one is also navigated in the same way and has the same basic options and features.

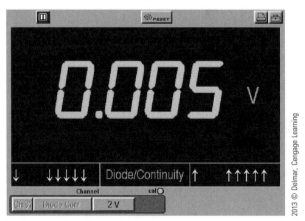

5. Press No until you return to the Digital Meter menu.

Vantage Pro Specific

The Vantage Pro is able to take direct amperage measurements using the blue-colored amps shunt connector. This connector is protected by the 10-amp fuse located on the back of the Vantage Pro. To measure amps using the amps shunt, the circuit to be tested must be broken or opened and the meter placed in series to reconnect the circuit. A load is then placed on the circuit, and amperage is displayed on the screen.

1. Scroll down to highlight Amps (Internal) and press Yes.

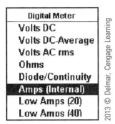

2. Scroll down to the scale button and set the amp meter to Auto just as you did when testing for resistance using the ohm meter.

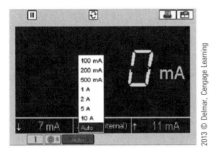

3. Remove the battery from the demo board. Simply connect the negative terminal of the battery to the demo board with the positive terminal of the battery hanging off the left side of the demo board. This will be the opening in the circuit.
4. Connect the Black alligator clip to the positive terminal on the demo board.
5. Connect the Yellow alligator clip to the positive terminal of the battery. This will reconnect or close your circuit.

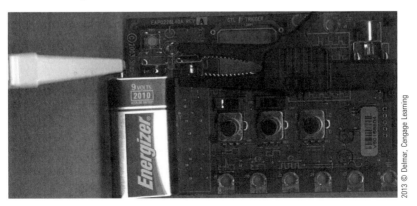

6. Push the power button to turn on the demo board and look at the amperage reading.

504 Chapter 14 Multimeter Operation

7. As with the other screens we have discussed, this one is also navigated the same and has the same basic options and features.

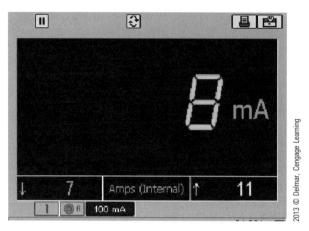

8. When you have finished, press No to return to the Digital Meter menu.

MODIS and Vantage Pro

1. Scroll down to highlight Low Amps (20) and press Yes.

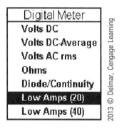

2. Scroll down to the scale button and set the amp meter to Auto, just as you did when testing for resistance using the ohm meter.

3. This test requires the use of the Low Amps probe.

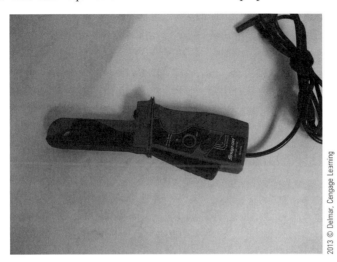

4. Connect the low amps probe to the tool. The black lead goes to the ground connector, and the red lead goes to channel one.
5. Turn the Low Amps probe on and set it to the 20 amp setting.

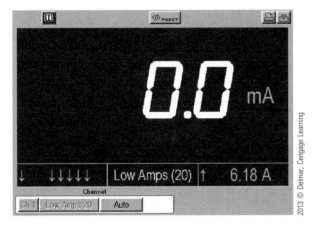

6. Press the zero button to zero the reading on the meter.
7. Keep the demo board built just the same as the last Internal Amps test. Either use a separate jumper wire, or build a jumper using the Yellow/Black test lead. Using the yellow lead, connect one end to the battery and the other to the positive terminal on the demo board. This should complete or close the circuit.
8. Turn on the demo board.
9. Clamp the low amps probe around the jumper wire.

10. Take a reading from the meter.

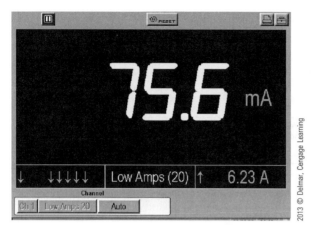

11. As with the other screens we have discussed, this one is also navigated the in the same way and has the same basic options and features.
12. The Low Amps (40) option works exactly the same but measures up to 40 amps. Be sure to set the low amps probe to the 40-amp setting when you are using this feature. Otherwise, the rest of the procedure remains the same.

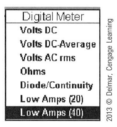

Digital Meter Summary

The Digital Meter found on the MODIS and Vantage Pro can be used just as any other handheld DMM. With a bigger color display built into a tool with other diagnostic options that will be needed later in the diagnostic procedure, there is a productivity reason to use this meter rather than stop the diagnosis to go and find another meter. The following charts summarize the Vantage Pro and MODIS Digital Meter specifications and were taken from the user manuals of the respective tools.

Option	Use
Volts DC	Measures direct current voltage (GND and CH1)
Volts DC-Average	Measures direct current and uses a filter to remove excess noise/hash on the signal (GND and CH1)
Volts AC rms	Measures the effective voltage rather than the Peak or Average voltage (GND and CH1)
Ohms	Measures electrical resistance (CH3 and CH4)
Diode/Continuity	Measures voltage drop across a diode or continuity (CH3 and CH4)
Low Amps (20)	Measures current from components like ignition coils, injectors, fuel pumps and parasitic draw using the Low Amp Probe (GND and CH1)
Low Amps (40)	Measures current from components like fans and electric motors using the Low Amp Probe (GND and CH1)

MODIS Digital Meter Tests

Option	Use
Volts DC	Measures direct current voltage
Volts DC-Average	Measures direct current and uses a filter to remove excess noise/hash on the signal
Volts AC rms	Measures the effective voltage rather than the Peak or Average voltage
Ohms	Measures electrical resistance
Diode/Continuity	Measures voltage drop across a diode or continuity
Amps (Internal)	Measures current for component and parasitic draws under 10A when connected in series with the circuit being tested
Low Amps (20)	Measures current from components like ignition coils, injectors, fuel pumps and parasitic draw using the Snap-on® Low Amp Probe
Low Amps (40)	Measures current from components like fans and electric motors using the Snap-on® Low Amp Probe

Vantage Pro Digital Meter Tests

Power Graphing Meter

The PGM resembles and functions much like the lab scope when it comes to navigation. The big difference is what is being displayed and how it is being calculated. Lab scopes plot a voltage reading over a specified time period. Sometimes we use accessories like Low Amps Probes or Transducers so that we can view amps and pressures on the lab scope. But in reality, the input from these accessories is still a voltage, and a calculation is done to display amps or pressures. The lab scope is still plotting a voltage over time. The PGM does not average the samples it collects, as a typical meter would. It instead samples and plots individual values as they change, always keeping track of the minimum and maximum. This type of sampling works well with repetitive signals such as duty cycles, pulse widths, and frequencies. The PGM is looking for changes, so as soon as a high-speed glitch changes the repetitive signal, the PGM displays that change and the glitch appears on the meter. The PGM is capable of displaying a 300-microsecond glitch 100% of the time. This type of glitch would be impossible to detect using an averaging meter and difficult to find with a lab scope.

If we set the Lab Scope and PGM to the exact same settings and use both to measure a voltage, then the two look almost identical. In this example, a frequency (square wave) signal is being looked at using a voltage setting on both the Lab Scope and PGM. There are no noticeable differences in the actual signal, so what distinguishes the Lab Scope from the PGM?

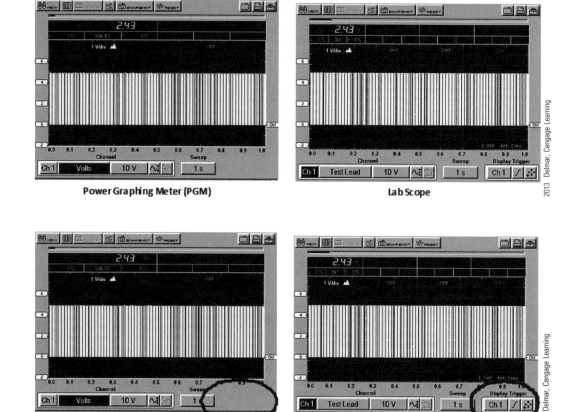

The first main difference is the absence of the trigger option on the PGM. The PGM samples and displays data as fast as it receives it. This is similar to the lab scope with the trigger set to "None," but lab scopes have the ability to trigger from a very specific point and sync other channels (waveforms) to that point. This makes it easier to view relationships with multiple channels displayed on the screen.

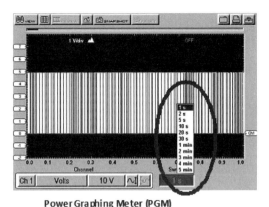 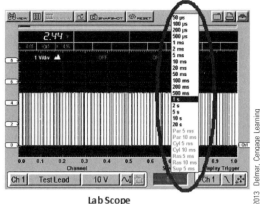

Power Graphing Meter (PGM)   **Lab Scope**

The next difference is the sweep rate abilities between the two. The PGM's lowest sweep rate is one second, whereas the lab scope can use a sweep rate down to 50 microseconds. If you need to view minute details of a signal, you will need to use the lab scope. The longest sweep rate for the PGM is five minutes, whereas the longest for the lab scope is 20 seconds. Sweep rates also determine the total length time captured in the data buffer. A frame of data is one screen. If the buffer can hold 262 frames of data, then to find the total amount of time than can be captured by the tool in a single recording you would have to multiply the number of frames (262) by the sweep rate, which is the amount of time displayed on one screen or frame. The maximum amount of time for the lab scope would be 262 frames multiplied by the longest available sweep rate of 20 seconds, for a total recording of about 87.3 minutes or not quite an hour and a half. Using the same formula for the PGM, we would multiply 262 frames by the longest sweep available, which is five minutes. This equates to 21.8 *hours* of *reviewable* recording time. An overnight parasitic draw test is a common example of where this is used. Hook up your low amps probe to one of the battery wires (be sure the auto power off feature on the amp probe is disabled), connect your tool to external power, set the PGM to a three-minute sweep (gives you about a 13-hour recording) or longer if needed, and then let it record overnight. Any change in the amps draw will be clearly shown by morning. If we are looking for a glitch and want to find it quickly, then the longer sweep times allow more data to be viewed on the screen at a given time. As shown in the example at the beginning of this section, the PGM will typically make finding a high-speed glitch easier but the lab scope should then be used to gather more details about the glitch if necessary.

The real difference between the lab scope and PGM is seen in the variety of available meter configurations. Again, the lab scope can only plot a voltage over time but not interpret the signal and calculate a different measurement such as frequency, pulse width, and duty cycle before they are displayed. These calculated values can be very beneficial in finding glitches. The following is another example of how the power of the PGM can be put to use to quickly find a high-speed glitch.

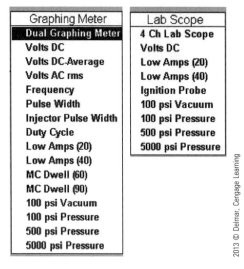

Assume there is an intermittent misfire on a vehicle and we want to be sure the fuel injectors are receiving a signal to fire. When can set up or lab scope to measure our voltage over time and collect a bunch of common fuel injector signals and then capture that data and look through it to see if there are any missing injector firings. We can do the same basic thing using the PGM by setting it up to measure voltage. Remember the earlier example where the screens on both the lab scope and PGM looked basically the same. Since an injector firing is a repetitive event it could be looked at by using the frequency configuration in the PGM. Since the PGM will be on the constant look out for changes in that frequency and display any such changes, if the injector doesn't fire it will be captured and displayed. Both tools have the ability to capture and display the non-firing injector, but which one can do it in a way that is easier for the eye to see? Ultimately, the technician is required to make a diagnostic decision based on what he or she sees. So what the tool is capable of capturing is only half the battle; the other half is displaying it so that the technician can easily spot these high-speed, intermittent events. To illustrate this example, I will use the PGM and channels one and two. Both channels will connected to the same fuel injector signal connector on the demo board. Channel one will be set up to measure voltage and be used as our lab scope because when measuring DC voltage there is little difference in the display. Channel two will be set to measure Frequency and illustrate the PGM's capabilities because the lab scope can not measure and display a frequency. The screenshot shows the results.

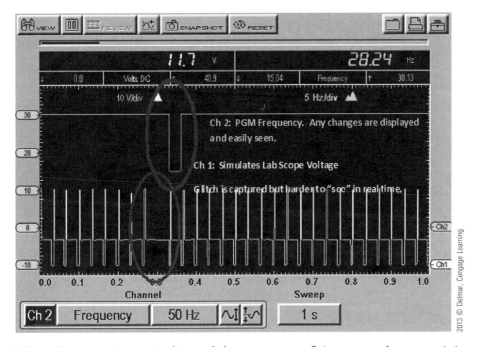

When this screen is running live and the patterns are flying across the screen, it is much easier to spot a single drop out than it is to spot a single missing peak amongst a steady stream of peaks. What would be easier for the eye to see: the spot where a missing tree used to be in a forest, or a single tree in the middle of an empty field?

Power Graphing Meter Connections

Even though the color coding makes connecting the test leads to the different channels an intuitive process, the ground connection requires a little explanation. As you will notice, on the channel one yellow/black test lead there are two ground connections. The purpose of the double-ended ground connection is to allow it to be piggy-backed by other ground connections from other test leads, such as the channel two green/black test lead and the Amp Current Probe, which also has its own ground connection. The double-ended ground connection is called the flying ground because it hangs off the side of the device and all other ground connections are connected to it. Only the single-capped ground connection is physically connected to the diagnostic tool, in order to protect the diagnostic tool from damage. If all the grounds were stacked on top of the tool, this would create a leverage point such that, if the tool were to fall, it could and hit this protruding stack of connections and severe damage could result. See the Lab Scope section for an illustrations of this.

Power Graphing Meter Configuration

Because the Power Graphing Meter and Lab Scope are very similar, many of the configuration settings are the same. Please review the Lab Scope section and its configurations to familiarize yourself with the PGM. No matter what specific PGM test (Volts, Frequency, Duty Cycle, etc.) you select, the display-screen navigation, features, and options will be the same. The following PGM settings are identical to the Lab Scope and are discussed in greater detail in that section. Please review the Lab Scope section to become a power user of the PGM.

Power Graphing Meter and Lab Scope Similar Configurations

- Data Collection and Review
    - Freeze/Run
    - Review
    - Zoom
    - Cursors
- "Snapshot" Data collection Feature
- File Folder
    - Save Movie
    - Save Frame
    - Save Preset
    - Save Image
- Tool Box icon
    - Units
    - Scales Display
    - Inverse Colors
    - Save data
- Channel Configuration
    - Displayed
    - Inverted
    - Coupling AC
    - Peak Detect
    - Filter
    - Auto Find
- Zero Offset Feature

Please review the Lab Scope section for an in-depth review of all the functions just listed and their diagnostic advantages.

Task: You will not need to connect the demo board for this part, but your instructor may have you take Screenshots (Save Images) of the following screens as proof you are navigating the tool and exploring the different features. The easiest way to do this is to program the

S-button (Brightness/Contrast button for MODIS) to "Save Image." After collecting the screenshots, you may be asked to put those pictures into a PowerPoint or Word file and print that out to turn in for credit.

1. From the Main Screen, select Graphing Meter and then scroll over to highlight Dual Graphing Meter and press Yes.

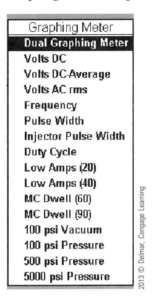

You screen may look different due to the various options selected or not. Focus on the particular feature being explained, and, by the end, your screen will match the one in the book.

2. Your tool may have a different number of channels turned on or off. At this time, however, that will not matter, so ignore this difference.

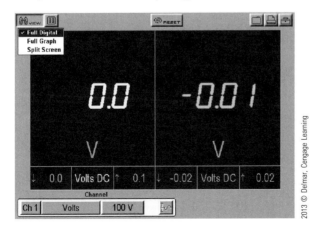

3. With the "View" button highlighted, press Yes.

4. Scroll down to "Full Digital" and press Yes. This will place a check mark next the viewing option and activate it. You will see a side-by-side view of a digital meter or essentially two multimeter screens at the same time.

5. Scroll down to "Full Graph" and press Yes. This will place a check mark next the viewing option and activate it. This will now remove any numeric digital information from the screen, which will now only display the graphs.

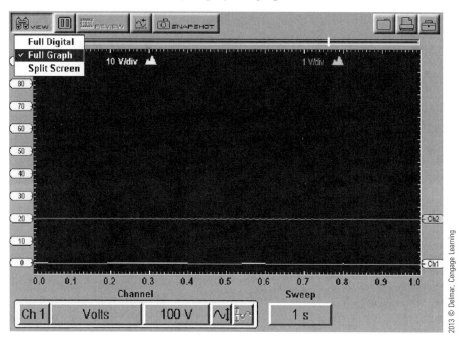

6. Scroll down to Split Screen and press Yes. This will place a check mark next the viewing option and activate it. This option allows you to view both the digital and graphed data on the same screen for both channels. We will leave the view in Split screen for this demonstration.

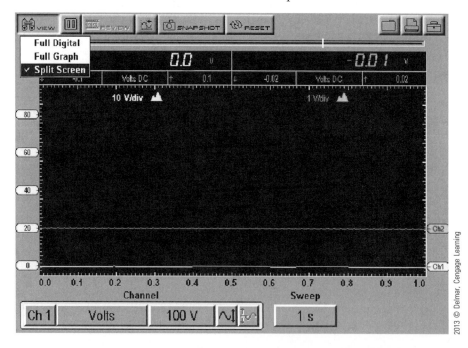

7. Press No to exit the View menu options.
8. To reset the captured Minimum and Maximum values, scroll over to "Reset" and press Yes. A dialogue box will appear, telling you that the values are being reset.

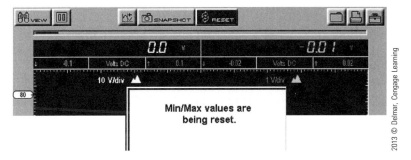

PGM Channel Configurations

Each channel on the PGM can be individually configured to show the best possible picture of the incoming signal.

1. Scroll over to the right and select the "Test Lead" button, which, in this case, is currently set to Volts and press Yes. This will open a dialogue box that will allow you to select the type of test you wish to perform. The test lead option is broken into many sub-settings that specify what is actually being measured. This selection will allow the PGM to automatically configure itself to give the best information possible. For example, when Volts is selected, the scale is set to voltage and is displayed as volts per division. But when switched to either the Low Amps 20 or 40 probe, the scale is changed to amperage and expressed as amps per division. The other possible selection is one of four pressure Transducer settings. This will help to configure the display to allow for the best waveform view and the most accurate information gathering.

Let's look at each option under the Test Lead feature.

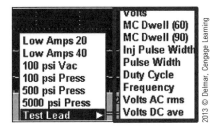

Volts: Measure direct current (DC) voltage. The unit of measure is displayed in volts or millivolts.

MC Dwell (60): This setting measures the duty cycle of the carburetor Mixture Control solenoid. The duty cycle is expressed as the dwell angle of a six-cylinder engine. 100% duty cycle equals 60 degrees. (360 degrees / 6 cylinders = 60°). The unit of measure is displayed in degrees

MC Dwell (90): This setting measures the duty cycle of the carburetor Mixture Control solenoid. The duty cycle is expressed as the dwell angle of a four-cylinder engine. 100% duty cycle equals 90 degrees. (360 degrees / 4 cylinders=90°). The unit of measure is displayed in degrees.

Inj Pulse Width: Injector Pulse Width measures the on time of a fuel injector pulse. The unit of measure is displayed in milliseconds.

Pulse Width: This feature measures the on time of various other components. The unit of measure is displayed in milliseconds.

Duty Cycle: The feature measures the on time of various components and displays the reading as a percentage of on time, compared to the complete cycle time. 100% is on all the time. The unit of measure is displayed as a percentage. An EGR valve and a Canister Purge Solenoid are typical examples of what is measured using this feature.

Frequency: This feature measures the number of times a signal repeats itself in a second. The unit of measure is displayed in hertz or kilohertz.

Volts AC rms: The Volts Alternating Current (root mean square) feature calculates the effective alternating current instead of the peak or average current. This is the best measurement of AC current.

Volts DC ave: Volts DC average will average the sample before it is displayed. This feature acts as a filter, removing electrical noise from the signal.

2. **Low Amps (20) and (40).** This feature is used to measure amperage in a circuit.

3. **Pressure Transducers** are used to measure pressure over time.

   100 psi Vac: This setting will measure vacuum up to 20 inches of mercury (20 inHg) and is commonly used to measure engine vacuum.

   100 psi Press: This setting will measure pressure up to 100 pounds per square inch (100 psi) and is commonly used to measure fuel pressure, engine oil pressure, and some transmission pressures.

   500 psi Press: This setting will measure pressure up to 500 pounds per square inch (500 psi) and is commonly used to measure transmissions, engine compression, and air conditioning systems.

   5000 psi Press: This setting will measure pressure up to 5000 pounds per square inch (5000 psi) and is commonly used to measure ABS braking systems, power steering, and other heavy duty hydraulic systems.

PGM Channel Configurations **517**

4. Scroll over to highlight the Scale button and press Yes. This will open a dialogue box of the available scale settings. As stated before, the scale options will change depending on the selected test probe.

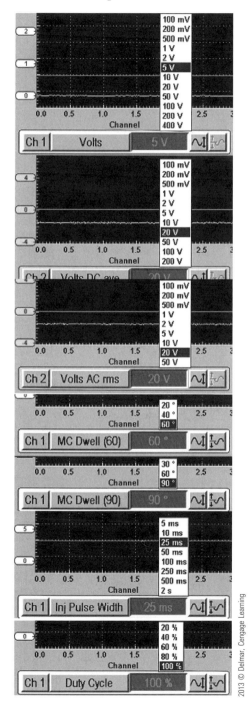

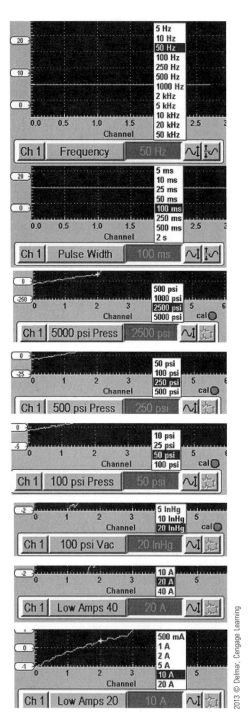

5. Scroll over to the right to highlight the Threshold button and press Yes. The Threshold button is only used when calculating measurements such as duty cycle, frequency, or pulse width. This option changes the reference point on the waveform.

6. This will open a dialogue box with two choices: Auto Threshold Select (ATS) or Manual Threshold Select (MTS).

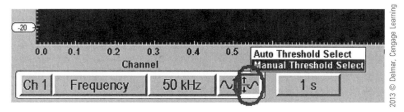

7. Pressing Yes with Auto Threshold Select (ATS) highlighted will bring up a dialogue box that explains that the signal is being evaluated and a threshold is being set. This feature automatically sets the threshold level in the middle of the minimum and maximum value that the waveform travels.

8. Scrolling down to highlight Manual Threshold Select (MTS) and pressing Yes will open up a dialogue box that allows the user to set the threshold. Typically, this is not used unless the initial waveform is not showing the information correctly and needs to be tweaked.

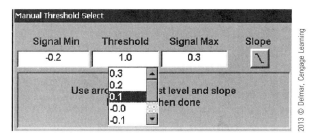

9. The minimum and maximum values will be displayed in the MTS box as references, but this may take a few seconds, so be patient. The Threshold must be set somewhere between the minimum and maximum values. The last option is the slope button, which can be changed by scrolling over to highlight it and pressing Yes. By default, it is set to the falling side, but this can be changed if desired.

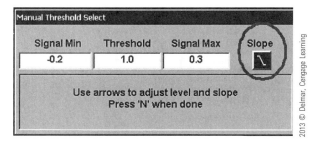

Practice

<u>Task:</u> Your instructor may have you take Screenshots (Save Images) of the following screens as proof you are navigating the tool, making the connections, and capturing readings. The easiest way to do this is to program the S-button (Brightness/Contrast button for MODIS) to

"Save Image." After collecting the screenshots, you may be asked to put those pictures into a PowerPoint or Word file and print that out to turn in for credit.

1. **Task:** Set up the demo board to see the difference between a Volts DC and a Volts DC ave (average) signal.

    a. Configure both Channels 1 and 2 to the following settings:
        - Channel 1 = Volts
        - Channel 2 = Volts DC ave
        - Both Channels:
            - 1 Volt Scale
            - 1 Second Sweep
            - No Peak Detect
            - No Filter
    b. Connect the demo board as shown.
        - Black = Ground
        - Yellow = Variable Voltage
        - Green = Variable Voltage
        - Glitch = On
        - Voltage dial all the way counterclockwise

    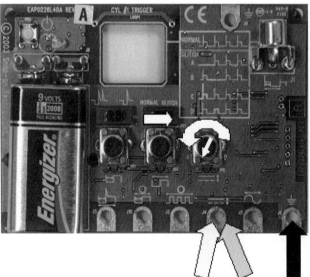

    c. Turn on the demo board and view the signals.
    d. Use the zero offset to separate the signals if necessary.
    e. Place the board near a fluorescent light or some other RFI source to better see the difference.

f. Pause, Zoom, and Review until you find the glitch and then save a screenshot.
g. Your screen should look similar to this.

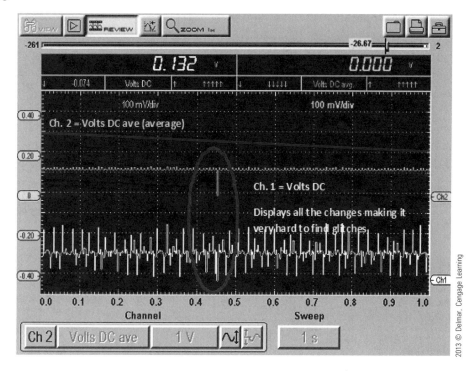

1. **Task:** Set up the demo board to see the difference between a Volts DC and an Inj Pulse Width signal.
   a. Configure both Channels 1 and 2 to the following settings:
      - Channel 1:
        - Volts
        - 100 Volt Scale
      - Channel 2:
        - Inj Pulse Width
        - 10ms Scale
        - Peak Detect On
      - Both Channels:
        - 1 Second Sweep
        - No Filter
   b. Connect the demo board as shown.
      - Black = Ground
      - Yellow = Fuel Injector
      - Green = Fuel Injector

522 Chapter 14 Multimeter Operation

- Glitch = Off
- Injector Switch = Right (single peak)
- All dials straight up at the 12 o'clock position

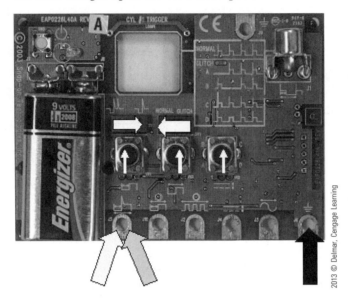

c. Turn on the demo board and view the signals.
d. Use the zero offset to separate the signals if needed.
e. Your screen should look similar to this.

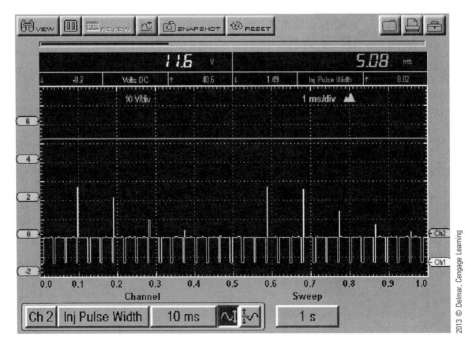

f. Now quickly turn the Injector dial (far left dial) back and forth and watch the pulse width change. Can you see it change in the waveform in channel one? Can you see it change in the waveform in channel two?

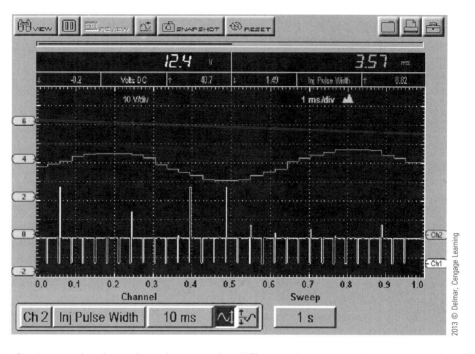

1. **Task:** Set up the demo board to see the difference between a Frequency and a Duty Cycle signal.

    a. Configure both Channels 1 and 2 to the following settings:
    - Channel 1:
        - Frequency
        - 100 Hz Scale
        - Filter
    - Channel 2:
        - Duty Cycle
        - 100% Scale
        - Filter
    - Both Channels:
        - 1 Second Sweep

    b. Connect the demo board as shown.
    - Black = Ground
    - Yellow = Variable Frequency & Duty
    - Green = Variable Frequency & Duty

- Glitch = Off
- Injector Switch = Right (single peak)
- All dials straight up at the 12 o'clock position

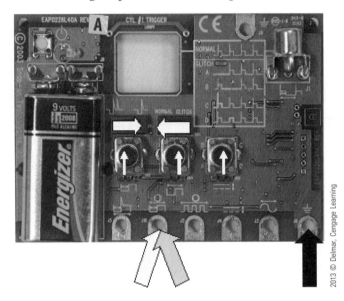

c. Turn on the demo board and view the signals.
d. Use the zero offset to separate the signals if needed.
e. Your screen should look similar to this.

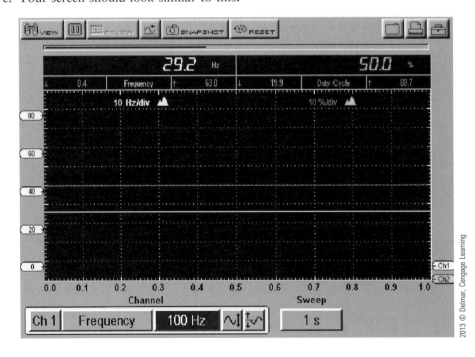

f. Slowly rotate the Frequency dial (middle dial) clockwise to its maximum position. How did this affect the Duty Cycle? Your screen should look like this.

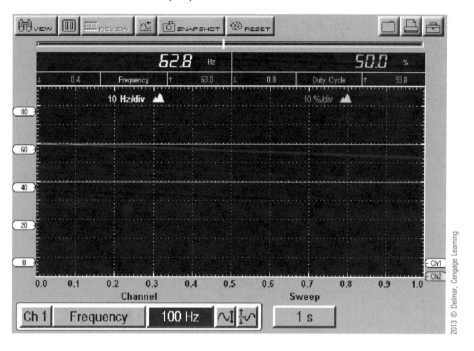

g. Slowly rotate the Frequency dial (middle dial) counterclockwise to its minimum position. How did this affect the Duty Cycle? Your screen should look like this.

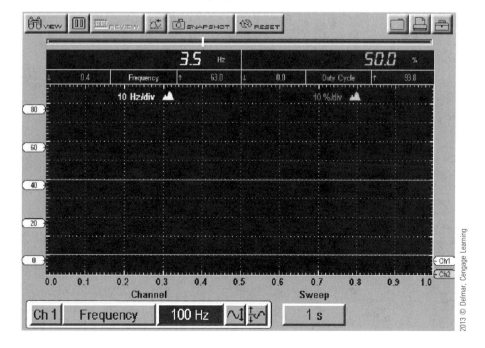

526 Chapter 14 Multimeter Operation

    h. Return the Frequency dial (middle dial) to the 12 0'clock position.

    i. Slowly rotate the Duty Cycle dial (left Inj Pulse width dial) clockwise until the Duty Cycle is under 10%. How did this affect the Frequency? Your screen should look like this.

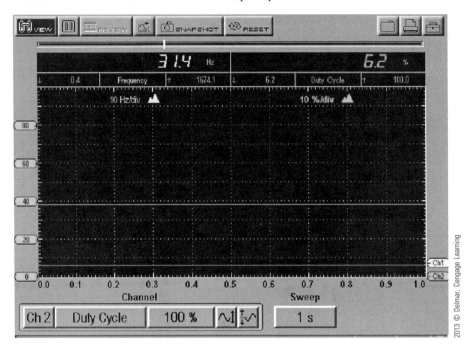

    j. Slowly rotate the Duty Cycle dial (left Inj Pulse width dial) counter clockwise until the Duty Cycle is over 90%. How did this affect the Frequency? Your screen should look like this.

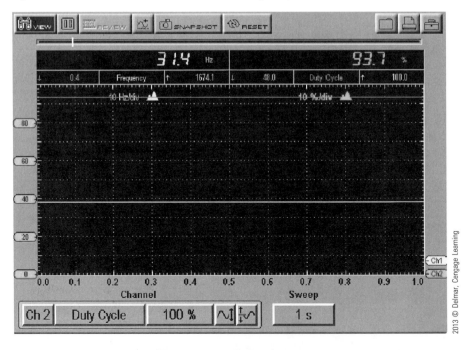

k. Explain the relationship between Frequency and Duty Cycle. Use your screenshots to justify your answer.
l. Turn the Glitch switch "On" and capture the glitch using the procedures we have learned about. Your screen should look like this.

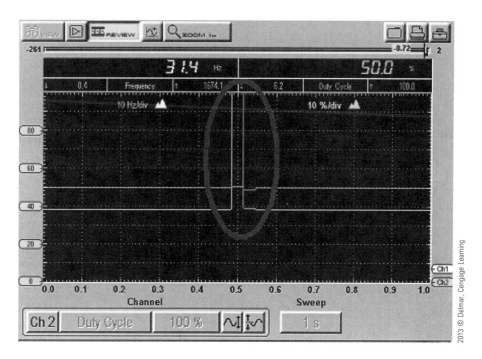

Summary

The overall diagnostic capabilities of your diagnostic tool are enhanced by having the Digital and Graphing Meters at your disposal and conveniently packaged with your lab scope. The larger color display found on the MODIS or Vantage Pro is much easier to read when compared to a normal DMM display. Both meters continually update and display the minimum and maximum values, and these can be rest by the user at any time. This can be helpful for confirming an abnormal reading that may have been too fast to see on the screen or was inadvertently missed while displayed.

The Digital Meter is similar to any common handheld digital multimeter (DMM) that you have used. It is capable of taking a variety of readings, including AC and DC Voltage, Resistance, Continuity, Diode testing, and Amps (with the use of a low amps probe).

The Power Graphing Meter (PGM) is a unique piece of testing equipment as it reports only the changes to the signal it is sampling and not an average of the incoming signals. The graphing capabilities allow signals to be viewed over time, enabling the eye to more easily see the changes. The real diagnostic advantage of the PGM is that it does not average the samples before it displays them. The PGM instead samples and plots individual values as they change, always keeping track of the minimum and maximum. This type of sampling works well with repetitive signals such as duty cycles, pulse widths, and frequencies. It also collects more common measurements, such as Amps, Pressure, and Vacuum (with the appropriate accessories). The PGM is always looking for changes, so as soon as a high-speed glitch changes the repetitive signal, the PGM displays that change, and the glitch appears on the meter. The PGM is capable of displaying a 300-microsecond glitch (300 millionths of a second) 100% of the time. This type of glitch would be impossible to detect using an averaging meter and difficult to find with a lab scope.

Another advantage of the PGM when compared to the lab scope is that the longest sweep rate for the PGM is five minutes, whereas the longest for the lab scope is 20 seconds. Sweep rates also determine the total length time captured in the data buffer that can be reviewed or saved and reviewed later. The maximum amount of time for the lab scope would be 262 frames multiplied by the longest available sweep rate of 20 seconds for a total recording of about 87.3 minutes, or not quite an hour and a half. Using the same formula for the PGM, we would multiply 262 frames by the longest sweep available, which is five minutes. This equates to 21.8 *hours* of *reviewable* recording time useful for capturing parasitic draws or other intermittent problems. With the vast amount of information that can be gathered using a PGM, it is important for you to be able to interpret this data and save it for later reference. Because most PGM data is displayed as a pattern or picture, the first step in interpreting that picture is to extract actual numerical data or values from the measurements that you are collected with the PGM. After this, you can save and archive this information in either a Movie or Snapshot file for future reference, or easily send it electronically to a colleague for a second opinion or post it to an Internet forum for even more different opinions. With all of these features and capabilities, the PGM is definitely more than just a meter, and it offers more capabilities and features than some lab scopes on the market.

Review Questions

1. Technician A states that the Digital Meter is similar to a common handheld digital multimeter (DMM). Technician B states that the Power Graphing Meter (PGM) is capable of capturing a 300 microsecond glitch 100% of the time. Who is correct?
   a. Tech A only
   b. Tech B only
   c. Both Tech A and Tech B
   d. Neither Tech A nor Tech B

2. Technician A states that most DMMs will average the collected samples before displaying a reading. Technician B states that the PGM does not average the collected samples but continuously looks for changes and displays the changes in the readings. Who is correct?
   a. Tech A only
   b. Tech B only
   c. Both Tech A and Tech B
   d. Neither Tech A nor Tech B

3. Technician A states that there are specific channels that must be used on each tool when testing resistance, diodes, and continuity. Technician B states that, in some situations, the PGM can more easily display a high-speed glitch than a lab scope. Who is correct?

   a. Tech A only
   b. Tech B only
   c. Both Tech A and Tech B
   d. Neither Tech A nor Tech B

4. Technician A states that the Digital Meter is capable of capturing and displaying the minimum and maximum values of a signal. Technician B states that the PGM has lower sweep rates than the lab scope to provide more detailed analysis of waveforms. Who is correct?

   a. Tech A only
   b. Tech B only
   c. Both Tech A and Tech B
   d. Neither Tech A nor Tech B

5. Two technicians are discussing the screenshot below. Technician A states that this is a live reading. Technician B states that increasing the voltage scale could lower the accuracy (lose decimal places) of the meter. Who is correct?

   a. Tech A only
   b. Tech B only
   c. Both Tech A and Tech B
   d. Neither Tech A nor Tech B

6. Two technicians are discussing the screenshot below. Technician A says this meter is calibrated to read zero ohms. Technician B says that this meter's scale is set to the 2 kilohms setting. Who is correct?

   a. Tech A only
   b. Tech B only
   c. Both Tech A and Tech B
   d. Neither Tech A nor Tech B

7. Two technicians are discussing the screenshot below. Technician A states this meter is reading an infinite ohm (open) signal. Technician B states that, to correct the reading on the screen, press Yes and choose a more appropriate scale. Who is correct?

   a. Tech A only
   b. Tech B only
   c. Both Tech A and Tech B
   d. Neither Tech A nor Tech B

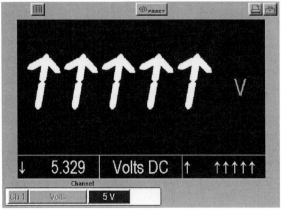

8. Technician A states that the longer the sweep rate is set for, the longer the total Movie that can be captured and saved. Technician B states that the PGM can record longer length Movies than the lab scope. Who is correct?

   a. Tech A only
   b. Tech B only
   c. Both Tech A and Tech B
   d. Neither Tech A nor Tech B

9. Two technicians are discussing the screenshot below. Technician A states this is a screenshot from the Digital Meter. Technician B states that the PGM always displays a graph of the signal. Who is correct?

   a. Tech A only
   b. Tech B only
   c. Both Tech A and Tech B
   d. Neither Tech A nor Tech B

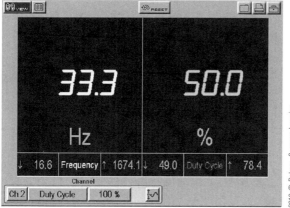

10. Two technicians are discussing the screenshot below. Technician A states that, in order to remove the Grid and Scales from this screen, the display options must be accessed through the Toolbox icon to the far upper right of the screen. Technician B states that the Zero Offset button is currently highlighted in blue. Who is correct?

   a. A. Tech A only
   b. B. Tech B only
   c. C. Both Tech A and Tech B
   d. D. Neither Tech A nor Tech B

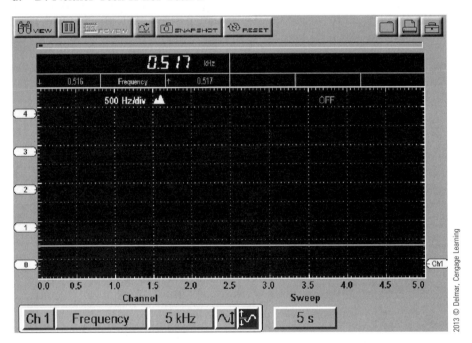

11. Technician A states that Peak Detect will only capture fast-acting, high-value events. Technician B states that Filter should be used when viewing small value scales such as 2 volts or less. Who is correct?

   a. Tech A only
   b. Tech B only
   c. Both Tech A and Tech B
   d. Neither Tech A nor Tech B

12. Two technicians are discussing the screenshot below. Technician A states that Channel one is set to a 10-volt scale. Technician B states that pressing Yes now will display Channel two. Who is correct?

   a. Tech A only
   b. Tech B only
   c. Both Tech A and Tech B
   d. Neither Tech A nor Tech B

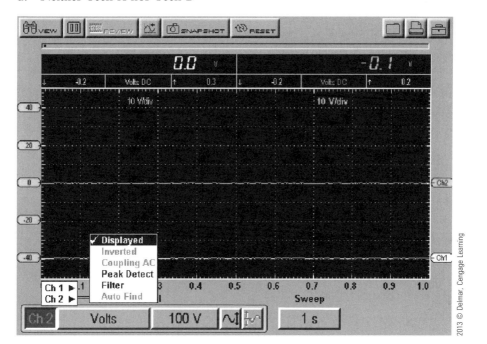

13. Two technicians are discussing the screenshot below. Technician A states that pressing Yes now will Zoom out and allow more time to be displayed. Technician B states that the current view is set to 2x zoom. Who is correct?

   a. Tech A only
   b. Tech B only
   c. Both Tech A and Tech B
   d. Neither Tech A nor Tech B

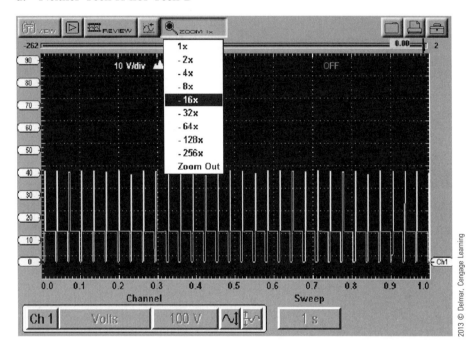

14. Two technicians are discussing the screenshot below. Technician A states that, when Review is activated, the Right and Left Buttons will scroll the data at multiple frames at a time, while the Up and Down arrows will scroll the data at one frame at a time. Technician B states that the white vertical bar (circled in blue) on the screen represents the zoom line where the pattern would be centered if zoomed back to a lower level. Who is correct?

   a. Tech A only
   b. Tech B only
   c. Both Tech A and Tech B
   d. Neither Tech A nor Tech B

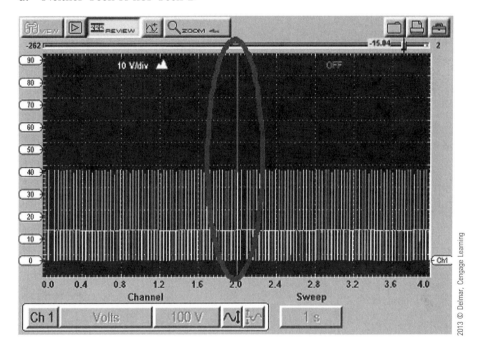

15. Two technicians are discussing the screenshot below. Technician A states that pressing the Left or Right arrow buttons will move Cursor 2. Technician B states that the value at Cursor 1 is 28.76 volts. Who is correct?

   a. Tech A only
   b. Tech B only
   c. Both Tech A and Tech B
   d. Neither Tech A nor Tech B

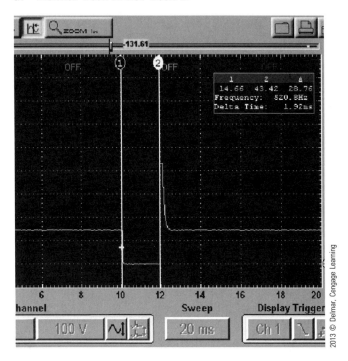

Index

AC/DC power adapter, 7, 8
AC/DC power input, 30, 31
Accessories
   features/benefits, 378–380
   Parts catalog, 378
Adapters, 415, 416
Add Note
   data management feature, 244
   Lab Scope data, 436
Add program
   MODIS, 71
   Solus Pro, 19
   Vantage Pro, 44
Airbag Functional Tests, 299–300
Amp fuse, 59
Amps, 503
Antilock Brakes system (ABS) tests, 298–299
Area of interest
   bad fuel injector, 469
   finding glitches, 464–465
   bad wheel speed sensor, 472
Ask-A-Tech, 274, 275, 276, 277
Auto, 422, 424
Auto Find, 411–412
Auto Threshold Select (ATS), 519
Aux port
   MODIS, 56, 57
   Vantage Pro, 30, 31
A-Z Index, 380–383

Back-door data, 87, 88
Backlight drop, 38
Backlight off, 39
Backup to CF, 19
Batteries
   MODIS, 59
   Solus Pro, 8
   Vantage Pro, 31
   Waveform Demonstrator Board, 489
Battery charger, 58
Battery indicator, 55, 56
BCM. *See* Body Control Module (BCM)
Bitmap format (BMP)
   MODIS, 67
   Solus Pro, 14
   Vantage Pro, 40
BMP, file type, 251
Body Control Module (BCM)
   description, 96
   functional tests, 301–302
Body Systems, 96
   Brightness/contrast button
   MODIS printers, 68, 69
   navigation example, 10–11
   Solus Pro, 9
   Vantage Pro, 33, 34

Cables
   channel configurations, 415, 416
   MODIS, 67, 74
   PC connections, 19
   scanner connection, 97
"Calibration Identification," 217

"Calibration Verification Number" (CVN), 217
Capture/Freeze data
   bad fuel injector, 466
   bad wheel speed sensor, 469
   finding glitches, 462
CF memory card slot, 6
CF2, 30
"Ch 1," 421, 425
Channel 1
   color coding, 406
   MODIS, 56, 57
   Vantage Pro, 30
Channel 2
   color coding, 406
   configurations, 407
   MODIS, 56, 57
   Vantage Pro, 30, 31
Channel 3
   color coding, 406
   MODIS, 56, 57
Channel 4
   color coding, 406
   MODIS, 56, 57
Channels
   lab scopes, 394–395
   MODIS connectors, 56, 57
Charge battery, 73–74
"Check Engine Light," 86
Chrysler systems, 97, 97
CKP Variation Learn Test, 287
Clear, 129–130
Clear Codes, 234–235

Clear Emission Related Data, 211
Code Menu
　Clear Codes, 234–235
　Diagnostic Trouble Codes (DTCs), 228–229
　DTC Status, 236–237
　Freeze Frame/Failure Record, 235–236
　overview, 228
　review questions, 238–240
　Trouble Codes, 230–234
Code Tips, Troubleshooter features, 322, 323–327
Coil in Cap (CIC), 448
Coil on Plug (COP), 448–449
Coil over Plug (COP), 37, 65
Color theme
　MODIS, 69
　Solus Pro, 17–18
　Vantage Pro printer, 43
Common ground, 30
Compact Flash (CF) card, 30
　connector port, 7
　MODIS, 55
Component
　data management feature, 244
　Lab Scope data, 436
Component Identification (CID), 214
Component Test Meter (CTM)
　MODIS overview, 54
　overview, 342
　Vantage Pro, 28
　Vantage Pro options, 35
Component Testing
　compression tests, 378
　fluid pressure tests, 377–378
　fuel injector test, 357–372
　knock sensor, 347–356
　and Lab Scope, 347
　power user tests, 375–377
　review questions, 387–390
　Vantage Pro options, 35
Component Testing Menu
　component tests, 345–373
　navigating, 342–345
Compression tests, 378
Condition
　data management feature, 244
　Lab Scope data, 436
Connect to PC
　data storage device, 250
　Solus Pro, 19–21
Connector(s)
　channel configurations, 416
　fuel injector test, 358
　knock sensor test, 348, 352–353
　Waveform Demonstrator Board, 490
Connector ports
　Solus Pro, 7–8
　Vantage Pro, 29–31
Contrast button
　MODIS, 68, 69
　Solus Pro, 9–11
　Vantage Pro, 33, 34
Control buttons
　MODIS, 59
　Solus Pro, 9
　Vantage Pro, 33
Copy function, 247
Coupling AC, 409
Current Probe, 365–371
Current Ramp Class Menu, 386
Current Ramp Test, 362–365, 368, 370, 371, 372
Cursor button
　graphing view, 148
　Lab Scope data, 441–443
Custom Data List
　data management, 267–270
　graphing view, 159
　printing scanner data, 141
　scan data, 117, 173–177
　summary, 177
Custom Data List (Version 9.4), 125
Custom Data View
　data management, 265–266
　graphing, 160–168
　PID List, 133–139
Custom Setup
　Code Menu, 232
　graphing view, 159
　printing scanner data, 140
　scanner data, 116
Custom Setup (Version 9.4), 124
"Cyl," 423, 429
"Cyl 5ms," 458–459
Cylinder, 445
Cylinder Polarity, 37

Data item, 258
Data Link Connector (DLC)
　Global OBDII, 200
　standard pin, 86–87
Data management
　custom data list, 267–270
　custom data view, 265–266
　data storage devices, 249–251
　edit feature, 244–248
　features, 243–244
　file types, 251–253

Index **539**

graph properties, 262–264
Lab Scope navigation, 272–277
LS(M) review, 270–271
main screen option, 11
overview, 242
review questions, 278–282
reviewing SC(M), 258–261
ShopStream Connect management, 253–255
ShopStream Connect navigation, 255
Data memory buffer, 149–150
Data menu (MT2500), 13–114
Data menu (Version 9.4), 121–123
Data scan, 322–323, 331–332
Data storage devices, 249–251
Date
    MODIS, 67–68
    printer setting, 15
    Vantage Pro printer, 41, 42
DC Voltage Test, 351–354
Delete feature, 247
Demo board. *See* Waveform Demonstrator Board
Diagnostic tools
    certification program, 1
    training programs, 2
Diagnostic Trouble Codes (DTCs), 228–229
Dials, 489–490
Different Graph views, 258
Digital KV View, 457
"Digital" Lab Scope configurations, 397, 405
Digital Meter
    MODIS navigation, 497–501
    Multimeter overview, 485
    resistance testing, 501–507

system navigation, 490–495
Vantage Pro navigation, 496–497, 500–501
Digital Meter box, 491–492
Digital multimeter (DMM)
    common use, 392
    Multimeters, 485, 490
Digital Volt Ohm Meter (DVOM), 56
Direct, 65
Direct ignition type, 37
"Display As,"
    Lab Scope configurations, 397–398
    MODIS, 64
    Vantage Pro, 37
Display control, 59, 60
Display Current Data, 207–209
Display Freeze Frame Data, 209–210
Display Permanent Trouble Codes, 219
Display Trigger, 459
Display Trouble Codes, 210
Displayed, 407
DTC Status, 236–237
DTCs Detected Last Drive Cycle, 215–216
Duty Cycle, 516

Easy scroll
    MODIS legacy software, 73
    Solus Pro, 19
    Vantage Pro, 45, 46
"ECU's Acronym and Tex Name," 218
Edit, 243
Edit feature, 244–249
Electron Class lessons, 384–385
Engine system (MT2500), 113
Engine system (Version 9.4), 121–126

Environmental Protection Agency (EPA), 90, 201–202
Exit, 258

Failed This Ignition, 233, 234
Falling edge, lab scopes, 394
Fast Forward, 259
Fast-Track Data Scan, 322–323, 331–332
Fast-Track Troubleshooter
    features, 321–323
    functional tests, 304–307
    MODIS overview, 54
    overview, 320–321
    research, 333–335
    review questions, 336–339
FGA Demo, 70
File type(s)
    data management, 251–253
    Lab Scope presets, 431
    MODIS, 67
    Solus Pro, 14
    Vantage Pro, 40
Filter
    channel configurations, 410–411
    MODIS, 59–60
Fix Line, 117, 119, 120
Flexible Gas Analyzer, 70, 71
Fluid Pressures Tests, 377–378
Flying ground, 394
Folder Icon, 435
Ford system module, 96
Frame Number, 259
Freeze button, 439
Freeze Frame/Failure Record, 235–236

Index

Freeze/Run
  graphing view, 145–146
  MODIS, 68
  PID List, 128
  Solus Pro, 16
  Vantage Pro printer, 43
Frequency, 516
Frequency Connector, 490
Frequency Dial, 489
Front-door data, 87, 88
Fuel Injector Connector, 490
Fuel injector glitches, 466–469
Fuel Injector Test, 357–372
"Full Code List," 115, 231
"Full Code List" (Version 9.4), 123
"Full Digital," 513
Full Graph, 514
"Full PID List"
  graphing view, 158
  printing scan data, 139
  scan data, 115, 117
"Full PID List," (Version 9.4), 123
Full Screen
  Code Menu, 231
  graphing view, 158
  scan data, 139
"Full Screen" (Version 9.4), 123
Functional Test Menu, 286
Functional Tests
  ABS test, 298–299
  airbags, 299–300
  BCM tests, 301–302
  CKP Variation Learn Test, 287
  four categories, 285
  Idle Air Control Test, 289–295
  Injector Balance Test, 287–288

IPC, 302–304
Output Controls, 288–289
overview, 284
researching, 307–314
review questions, 315–318
training, 304–307
transfer case, 300–301
transmission tests, 297–301
Fuses, 28–29

Gas bench setup options, 71
Gases, 62, 63
General Motor system module, 96
Generic Function
  Global OBDII, 201–202, 2222–223
  scanner procedure, 91
Glitches
  bad fuel injector, 466–469
  bad wheel speed sensor, 469–473
  Lab Scopes, 462–464
Global OBDII
  Clear Emission Related Data, 211
  description, 200–201
  Display Current Data, 207–209
  Display Freeze Frame Data, 209–210
  Display Permanent Trouble Codes, 219
  Display Trouble Codes, 210
  DTCs Detected Last Drive Cycle, 215–216
  Generic Functions, 222–223
  In-Use Performance Tracking, 218
  MIL status, 206–207
  Non-Cont. Monitored Systems, 213–215
  OBD Health Check, 219–221
  Oxygen Sensor Monitoring, 212–213

Read Vehicle Identification, 217–218
readiness monitors, 204–205
Request Control On-Board System, 216
review questions, 224–226
scanner procedure, 90, 91
scanner software, 200
system navigation, 202–204
vehicle connection, 94
Graph(s)
  clearing data memory buffer, 149–150
  Custom Data View, 160–168
  data view, 141–146
  MODIS, 70
  PID Sort, 150–152
  PID scaling options, 169
  properties, 262–264
  reviewing, 147–147
  Save/print/utilities, 158–159
  scan data, 141–159
  Snapshot button, 158
  Solus Pro, 18
  Zoom example, 155–157
  Zoom features, 152–155
Graphic Meter, 513
"Grid," 400
Ground
  Lab Scopes, 394–395
  Waveform Demonstrator Board, 490
Ground clamp
  channel configurations, 416
  Single Cylinder Ignition pattern, 449
Ground (Zero) point, 393–394

HDS units, 72
"History Codes," 233

Index **541**

Horizontal axis, 393
How To Menu, 383–385

Idle Air Control Test, 289–295
Ignition Ground, 489
Ignition Probe, 404, 412, 416
Ignition Scope
   configuration, Lab Scopes, 443–448
   MODIS, 56
   Movies/Snapshots, 458–462
   multiple patterns, 449–458
   single pattern, 447–449
Ignition Scope Movie file, 252
Ignition Scope Preset file, 252
Ignition Scope Snapshot file, 252
Ignition system
   MODIS, 65
   Vantage Pro, 37–38
"Ignition System," 400
In-Use Performance Tracking, 218
"Individual Properties," 262, 263
Info/component testing, 62
Info function, 248
Information Tests, 285
Infrared printer port. *See* IRDA
Inj Pulse Width, 516
Injector Balance Test, 287–288
Injector Pulse Width Dial, 489
Injector Type Switch, 489
Instrument Panel Cluster (IPC) tests, 96, 302–304
Internal drive, 249
Inverse Colors, 402, 493
Inverted configuration, 408–409
IRDA, 55, 56
IS(C), file type, 252

IS(M), file type, 252
IS(P), file type, 252

Joint Photographic Experts Group (JPEG) format
   MODIS, 67
   Solus Pro, 14
   Vantage Pro, 40
JPEG, file type, 251
JPG, file type, 251

Keep Entries
   data management feature, 244
   Lab Scope data, 436
Knock Sensor
   component information, 347–350
   component testing, 347–348

Lab Scope(s)
   basics, 392–394
   channel configurations, 406–420
   channels/connections, 394–395
   display configurations, 396–405
   finding glitches, 462–464
   ignition configuration, 443–444
   knock sensor, 352
   MODIS connectors, 56
   Multimeter overview, 485–488
   navigating, 272–277
   need for, 392
   and PGM, 507–512
   power user tests, 375
   presets, 430–435
   review questions, 474–482
   specifications by sweep rate, 395
   trigger configuration, 421–430

Lab Scope data, 435–443
Lab Scope Movie file, 252, 270–271
Lab Scope Preset file, 252
Lab Scope Snapshot file, 252
"Last Test Failed," 234
LED Menu, 117–118
LED Menu (Version 9.4), 125–126
Left side panel, 57
Legacy software, 72–74
Load, 243
Location
   fuel injector test, 358–361
   knock sensor test, 348, 349–350
Low amps
   resistance testing, 504, 506
   PGM channel configurations, 516
Low amps probe
   channel configurations, 411, 414
   MODIS, 56
   resistance testing, 505
LS(C) file, 252, 431
LS(M), 252
LS(P), 252

Main menu
   MODIS, 61–63
   scanner, 106–107
Main screen options, 34–35
Make
   data management feature, 244
   Lab Scope data, 436
Malfunction Indicator Light (MIL), 86
"Manual," 438
Manual scale, 172
Manual snapshot, 178–180
Manual Threshold Select (MTS), 519

Manufacturer
  common modules, 96
  scanner flowchart, 91
  vehicle connection, 93–94
MC Dwell, 516
Memory indicator, 251
MIL status, 206–207
Mini USB
  connector port, 7
  MODIS connectors, 57–58
  Vantage Pro, 30, 31
MODIS (Modular Diagnostic Information System)
  battery, 59
  Code Menu, 228
  connect to PC, 74–75
  connectors, 55–58
  control buttons, 59–60
  CTM, 342, 374
  Custom data view, 136
  data management, 242
  diagnostic tool, 2–3, 4
  Digital Meter navigation, 497–501
  functional tests, 282
  gas bench set up options, 71
  Global OBDII, 200
  graphs, 143–144, 160
  Ignition Scope configuration, 443
  and Lab Scopes, 392, 398
  legacy software, 72–74
  Multimeters, 484
  platform overview, 54, 55
  printer, 67–69
  resistance testing, 504–507
  review questions, 78–84
  Single Cylinder Ignition pattern, 448
  system navigation, 60–61
  system tools, 71
  tool setup options, 63–64
  user capabilities, 1
  utility options, 63
Monitor Identification (MID), 214
Move function, 247
Movie
  data saving, 13
  MODIS, 66
  reviewing ignition scope, 458–462
  Vantage Pro, 39
MT2500 Scanner, text view, 113–121
Multimeter(s)
  Digital Meter navigation, 490–501
  MODIS menu, 62
  overview, 484–485
  PGM channel configurations, 515–519
  PGM configurations, 512
  PGM connections, 511
  PGM/Lab Scope, 507–511
  PGM navigation, 512–515
  practice exercises, 519–527
  resistance testing, 501–507
  review questions, 528–536
  Vantage Pro options, 35
  Waveform Demonstrator Board, 488–489
Multiple Ignition Scope patterns, 449–458
My Computer, 47
My data
  MODIS, 67
  Solus Pro, 14
  Vantage Pro, 40

National Coalition of Certification Centers (NC3), 1–2
New Vehicle ID, 373–374
No button
  Solus Pro, 9, 13
  Vantage Pro, 33
No-Start Basics option, 383–386
Non-Cont. Monitored Systems, 213–215
"None," 424
Normal, 423, 424
Normal/Glitch Switch, 489

OBD Health Check, 219–221
OBDI (On Board Diagnostics) connections
  scanners, 97, 98, 99, 100
  steps, 100–102
OBDII. See also Global OBDII
OBDII adapter, 102, 105
OBDII connections, 102–107
OBDII Data Link Connector, 86
OBDII Training Mode, 203
OEM room, 200–201
Ohms Meter, 497–500
On Board Diagnostics (OBDI). See Global OBDII, OBDI, OBDII
Operation
  fuel injector test, 358
  knock sensor test, 348, 349
Output Controls, 288–289
Oxygen Sensor Monitoring, 212–213

"Parade"
  Ignition Scope configurations, 444
  multiple ignition patterns, 455
Parameter identifications (PIDS). See also PIDS List, PID scaling options, PID Sort, PID Trigger
  computer data, 188
  scan data, 112
Peak, 394
Peak Detect, 410

Index **543**

"Pending Codes," 233
Personal computer (PC)
   data storage device, 250
   MODIS connection, 74–75
   Solus Pro connection, 19–21
   Vantage Pro connection, 46–47
PID List
   custom data views, 265
   MODIS, 70
   Solus Pro, 18
   viewing, 126–133
PID scaling options, 169–173
PID Sort, 130–131, 150–152
PID Trigger
   capturing technique, 180–186
   summary, 186–188
Plastic screen, 8
Play, 258
Plugs, 416
Polarity, 451–454, 455–456
Power button
   MODIS, 59, 60
   MODIS navigation, 61
   navigation example, 10
   Solus Pro, 9
   Vantage Pro, 33, 34
Power Graphing Meter (PGM)
   channel configurations, 515–519
   configurations, 512
   connections, 511
   knock sensor, 352
   and Lab Scope, 507–511
   Multimeter overview, 485, 487–488
   navigation, 512–515
   power user tests, 375
   practice exercises, 519–527

Power input, 55, 56
Power management
   MODIS, 65–66
   Utilities menu, 12, 13
   Vantage Pro, 38–39
Power User, 2
Power User Tests, 375–377
Powertrain Control Module (PCM),
Powertrain Control Module (PCM)
   functional tests, 284
   Global OBDII, 200
Presets, 430–435
Pressure Traducers, 376–378, 516
"Previous vehicles," 92
Print
   Code Menu, 231–232
   graphing view, 158–159
   scanner data, 115
   scanner data, 139–140
   Version 9.4, 123–124
Print data, 117
"Print Frame," 117
Print list
   MODIS, 68, 69
   Solus Pro, 17
Print page
   MODIS, 69
   Solus Pro, 17
   Vantage Pro printer, 43
Printer
   Digital Meter navigation, 493
   MODIS, 67–69
   Solus Pro, 14–16
   Vantage Pro options, 41–43
Protective sliding cover, 30, 31

Raising edge, 394
"Ras 5ms.," 60
Raster, 445–446
Read Vehicle Identification, 217–218
Readiness Monitors, 3, 204–206
Red brick scanner, 6
Replaceable screen cover, 33
Request Control On-Board System, 216
"Reset"
   Lab Scope configurations, 397
   PGM navigation, 515
Reset button, 492
Reset Tests, 285
Resistance testing, 501–507
Review, 128, 129, 147–148
Review button
   bad fuel injector, 467–468
   bad wheel speed sensor, 471
   finding glitches, 464
   Lab Scope data, 439, 440, 441
Review questions
   Code Menu, 238–240
   component testing, 387–390
   data management, 278–282
   Fast-Track Troubleshooter, 336–339
   functional tests, 315–318
   Global OBDII, 224–226
   Lab Scopes, 474–482
   MODIS, 78–84
   multimeters, 528–536
   scan data, 190–191
   scanner, 108–109
   Solus Pro, 23–25
   Vantage Pro, 49–52

Rewind, 258

Run
  graphing view, 145
  MODIS, 68, 75
  Solus Pro PC program, 20
  Vantage Pro PC program, 47

Run button, 128

S-button, 9
  MODIS printers, 16, 17, 68
  Vantage Pro, 33, 42

S-button popup
  MODIS, 69
  Vantage Pro printer, 43

Save
  graphing view, 158
  Lab Scope data, 436
  scan data, 115, 139
  Version 9.4, 123

"Save As," 512–513

Save data
  Code Menu, 232–234
  data management feature, 243–244
  graphing view, 159
  MODIS, 66
  printing scanner data, 140–141
  scan data, 116–117
  utility menu, 13–14
  Vantage Pro, 39–40
  Version 9.4, 124–125

"Save Data"
  Lab Scope configurations, 403
  Lab Scope data, 437

Save dialogue box, 433–435

Save frame
  graphing view, 158
  Lab Scope data, 437

MODIS, 68
  printing scanner data, 139
  Solus Pro, 16
  Vantage Pro printer, 43

Save image
  graphing view, 158
  Lab Scope data, 437
  MODIS, 68
  Solus Pro, 16, 17
  Vantage Pro, 42, 43

Save Movie
  graphing view, 158
  printing scan data, 139

"Save Movie," 435–437

Saved data
  MODIS menu, 62, 63
  Vantage Pro options, 35–36

SC(M) file, 251

SC(P) file, 252

Scale
  PGM channel configurations, 517–518
  SC(M), 258

Scale button, 417

Scales Display, 401

Scanner
  Code Menu, 228
  Global OBDII software, 200
  main screen option, 11
  MODIS menu, 62
  OBDI connection steps, 100–107
  OBDI connections, 97–100, 102
  overview, 86–87
  repair scenario, 88–89
  review questions, 108–109
  systematic procedure, 89–91
  vehicle connection steps, 91–97

Scanner button, 121–123

Scanner data
  cable connector, 7, 8
  MT2500, 112–113
  overview, 112
  review questions, 190–191
  Version 9.4, 121

Scanner Demo, 94

Scanner module™, 56

Scanner Movie file, 251–252

Scanner Movie SC(M) file, 258–261

Scanner Snapshot file, 252

Scanner units
  MODIS, 64
  Version 9.4, 124

Scanner view
  MODIS, 70
  Solus Pro, 18

Scope
  MODIS menu, 62
  Vantage Pro options, 35

Screen cover, 33

Screenshot, 14

Secondary Clip-on Wire Adapter, 448, 450

Secondary Coil Adapter, 448

Secondary Ignition, 489

Select all function, 247–248

Serial port, 55, 56

"Service Vehicle Soon (SVS) Light," 86

Setup function, 248

"Shared Properties," 262, 263

Shop info
  MODIS, 75–76
  PC program, 20–21
  Vantage Pro PC program, 47

Index **545**

ShopStream Connect, 20–21
    data management advantages, 253
    management, 254–255
    MODIS, 74–75
    navigating, 255–257
    Vantage Pro, 28, 47, 48
Shunt amps, 30, 31
SIA 2000 Ignition scope adapter, 449, 450–452
Side CF card slot, 57
Signal signature, 392
Sine Wave Connector, 490
Single Cylinder Ignition, 447–449
"Single Cylinder Ignition," 452
Single Cylinder Ignition pattern, 448
Snap-on
    accessories, 378–380
    Parts catalog, 378
    PC program, 20
    training program, 1, 2
Snap-On Certification Manual, 3
Snap-On Certification Tests, 3–4
Snap-On Certified Diagnostic Training Centers, 2
Snapshot
    data capture, 178
    data saving, 13
    graphing view, 158
    Lab Scope data, 438–439
    MODIS, 66–67
    reviewing ignition scope, 458–462
    Vantage Pro, 39–40
Snapshot button, 158
Society of Automotive Engineers (SAE), 228, 229
Solus Pro
    battery compartment, 8
    CF memory card slot, 6

Code Menu, 228
connector ports, 7–8
control buttons 9
data management, 242
diagnostic tool, 2, 4
functional tests, 282
and Global OBDII, 200
graphs, 142–143, 160
main menu options, 11
PC connection, 19–20
platform, 6–7
platform overview, 6
review questions, 23–25
"S" button, 9, 16, 17
system navigation, 10–22
viewable PIDS, 136
Solus Ultra, 6
Sort
    graphing PID, 150–152
    PID List, 130–131
Split Screen, 514
Stand By, 38, 39
Standard, 65
Standard distributor system, 37
Start Communication, 203–204
"Sup 5ms.," 461
Superimposed, 446–447
Sweep
    defined, 393
    ignition scope movies/snapshots, 458, 460, 461
    Lab Scope configurations, 403–405
    multiple ignition patterns, 457
    SC(M), 258
Switches, 489
Symptoms, 322, 327–329

System
    flowchart, 91
    scan data, 113
    tools, 18–19
    vehicle connection, 96
System Info
    MODIS, 76
    PC connection, 21–22
    Vantage Pro PC program, 48
System restore
    MODIS, 71
    Solus Pro, 19
    Vantage pro, 44

Tech notes, 348
Technical Assistance, 322, 330–331
Technical colleges, 1, 2
Technical Service Bulletins, 330
"Test Failed Since Code Cleared," 234
Test Identification (TID), 214
Test Lead, 412, 413
"Test Lead," 515
Test Tips, 386
Tests & Procedures, 322, 329–330
Text, 70
Text view
    MT2500, 113–121
    Solus Pro, 18
    Version 9.4, 121–126
Threshold, 518
Thumb pad
    MODIS, 59, 60
    Solus Pro, 9
    Vantage Pro, 33

Time
- MODIS, 68
- printer setting, 15–16
- Vantage Pro PC program, 49
- Vantage Pro printer, 42

Time zone, 16
Timer, 12
Toggle Tests, 285
Tool help
- MODIS legacy software, 72–73
- Vantage pro system tools, 44

Tool setup
- MODIS options, 63–65
- Utility menu, 12, 14, 16
- Vantage Pro, 36, 37, 39

Tool setup menu, 17–18
Tool setup options, 39–43
Top CF card, 249
Top CF Card slot, 55
TPS, 266–267
Transfer Case functional test, 300–301
Transmission functional tests, 297–301
Trigger configuration, 421–430
"Trigger Display," 400–401
Trigger Position button, 430
Trigger Position Cursor, 426–429
Trigger Slope icon, 425–426
Trouble Codes, 30–234
Turn off, 38, 39

Units
- MODIS tool setup options, 64
- Vantage Pro, 36–37

"Units," 398
Update from CF, 19
Update scanner module
- MODIS, 71
- Solus Pro, 19

USB port
- connector port, 7
- MODIS connectors, 55, 56
- Vantage Pro, 30, 31

USB storage device, 250
Utilities
- MODIS menu, 62, 63
- Vantage Pro, 45, 46
- Vantage Pro options, 36–37

Utility Menu, 11–14

Vantage Pro
- batteries, 31, 32
- connect to PC, 46–47
- connectors, 29, 30–31
- control buttons, 33
- CTM, 342
- data management, 242
- diagnostic tool, 2, 4
- Digital Meter navigation, 496–497, 500–501
- Ignition Scope configuration, 443
- and Lab Scopes, 392
- Multimeter, 484
- platform features, 28, 29
- platform overview, 28
- resistance testing, 502–507
- review questions, 49–52
- Single Cylinder Ignition pattern, 448
- system navigation, 34–43

system tools, 44
Utilities, 45, 46
Variable Control Tests, 285
Variable Frequency/Duty Connector, 490
Variable Voltage Connector, 490
"Vehicle Comm," scanner connection, 92
"Vehicle Identification Number," 217
Version 9.4, 121
Vertical axis, 393
VGA output, 58
"View," 513, 515
Vin character, 94–95
Vin number, 217
VIN specific details, 200
VIN specific software database
- scanner procedure, 90–91
- vehicle connection, 91, 93

Voltage Dial, 490
Voltage Scale, 494–495
Voltage Settings, 495
Volts, 515
Volts AC rms
- Digital Meter navigation, 495
- PGM channel configurations, 516

Volts DC, 495
Volts DC ave, 516
Volts DC-Average, 495

Wasted spark, 37, 65
Waveform Demonstrator Board, 488–489
Windows CE program
- MODIS overview, 54
- Vantage Pro, 28

Year
    data management feature, 244
    Lab Scope data, 436
Yes button
    Solus Pro, 9
    Vantage Pro, 33
Yes-No button, 59, 60

Zero Offset button, 418–419

Zoom
    graphing, 152–155
    graphing example, 155–157
    Lab Scope data, 439, 440–441
    PID List, 131–133
Zoom In
    bad wheel speed sensor, 473
    finding glitches, 465
    SC(M), 259

Zoom Out
    bad fuel injector, 466–467
    bad wheel speed sensor, 470
    finding glitches, 463–464
    SC(M), 259

	DATE DUE		